Water Pollution and Management Practices

Anita Singh • Madhoolika Agrawal • Shashi Bhushan Agrawal
Editors

Water Pollution and Management Practices

Springer

Editors
Anita Singh
Centre of Advanced Study in Botany
Institute of Science, Banaras Hindu University
Varanasi, Uttar Pradcsh, India

Madhoolika Agrawal
Centre of Advanced Study in Botany
Institute of Science, Banaras Hindu University
Varanasi, Uttar Pradesh, India

Shashi Bhushan Agrawal
Centre of Advanced Study in Botany
Institute of Science, Banaras Hindu University
Varanasi, Uttar Pradesh, India

ISBN 978-981-15-8357-5 ISBN 978-981-15-8358-2 (eBook)
https://doi.org/10.1007/978-981-15-8358-2

This Springer imprint is published by the registered company Springer Nature Singapore Pte Ltd.
The registered company address is: 152 Beach Road, #21-01/04 Gateway East, Singapore 189721, Singapore

Dedicated to our beloved parents

Preface

In order to understand the meaning and to communicate successfully in groups, it is necessary to have complete knowledge about a particular problem. The main objective of the present edited book is also to provide a holistic view of every aspect of the topic entitled "**Water Pollution and Management Practices**". The concentrations of pollutants are increasing in the water system day by day. These pollutants include organic contaminants such as pesticides, solvents, and petroleum products and inorganic pollutants such as heavy metals, nonmetals, metalloids, and simple soluble salts. Water pollutants have both natural (fires, volcano eruptions, soil or rock erosion, biodegradation) as well as anthropogenic (air and soil pollution, herbicides) origins. Above safe limits, these pollutants may have a toxic impact on aquatic systems, including various life forms, and upon transfer to terrestrial systems, they can negatively affect biotic and abiotic components of the ecosystems. In plants, pollutants not only affect the biochemical and physiological activities but also disturb the signalling pathway.

The book gives comprehensive and concise knowledge about water pollutants and recent advancements in the field of sustainable technologies to reduce the level of pollutants as well as to reduce the burden on human health. The major water pollutants dealt in this book are water-borne microbial contaminants, nitrogenous wastes, industrial effluents, textile wastes, agricultural runoff, dyes, and antibiotics and their consequent effects on aquatic and terrestrial ecosystems. Microbial contamination of drinking water is suggested to be one of the most important issues of human health risk during present times. Further, to monitor the groundwater contaminants, an application of artificial intelligence is discussed in detail, which can help in developing prediction models for globally concerning contaminants, including arsenic, fluoride and nitrate. Hybrid ML–DL models, could be especially useful under the many-dimensional nature of environmental data and can be used as a decision-support system to create proactive environmental-management policies. The expansion of urban and cultivated areas has been found to cause evident alterations in the natural landscape and natural processes that interfere with hydrological cycles and influence the fluxes of discharge and runoff on various scales and thus affect the quality of water adversely. For reducing river pollution, the role of riparian ecotone in scrubbing the pollutants, mainly sediments, nitrogen and phosphorus, was found to be significant; therefore, extensive riparian ecotone–based studies are recommended so that riparian ecotone can be used as mean to manage the degrading rivers. There are multiform determinants such as transparent exopolymeric particles (TEP), diatom-dominance-diversity linkages, microbial extracellular enzymes (EE), ecological response index (ERI) and ecosystem feedbacks. They can be used as novel tools to assess the impact of human activities on ecological functioning and assimilation capacity of the river.

Few of the chapters also deliberated upon the current scenario of wastewater application and its consequences on agricultural system. The control measures to reduce contaminants in water, which reduce their utility in agriculture, are discussed in detail to find out sustainable solutions. In this regard, several physico-chemical and biological methods and their limitations are discussed. Some of the important low-cost and easy control methods include precipitation, adsorption, membrane based filters, etc. The biological control methods including bioremediation and phytoremediation have also been discussed particularly for reducing arsenic contamination. For heavy metals, which are major components of industrial waste water, technologies using nanoparticles, ion-exchange, adsorption, membrane filtration, chemical precipitation, coagulation–flocculation and electrochemical treatments have been suggested. Adsorption on biomass-derived bio-sorbents can provide the capability to treat wastewater on a large scale and has been detailed in this book. Various issues discussed in the book set the stage for understanding and assessing the current and possible future scenarios of management of river pollution.

Overall, the information compiled in this book has covered the in-depth knowledge about advanced technologies to sort out the problem of water pollution sustainably in the future. Finally, the book is in its present form only because of the concerted efforts of the contributors, who are experts from India and the other parts of the world in their respective fields.

Varanasi, India

Anita Singh
M. Agrawal
S. B. Agrawal

Acknowledgement

We are very grateful to the Head, Centre of Advanced Study in Botany, Institute of Science, Banaras Hindu University, Varanasi, for providing necessary facilities. The Banaras Hindu University is also thankfully acknowledged for providing financial assistant to Dr. A. Singh as PI under IoE seed grant scheme (Dev. IoE Scheme No. 6031). All editors thankfully acknowledge the contributions of all the authors and their efforts and time dedicated to this book. Editors also gratefully acknowledge the team at Springer Nature who made possible the proposed book in its present form.

Contents

Editors and Contributors

About the Editors

Anita Singh is an assistant professor at Centre of Advanced Study in Botany, Institute of Science, Banaras Hindu University, Varanasi, UP, and India. She has completed her M.Sc. and Ph.D. at Banaras Hindu University. Her expertise falls under the domain of sustainable agriculture. Dr. Singh has worked on metal remediation, impact of abiotic stress on plant physiology and biochemistry, and assessment of florescence transients. She has 40 publications to her credit. She is a recipient of several awards including the International Green Talent Award presented by BASF, Germany, and the International Alice J. Murphy Outstanding Achievement Award. Dr. Singh has edited some informative books published by Springer Nature, IGI, and John Wiley & Sons Ltd. and she has also reviewed several papers published in the reputed journals.

Madhoolika Agrawal is a professor at Centre of Advanced Study in Botany, Institute of Science, Banaras Hindu University, Varanasi, India, since 1999. She has worked on environmental pollution and global climatic change effects and their management. Dr. Agrawal is a recipient of the US Fulbright Fellowship and has worked at USDA, Beltsville, Maryland, USA, and INSA–Royal Society exchange fellowship at Lancaster University, UK. She is a fellow of the Indian National Science Academy, the National Academy of Sciences, and the National Academy of Agricultural Sciences, India. Dr. Agrawal has worked on various projects funded by the Government of India and international agencies. She has published more than 250 research papers, co-edited a book (Lewis Publishers, Boca Raton, USA), and co-authored a book (Springer).

Shashi Bhushan Agrawal is a professor of ecology at Centre of Advanced Study in Botany, Institute of Science, Banaras Hindu University. He has completed his M.Sc. at Agra University and Ph.D. at Banaras Hindu University. Professor Agrawal has worked in the area of stress biology and conducted long-term field experiments to ascertain the effects of abiotic stress factors on plants. He has worked as a visiting scientist at USDA, Beltsville, Maryland, USA. He edited two books published by SPB Academic Publishing and CRC/Lewis. Professor Agrawal has published more than 160 research papers and is serving in the editorial board of several international journals. He has successfully completed several research projects funded by ICAR, MOEn, DST, UGC, UP-CST, and CSIR.

Contributors

Saja H. Abdul-Ameer The Department of Biology, College of Science for Women, University of Baghdad, Baghdad, Iraq

Madhoolika Agrawal Centre of Advanced Study in Botany, Institute of Science, Banaras Hindu University, Varanasi, Uttar Pradesh, India

Halah H. Al-Haideri The Department of Biology, College of Science for Women, University of Baghdad, Baghdad, Iraq

Surabhi Awasthi CSIR – National Botanical Research Institute, Council of Scientific and Industrial Research, Lucknow, Uttar Pradesh, India

Reshu Chauhan CSIR – National Botanical Research Institute, Council of Scientific and Industrial Research, Lucknow, Uttar Pradesh, India

Parmita Chawley Department of Environmental Science and Engineering, Indian Institute of Technology (Indian School of Mines), Dhanbad, Jharkhand, India

G. S. Dangayach Department of Management Studies, MNIT, Jaipur, Rajasthan, India

Sanjay Dwivedi CSIR – National Botanical Research Institute, Council of Scientific and Industrial Research, Lucknow, Uttar Pradesh, India

Kaushik Gautam Centre of Advanced Study in Botany, Institute of Science, Banaras Hindu University, Varanasi, Uttar Pradesh, India

Fikrat M. Hassan The Department of Biology, College of Science for Women, University of Baghdad, Baghdad, Iraq

Sheeja Jagadevan Department of Environmental Science and Engineering, Indian Institute of Technology (Indian School of Mines), Dhanbad, Jharkhand, India

Deepa Jaiswal Ganga River Ecology Research Laboratory, Environmental Science Division, Centre of Advanced Study in Botany, Institute of Science, Banaras Hindu University, Varanasi, India

Nor Rohaizah Jamil Department of Environment, Faculty of Forestry and Environment, Universiti Putra Malaysia, Serdang, Malaysia

Richa Kothari Department of Environment Science, Central University of Jammu, Jammu, India

Brijesh Kumar Mishra Department of Environmental Science and Engineering, Indian Institute of Technology (Indian School of Mines), Dhanbad, India

Monika Department of Environment and Sustainable Development, Institute of Environment and Sustainable Development, Banaras Hindu University, Varanasi, Uttar Pradesh, India

Indrani Mukherjee Integrated Science Education and Research Centre (ISERC), Institute of Science, Visva-Bharati, Santiniketan, Birbhum, West Bengal, India

Jitendra Pandey Ganga River Ecology Research Laboratory, Environmental Science Division, Centre of Advanced Study in Botany, Institute of Science, Banaras Hindu University, Varanasi, India

Usha Pandey Department of Botany, Faculty of Science and Technology, Mahatma Gandhi Kashividyapith University, Varanasi, India

Punyasloka Pattnaik Department of Management Studies, MNIT, Jaipur, Rajasthan, India

Binh Thai Pham Institute of Research and Development, Duy Tan University, Da Nang, Viet Nam

Shakti Rath Faculty of Dental Sciences & Nursing, Central, Research Laboratory, Institute of Dental Sciences, Siksha 'O' Anusandhan (Deemed to be University), Bhubaneswar, Odisha, India

Arijit Roy Indian Institute of Remote Sensing, Indian Space Organisation, Dehradun, India

Rajesh Kumar Sharma Centre of Advanced Study in Botany, Institute of Science, Banaras Hindu University, Varanasi, Uttar Pradesh, India

Shikha Sharma Indian Institute of Remote Sensing, Indian Space Organisation, Dehradun, India

Zakariya Nafi Shehab Department of Environment, Faculty of Forestry and Environment, Universiti Putra Malaysia, Serdang, Malaysia

Ataollah Shirzadi Department of Rangeland and Watershed Management, Faculty of Natural Resources, University of Kurdistan, Sanandaj, Iran

Anita Singh Centre of Advanced Study in Botany, Institute of Science, Banaras Hindu University, Varanasi, Uttar Pradesh, India

Pooja Singh Department of Science Institute of Computer Science and Technology, SHEPA, Varanasi, India

Prince Kumar Singh Centre of Advanced Study in Botany, Institute of Science, Banaras Hindu University, Varanasi, Uttar Pradesh, India

Rajeev Pratap Singh Department of Environment and Sustainable Development, Institute of Environment and Sustainable Development, Banaras Hindu University, Varanasi, Uttar Pradesh, India

Sushant K. Singh Artificial Intelligence & Analytics, Health Care and Life Sciences, Virtusa Corporation, New York, NY, USA

Umesh Kumar Singh Department of Environmental Science, School of Earth, Biological and Environmental Sciences, Central University of South Bihar, Gaya, Bihar, India

Sonalika Sonal Department of Environmental Science and Engineering, Indian Institute of Technology (Indian School of Mines), Dhanbad, India

Sudhakar Srivastava Institute of Environment and Sustainable Development, Banaras Hindu University, Varanasi, India

Rudra Deo Tripathi CSIR – National Botanical Research Institute, Council of Scientific and Industrial Research, Lucknow, Uttar Pradesh, India

Krishna Yadav Department of Environmental Science and Engineering, Indian Institute of Technology (Indian School of Mines), Dhanbad, Jharkhand, India

Chapter 1
Microbial Contamination of Drinking Water

Shakti Rath

1.1 Introduction

Water is one of the most essential components of our life and the availability of freshwater is limited on earth. In the last three decades, there has been a continuous decline in the availability of potable water all over the world due to increasing water pollution and global warming (Bi et al. 2018). The "Safe Drinking Water Act" coined the term "contaminant" which means any type of waste substance in the physical, chemical, or biological form found in the drinking water (Allaire et al. 2018). Drinking water is never 100% free from impurities. It naturally contains contaminants, mostly chemicals and microorganisms due to anthropogenic activities (Fawell and Nieuwenhuijsen 2003). However, if the level of the contaminants increases above the safe limit, then the water becomes unfit for drinking. It has also been well established that water has always remained a major source of infection and morbidity among mankind from time immemorial (Edokpayi et al. 2018).

1.1.1 Types of Drinking Water Contaminants

Water contaminants can be classified into four major categories,

S. Rath (✉)
Faculty of Dental Sciences and Nursing, Central Research Laboratory, Institute of Dental Sciences, Siksha 'O' Anusandhan (Deemed to be University), Bhubaneswar, Odisha, India

A. Singh et al. (eds.), *Water Pollution and Management Practices*,
https://doi.org/10.1007/978-981-15-8358-2_1

1.1.1.1 Physical Contaminants

These waste products hamper the appearance of the water and alter its physical properties. For example, human and animal feces, household sediments in the lakes, soil erosion from the rivers and small streams (Fawell and Nieuwenhuijsen 2003).

1.1.1.2 Chemical Contaminants

These waste products occur naturally or are human-made such as industrial wastes, bleaching agents, insecticides, and pesticides used in farming and pharmaceutical drugs (Barrett 2014).

1.1.1.3 Biological Contaminants

These waste products mainly comprise microbes such as viruses, protozoa, parasites, bacteria, and their byproducts (Fawell and Nieuwenhuijsen 2003).

1.1.1.4 Radiological Contaminants

These are the harmful waste products formed from the protons and neutrons of an unstable atom which emits harmful ionizing radiation like uranium and plutonium (Fawell and Nieuwenhuijsen 2003).

1.2 The Problem Scenario

For any nation, the most significant aspect, as well as a considerable successful public health practice, is the delivery of safe drinking water to its citizens. However, lack of knowledge about water contamination, risk factors, and incompetent staff working in the drinking water supply systems cause unavoidable outbreaks of waterborne diseases in many countries worldwide (Bain et al. 2014). There is an increase in concern about the old drinking water supply systems that are more prone to the damage and intrusion of pathogenic microorganisms from various sources (Ashbolt 2015). Additionally, the improper industrial waste disposal and sanitation system increase the intrusion of unwanted contaminants to the water bodies that increases the rates of life-threatening waterborne infections/disease. There are several types of waterborne diseases (Fig. 1.1). Further, water bodies, quality of drinking water, and sanitation systems are also seriously affected by droughts and floods. For example, flooding can directly spread fecal contaminants which can lead to an outbreak of diseases like waterborne diarrhea, cholera, and hepatitis (Luby

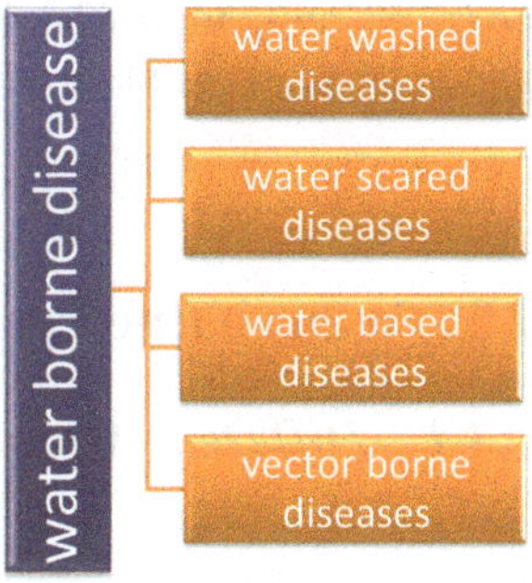

Fig. 1.1 Classification of waterborne diseases

et al. 2015). Usually, even in urban regions, cases of gastrointestinal diseases due to drinking water are not well quantified, reasons being surveillance insensitivities and lack of specific epidemiology studies (Levallois and Villanueva 2019).

A WHO report says that the most waterborne pathogens get introduced into the human body through drinking water and start infecting the gastrointestinal tract. In most underdeveloped and developing countries, 75% of the total diseases originated from water contamination and among them waterborne diarrhea and malaria being the topmost killers. Currently, approximately 1.1 billion people across the world are devoid of safe drinking water. Nearly 2.5 billion people live in unhygienic conditions with improper sanitation. Every year nearly 2.2 million deaths occur due to lack of water hygiene and sanitation-related issues with a yearly loss of more than 80 million years of life expectancy "Disability Adjusted Life Years (DALY's)" (WHO 2012).

Waterborne diarrhea accounts for 2195 deaths of children in 1 day which is much higher than deaths occurring due to AIDS, malaria, and measles collectively. It accounts for 1 in 9 child deaths worldwide, which makes it the 2nd largest killer of children worldwide. By the end of 2017, nearly 8% of children (under the age of 5 years) were killed by diarrhea (UNICEF 2019). In India, waterborne infections like diarrhea, cholera, typhoid, malaria, and hepatitis have accounted for 10,738 deaths in-between 2012–2017 (UNICEF 2019). Of this, acute diarrhea leads the chart followed by hepatitis, typhoid, and cholera. According to the World Health Organization (WHO) 2017, data, around the world, at least 219 million cases of malaria were recorded from 87 countries. More than four million people die every year because of malaria and the majority of them of children below the age of 5, mainly belonging to the countries of Africa. India accounts for 4% of the total global malaria cases. Further, soil-transmitted helminth infestations are one of the most common diseases worldwide which mostly affect the people living in slums and unhygienic conditions. Water and soil contamination occur by helminth egg present in human feces. Few major species are the hookworms (*Ancylostoma duodenale* and *Necator americanus*), the whipworm (*Trichuris trichiura*), and the roundworm (*Ascaris lumbricoides*). Nearly one-fourth populations of worldwide are infected with soil-transmitted helminth. Mostly, the people of tropics and subtropics are affected.

Over 267 million preschool-age children and over 568 million school-age children live in areas where these parasites are intensively transmitted (WHO 2017).

1.3 Waterborne Pathogens

1.3.1 Different Microorganisms Causing Diseases

There are over 500 waterborne pathogens of potential concern that have been identified by various regulatory authorities. They belong to various groups of microorganisms such as bacteria, viruses, fungi, and parasitic protozoa. WHO conducts repeated epidemiological studies in almost all the countries with the help of the local government and non-governmental organizations (NGOs) and estimates the risk factors, diseases, and organisms. The major waterborne enteric pathogens consist of Gram-negative bacteria such as *Escherichia coli, Salmonella* sp., *Shigella* sp., *Enterobacter* sp., and *Vibrio* sp., which are spread through the orofecal route, mainly in regions where sanitation facilities and sewage disposal are poor. This also leads to foodborne diseases through pathogenic bacterial species of *Campylobacter, Salmonella, Arcobacter, Helicobacter*, and *Yersinia* (Pandey et al. 2014; Ramirez-Castillo 2015). Some of the most important waterborne pathogens identified in WHO surveillance are listed in Table 1.1.

1.3.2 Major Bacterial Pathogens

Escherichia coli It is one of the major microorganisms found in the gut region of almost all warm-blooded animals causing no harm in normal conditions. However, strains like "verocytotoxin-producing *E. coli* (VTEC), Shiga toxin-producing *E. coli* (STEC), or enterohemorrhagic *E. coli* (EHEC)" are the pathogens that lead to severe diarrhea and they are released into the water bodies through feces. Transmission through contact and water is well established and recorded in many articles (Falup-Pecurariu et al. 2019; Pires et al. 2019).

Salmonella Salmonella generally does not multiply in the environment but they can survive in different environmental conditions. *Salmonella* infection (salmonellosis) is a general bacterial disease that affects the human gastrointestinal tract. These bacteria colonize animal and human guts and are shed through feces to the environment. They are generally transmitted through contaminated water and food. The other severe forms of *Salmonella* infections are typhoid (enteric fever) and paratyphoid (Mu et al. 2019; Zha et al. 2019).

Yersinia enterocolitica It is another pathogenic bacterium affecting humans through water contaminated with swine manure. Both *Salmonella* and *Yersinia* are

Table 1.1 Waterborne pathogens and their significance in water supplies

Pathogen	Health significance	Persistence in water supplies	Resistance to chlorine	Relative infectivity	Important animal source
Bacteria					
Burkholderia pseudomallei	H	May multiply	Low	Low	No
Campylobacter jejuni, C. coli	H	Moderate	Low	Moderate	Yes
Escherichia coli – Pathogenic	H	Moderate	Low	Low	Yes
E. coli – Enterohemorrhagic	H	Moderate	Low	High	Yes
Legionella sp.	H	May multiply	Low	Moderate	No
Non-tuberculous mycobacteria	L	May multiply	High	Low	No
Pseudomonas aeruginosa	Moderate	May multiply	Moderate	Low	No
Salmonella typhi	High	Moderate	Low	Low	No
Other salmonellae	High	May multiply	Low	Low	Yes
Shigella sp.	High	Short	Low	High	No
Vibrio cholerae	High	Short to long	Low	Low	No
Yersinia enterocolitica	High	Long	Low	Low	Yes
Viruses					
Adenoviruses	High	Long	Moderate	High	No
Enteroviruses	High	Long	Moderate	High	No
Astroviruses	High	Long	Moderate	High	No
Hepatitis viruses	High	Long	Moderate	High	No
Hepatitis E viruses	High	Long	Moderate	High	Potentially
Noroviruses	High	Long	Moderate	High	Potentially
Sapoviruses	High	Long	Moderate	High	Potentially
Rotavirus	High	Long	Moderate	High	No
Protozoa					
Acanthamoeba sp.	High	May multiply	Low	High	No
Cryptosporidium parvum	High	Long	High	High	Yes
Cyclosporacayetanensis	High	Long	High	High	No
Entamoebahistolytica	High	Moderate	High	High	No
Giardia intestinalis	High	Moderate	High	High	Yes
Naegleriafowleri	High	May multiply	Low	Moderate	No
Toxoplasma gondii	High	Long	High	High	Yes
Helminths					
Dracunculusmedinensis	High	Moderate	Moderate	High	No
Schistosoma sp.	High	Short	Moderate	High	Yes

Table Source: https://www.who.int/water_sanitation_health/gdwqrevision/watpathogens.pdf

often shed by apparently healthy animals and are capable of causing severe disease in humans (Strydom et al. 2019; Younis et al. 2019).

Vibrio cholera It causes cholera, which is a gastrointestinal infection of the intestine. The major symptoms are watery diarrhea, vomiting, and leg cramps. Rapid loss of body fluids leads to dehydration and shock in affected patients. Lack of medical attention can lead to death within hours of infection. *V. cholerae* enters the human body via contaminated drinking water or by eating contaminated seafood. The contamination of food or water occurs when untreated sewage is released into the supply water system, affecting the potable water or any food washed in that water. There are many reports on outbreaks and epidemics of cholera across the world and it is a major killer in areas with unhygienic conditions and improper sanitation (Hsueh and Waters 2019; Rijal et al. 2019). *Shigella* causes infectious disease Shigellosis. Patients with shigellosis have diarrhea, fever, and stomach cramps from the inception of the infection. Shigellosis does not last more than a week. Infected persons may remain asymptomatic, but can still act as carriers. The spread of *Shigella* can be reduced by proper handwashing techniques and maintaining proper hygiene (Carias et al. 2019; Williams and Berkley 2018).

Viruses are host-dependent and their transmission via water and other environmental routes is scientifically established. The viruses transmitted via water are shed with feces and infect through the mouth (Lodder et al. 2010). Each virus has a specific recipient cell in the host body, which helps in its replication and infection. These viruses can spread from the intestinal cell to other parts of the body and can cause serious complications other than diarrhea. To date, no major outbreak or epidemic has been recorded for water-transmitted viruses, but they have the potential of rapid spread and causing an epidemic. For example, adenoviruses, enteroviruses, astroviruses, hepatitis A and E viruses, noroviruses, sapoviruses, and rotavirus (Gall et al. 2015; Bonadonna and La Rosa 2019).

1.3.3 Protozoa

Numerous protozoan pathogens, like microsporidia, amoebae, ciliates, flagellates, and apicomplexans, originating in human or animal feces are frequently found in waterbodies worldwide. The major protozoa causing waterborne infections are species of microsporidia, the amoeba, *Entamoeba histolytica*, *Giardia duodenalis* (*G. lamblia*), *Toxoplasma gondii*, and *Cryptosporidium* species (Shanan et al. 2015). Most of these protozoa, in contrast to microorganisms and viruses that multiply rapidly, need a living host for its lifecycle and generally, the incubation period is 2–3 weeks long. Accurate diagnosis of these protozoans remains a concern. Higher molecular techniques may be employed for their identification, which will enable the scientists to conduct epidemiological research relating these organisms and to come up with a strategy for their prevention and management (Omarova et al. 2018).

1.3.4 Helminths

Major helminths causing waterborne infestations are nematodes which include ascarids, pinworms, hookworms, strongylids, angiostrongylids, capillarids and guinea worms, flukes which include schistosomes and liver flukes, tapeworms together with the beef, red meat and fish tapeworms as well as cystic and alveolar hydatid tapeworms. Improper sanitation and contaminated water facilitate the transmission of these helminths both among animals and humans (Strunz et al. 2014). The lifecycles of the maximum of these helminths are very established and they can be easily controlled by breaking these cycles. Recent studies suggest that waterborne helminth infestations have increased due to global warming and change in the global rain cycle (Ribas et al. 2017). Temperature and rainfall primarily determine the seasonal prevalence and human infestations cycle of these helminthes. However, they are more prevalent in rainy season or areas with heavy rainfall. Contaminated water used for irrigation, washing ingredients, and drinks are the major sources of the infestations (Ribas et al. 2017).

1.4 Waterborne Infections and Diseases

1.4.1 Transmission of Waterborne Diseases

The major sources of water-borne diseases are contaminated drinking water supply systems that contain several contaminants (Fig. 1.2). Unprotected water bodies or groundwater are often contaminated by human or animal feces or agricultural effluents. Piped water sources are also contaminated from a variety of pollutants (Hartemann et al. 1986). Moreover, due to lack of maintenance, supply pipes can consume chlorine residues, making the decontamination process ineffective. Broken pipes or crossed connections with sewage pipes also add contaminants into a water system, creating an efficient pathway for pathogen transmission. Further, during the power cuts, a negative pressure is created in the supply system which pulls the contaminants into the water system from the outer environment. Concealed networks (holes made in existing water pipes by persons) create openings in water pipes that also allow entry of contaminants into the water system. These microorganisms reach to human system through the orofecal route (Hatami 2013). There have been numerous reports on outbreaks of orofecal diseases such as typhoid or cholera. There are several other ways, where fecal material can reach to the human digestive system through contaminated food. The microorganisms in feces can spread rapidly and cause an outbreak in a very short period (Rooney et al. 2004; Pons et al. 2015; Mari et al. 2019). Some of the most prominent water-borne diseases are listed in Table 1.2.

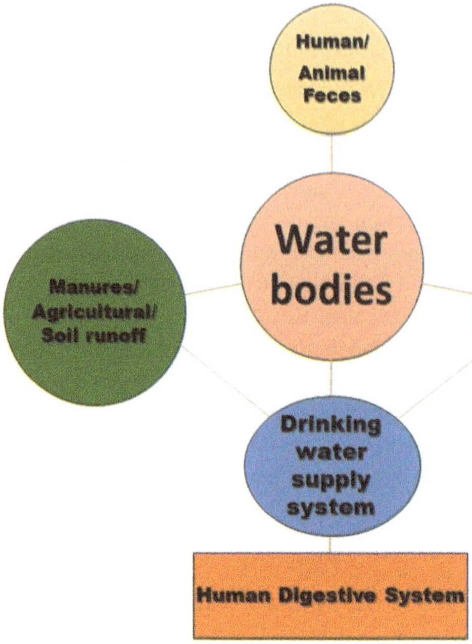

Fig. 1.2 Flowchart representing the source and transmission of water pollutants and pathogens

1.4.2 Biofilms in Water Distribution Systems

The existence of microbial biofilms is another major challenge encountered by authorities. Biofilms are microbial colonies with a layer of extracellular polymeric material protecting the biomass through environmental and shear tensions. Biofilms are a potential source of bacterial contamination and also affect the flavor and odor of the water and facilitate the corrosion of pipes (Chan et al. 2019). The biofilm contains bacterial colonies enveloped inside a polymeric slime that confirms adhesion to the particular pipe surface. Gradually, they become resistant to various disinfectants and antibiotics. Biofilms are found within pipes, tanks, business cooling towers, heat exchangers, hospital plumbing system, filtration media, etc. Because of their complex architecture, firm attachment, and their capacity to adapt to the environmental conditions, they may be very difficult to get rid of. These resistant bacteria could be an essential source of water contamination and, subsequently, leading food poisoning, diarrhea, and other outbreaks (Maes et al. 2019).

The newer technologies are mainly focusing on in situ differentiation of biomass and biofilms present in the water pipelines. In order to manage the biofilm formation in water distribution systems, their physiochemical structure must be correctly analyzed and appropriate chemical agents must be used to break the biofilm network. Present methods employed to control and remove biofilm include chlorine-chloramine disinfection, flushing involving water supply systems, and nutrient removal (Douterelo et al. 2016; Fish and Boxall 2018).

Table 1.2 Most common waterborne diseases

Disease	Major causative organism	Incubation period	Source/transmission	Symptoms	Geographical distribution
Traveler's Diarrhea Otherwise known as Aztec two-step, Delhi belly Hong Kong Dog, Montezuma revenge, Rangoon runs, Tourist trot	*Escherichia coli*	6–48 h	Drinking water with fecal contamination	Frequent loose motions, nausea, vomiting, abdominal cramps, bloating, fever, urgency, and malaise.	Very common under-developed countries of Asia, Africa, and South America
Cryptosporidiosis	*Cryptosporidium* sp.	2 days –3 weeks	Feces contaminated water, soil, food, and person to person contact	Frequent loose motions, nausea, vomiting, abdominal cramps, bloating, fever, loss of weight, and malaise.	Spread across the world Higher in tropical countries
Bacillary dysentery	*Shigella* sp.	1–3 days	Feces contaminated water, soil, food and person to person, sexual contact with an infected person	Diarrhea with blood and mucous, stomach ache, cramps, nausea, fever and vomiting	Spread across the world Mostly in Under-developed countries
Typhoid or enteric fever	*Salmonella* sp.	8–14 days	Feces contaminated water, soil, food and occasionally person to person contact	High fever, appetite loss, diarrhea, body aches, lethargy	All over the world, especially regions with poor sanitation facility
Cholera	*Vibrio cholerae*	12 h to 5 days	Contaminated water and seafoods	High fever, appetite loss, diarrhea, body aches, lethargy, abdominal cramps, dehydration, shocks	All over the world, especially regions with poor sanitation facility in coastal regions
Campylobacteriosis	*Campylobacter jejuni*	2–5 days	Ingestion of the organism via contaminated food or water, particularly raw or undercooked meats or raw milk, and from contact with pets, farm animals, and infected infants	Mild to severe bloody diarrhea, stomach ache, cramps, nausea and/or vomiting fever, headache, and muscle pain	All over the world, especially regions with poor sanitation

(continued)

Table 1.2 (continued)

Disease	Major causative organism	Incubation period	Source/transmission	Symptoms	Geographical distribution
Viral diarrhea	Enterovirus	2–3 days	Person to person via direct contact with virus shed from the gastrointestinal or upper respiratory tract	Diarrhea, running nose, body pain, vomiting, fever, mouth blisters	All over the world
Hepatitis	Hepatitis A & E virus	2–50 days	Feces contaminated water, soil, food, and person to person contact	Abdominal pain, nausea, vomiting clay-colored stool discharge, appetite loss, mild fever, dark urine, joint pain, jaundice, and severe itching	All over the world
Gastroenteritis	*Rotavirus*	2 days	Feces contaminated water, soil, food, and person to person contact	Diarrhea, black stool with pus and blood, high fever, body pain, vomiting, dehydration	All over the world, especially affects children
Dracunculiasis (Guinea worm disease)	*Dracunculusmedinensis* (nematode)	10–14 months	Consumption of larvae contaminated pond and stagnant water	Light fever, inflammation, nausea, vomiting, diarrhea, secondary bacterial infections, allergy, body rashes, dizziness, joint infection, and disorders	All over the world, especially regions with poor sanitation facility
Giardiasis	*Giardia intestinalis, G. lamblia or G. duodenalis*	1–14 days	Feces contaminated water, unhygienic sanitary facilities	Abdominal cramps, bloating, nausea and watery diarrhea	All over the world, especially regions with poor sanitation facility
Amoebic dysentery or amoebiasis	*Entamoeba histolytica*	2–4 weeks	Feces contaminated water, soil, food and person to person contact, sexual contact with an infected person	Fever, chills, diarrhea, abdominal cramps and diarrhea with blood and mucus	All across the world

1.5 Antibiotic-Resistant Bacteria in Water

1.5.1 Occurrence of Antibiotic-Resistant Bacteria in Wastewater

The spread of antibiotic-resistant bacteria (ARB) has been described as a global public health problem due to release of the antibiotics from hospital discharge. Every individual is at risk of getting infected with ARB bacteria, but some populations are at higher risk such as children, elderly people, and immune-compromised patients. If antibiotics become ineffective then most modern treatment procedures will become useless. In the last few decades, water bodies are detected harboring ARB and antibiotic-resistant genes (ARGs) in treated and untreated drinking water (Xi et al. 2009). ARB and ARGs are a serious medical concern as they are capable of transferring the ARGs to the human beings. As per the Center for Disease Control (CDC), US report, each year two million people get infected with ARB and nearly 25,000 people die of it. A large amount of antibiotics ARBs and ARGs have been repeatedly isolated from water bodies, animal wastes and manure, agricultural runoff, surface waters and sediments, municipal wastewaters, and even from treated drinking water (Chen et al. 2017). Hence, their presence becomes ubiquitous in the aquatic environment. The potential impact of the environmental presence ARB and ARGs in water bodies and water distribution systems is still not clear, but there is an increasing concern throughout the world regarding the rapid spread of ARB and ARGs in through water system (Fuentes et al. 2019).

Antibiotics and decontamination provide essential individual and public health protection, respectively, from water, sanitation, and health (WaSH)-related diseases ranging from dysentery to cholera. But there are growing concerns that even trace levels of antibiotics and intact DNA remnants from cell debris can promote antibiotic resistance by gene transfer to "downstream" bacterial populations, including pathogens (Khan et al. 2019). A recent European Commission review concluded that there is need of general safety measures to improve the effectiveness of wastewater treatment processes and efforts to control the use of antibiotics in animal husbandry and in human medical practices. It can reduce the spread of ARB (Pärnänen et al. 2019). A 2018 review by the Global Water Pathogen Project (GWPP) found that ARGs can be passed on and taken up by virtually all bacteria, so it is important to reduce general loads of bacteria as well as pathogens to reduce the potential spread of antimicrobial resistance. The GWPP chapter further concludes the known benefits of pathogen reduction through disinfection processes to balance the risk index (https://www.waterpathogens.org/).

1.5.2 Drinking Water and Antibiotic Resistance

For more than a century, microbial quality testing of drinking water has focused on the use of indicators for detecting waterborne pathogens. Adequate drinking water disinfection destroys or inactivates pathogenic and non-pathogenic bacteria, including ARB, most viruses, and degrades some antibiotics (Rath and Patra 2018). Compared to the wastewater, knowledge about the presence and public health significance of ARB, ARGs, and antibiotics in drinking water, including treatment plants, distribution systems, as well as the role of biofilms, is very less. Studies reported that ARB and ARG levels were higher in tap water than the source waters, indicating that there was regrowth of bacteria in drinking water during their distribution time. The investigators have hypothesized that chemical purification of water increases the chances of bacteria to be antibiotic-resistant and further leads to spread through drinking water supply systems. Such supply systems further turn into major reservoirs for these ABR bacteria as they form biofilms in the inner surface of the supply pipelines (Li and Gu 2019).

1.6 Intervening to Reduce Risk Caused Due to Water Contamination

Almost all the countries are working to formulate a standardized protocol to eliminate these organisms and to make a regulatory body to monitor the strict implementation of these protocols. However, only a few nations have been successful in achieving this target. Water Health Organization (WHO) has recommended risk-free preventive approaches to minimize the exposure of pathogenic microorganisms to the drinking water. Hence risk assessment becomes very essential in the formulation and implementation of treatment guidelines, supply, and safety of drinking water (Malik et al. 2012). Many such approaches are followed throughout the world, ranging from "risk scoring in sanitary inspections" and "risk matrices to quantitative microbial risk assessment (QMRA)" (Schijven et al. 2015). WHO needs to educate different countries regarding QMRA to facilitate its proper implementation for the supply of safe drinking water, its reuse, and even water recreation parks to effectively manage the risks linked with fecal pathogen-contaminated water. QMRA describes a four-step method for water disinfection, its reuse in domestic, industrial and recreational parks and lays out a clear guideline for controlling waterborne infections across the world (Ramírez-Castillo et al. 2015; Brouwer et al. 2018). Preventing diseases through reducing water contamination is a necessary intervention which could be made at every level of transmission, which includes:

- Prevention of contamination at the origin of water-basic level.
- Prevention of contamination while transmission using a proper filter and treating while it gets transported to the household.
- Treatment (chlorination) of storage tanks at home and prevention of further contamination during storage, such as from flies, wastes, and rainwater.
- Preventing while transportation of water through household pipes by using proper filters and purifiers.
- Boiling, purification, and filtration before consumption.
- Maintaining proper hygiene is also one of the most important components of prevention from the further transmission of diseases.
- More studies into the concentrations and removal of ARB and ARG are needed. When it comes to understanding the presence and human health significance of antimicrobial resistance in waterborne pathogens in environmental and treated waters, there are far more questions and research needs than there are answers. The GWPP and European Commission reviews both call for expanded science-based information and risk assessments into the relationships between aquatic antibiotic concentrations, the spread of antimicrobial resistance, and public health protection, all of which seems prudent.

1.7 Actions Taken by WHO and Other Organizations

According to the "Millennium Development Goal" target for drinking water, there was a plan of supplying 50% of the world population with safe drinking water. Till 1990, 76% of the total world population had access to safe drinking water, which increased to 89% in 2012. However, this distribution across the world is highly uneven (Fig. 1.3). By 2012, only 56% of the total world population had the highest level of access, supplied water facilities in their home. In 2004, WHO initiated "Water Safety Plans (WSPs)" that presents the guidelines for estimating the quality

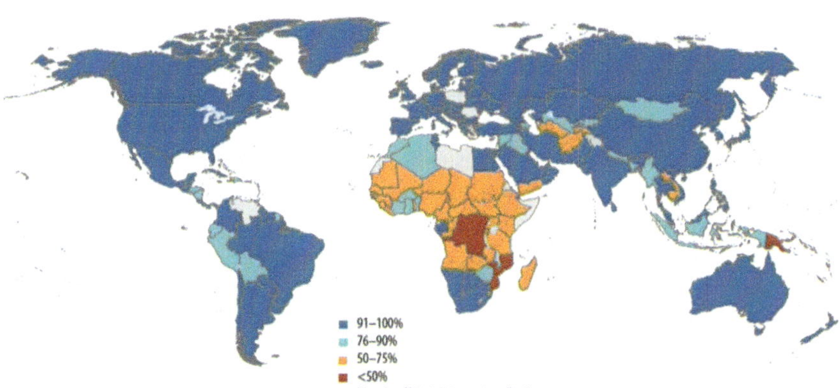

Fig. 1.3 Global coverage of improved drinking water 2012. (Source: https://apps.who.int/iris/bitstream/handle/10665/150112/9789241564823_eng.pdf?sequence=1)

of potable water. WSP was the most effective method for the evaluation and distribution of safe drinking water. But WSPs still require major support from the local governments for its effective implementation, monitoring, and data collection. WHO provides guidance and support to regulators and water suppliers with the help of local NGOs to effectively implement WSP, risk assessment, and monitoring the prevalence of waterborne infections/diseases. Currently, more than 50 countries have a national strategy established to scale up WSP implementation (WHO 2012). Figure 1.3 shows global coverage of improved drinking water. In 2015, WHO released a worldwide plan on antimicrobial resistance. ARBs and ARGs are a group of unusual and major contaminants of fecal wastes and water. The primary objective was to understand the role of water in transmission and increase of ARBs and ARGs and to take necessary steps to eliminate it. The plan was to highlight WASH and antimicrobial resistance and propose new directions for their threat evaluation control and management. Proper sanitation, hygiene, and safe water distribution in households, educational institutions, and health centers remain a priority; however, ensuring a check on the spread of ARBs and ARGs is also a concern for the regulatory authorities (Chen et al. 2017).

1.8 Conclusions

Microbial contamination of drinking water is one of the significant threats in developing countries and even in the most developed countries. Recently, it has been evident that there is an association between disinfection by-products and adverse health effects. It is essential to treat upcoming threats of diseases from contaminated water at the basic supply level, through proper sanitation and scientific waste management. Well-designed epidemiological data and studies are also needed so that microbes responsible for contamination can be identified. A major change is required in the current system of risk analysis and management of water safety and waterborne diseases. The existing water safety rules have lot of drawbacks, which make the implementation of newer WSPs difficult and prevents to assess the risk factors and to conduct an epidemiological study. Recently, the approach is practiced globally using principles from the food industry's Hazard Analysis Critical Control Point, described by the WHO as WSPs. Scientific and correct disposal of household, industrial, agricultural, and hospital wastes along with a proper sanitation system can increase the water safety levels, consequently decreasing the outbreak of waterborne diseases. By following the rules and regulations as well as by self-control, the risk from water contaminants can be reduced up to certain levels.

Conflict of Interests None

Funding None

References

Allaire M, Wu H, Lall U (2018) National trends in drinking water quality violations. Proc Natl Acad Sci U S A 115(9):2078–2083

Ashbolt NJ (2015) Microbial contamination of drinking water and human health from community water systems. Curr Environ Health Rep 2(1):95–106

Bain R, Cronk R, Wright J, Yang H, Slaymaker T, Bartram J (2014) Fecal contamination of drinking-water in low- and middle-income countries: a systematic review and meta-analysis. PLoS Med 11(5):e1001644. https://doi.org/10.1371/journal.pmed.1001644

Barrett JR (2014) Chemical contaminants in drinking water: where do we go from here? Environ Health Perspect 122(3):A80. https://doi.org/10.1289/ehp.122-A80

Bi W, Weng B, Yuan Z, Ye M, Zhang C, Zhao Y et al (2018) Evolution characteristics of surface water quality due to climate change and LUCC under scenario simulations: a case study in the Luanhe River basin. Int J Environ Res Public Health 15(8):E1724. https://doi.org/10.3390/ijerph15081724

Bonadonna L, La Rosa G (2019) A review and update on waterborne viral diseases associated with swimming pools. Int J Environ Res Public Health 16(2):166. https://doi.org/10.3390/ijerph16020166

Brouwer AF, Masters NB, Eisenberg J (2018) Quantitative microbial risk assessment and infectious disease transmission modeling of waterborne enteric pathogens. Curr Environ Health Rep 5 (2):293–304. https://doi.org/10.1007/s40572-018-0196-x

Carias C, Undurraga EA, Hurd J, Kahn EB, Meltzer MI, Bowen A (2019) Evaluation of the impact of shigellosis exclusion policies in childcare settings upon detection of a shigellosis outbreak. BMC Infect Dis 19:172. https://doi.org/10.1186/s12879-019-3796-7

Chan S, Pullerits K, Keucken A, Persson KM, Paul CJ, Rådström P (2019) Bacterial release from pipe biofilm in a full-scale drinking water distribution system. NPJ Biofilms Microbiol 5:9. https://doi.org/10.1038/s41522-019-0082-9

Chen Z, Yu D, He S, Ye H, Zhang L, Wen Y et al (2017) Prevalence of antibiotic-resistant *Escherichia coli* in drinking water sources in Hangzhou City. Front Microbiol 8:1133. https://doi.org/10.3389/fmicb.2017.01133

Douterelo I, Husband S, Loza V, Boxall J (2016) Dynamics of biofilm regrowth in drinking water distribution systems. Appl Environ Microbiol 82(14):4155–4168

Edokpayi JN, Rogawski ET, Kahler DM, Hill CL, Reynolds C, Nyathi et al (2018) Challenges to sustainable safe drinking water: a case study of water quality and use across seasons in rural communities in Limpopo Province, South Africa. Water 10(2):159. https://doi.org/10.3390/w10020159

Falup-Pecurariu O, Lixandru RI, Cojocaru E, Csutak K, Monescu V, Muhsen et al (2019) Shiga toxin producing *Escherichia coli*-associated diarrhea and hemolytic uremic syndrome in young children in Romania. Gut Pathog 11:46. https://doi.org/10.1186/s13099-019-0327-4

Fawell J, Nieuwenhuijsen MJ (2003) Contaminants in drinking water. Br Med Bull 68:199–208

Fish KE, Boxall JB (2018) Biofilm microbiome (re)growth dynamics in drinking water distribution systems are impacted by chlorine concentration. Front Microbiol 9:2519. https://doi.org/10.3389/fmicb.2018.02519

Fuentes MD, Gutierrez S, Sahagun D, Gomez J, Mendoza J et al (2019) Assessment of antibiotic levels, multi-drug resistant bacteria and genetic biomarkers in the waters of the Rio Grande River between the United States-Mexico Border. J Health Pollut 9(23):190912. https://doi.org/10.5696/2156-9614-9.23.190912

Gall AM, Mariñas BJ, Lu Y, Shisler JL (2015) Waterborne viruses: a barrier to safe drinking water. PLoS Pathog 11(6):e1004867. https://doi.org/10.1371/journal.ppat.1004867

Hartemann P, Newman R, Foliguet JM (1986) Epidemiology of infectious diseases transmitted by drinking water in developed countries. Epidemiol Public Health 34(1):59–68

Hatami H (2013) Importance of water and water-borne diseases: on the occasion of the world water day (march 22, 2013). Int J Prev Med 4(3):243–245

Hsueh BY, Waters CM (2019) Combating cholera. F1000 Res 8:F1000 Faculty Rev-589. https://doi.org/10.12688/f1000research.18093.1
https://apps.who.int/iris/bitstream/handle/10665/150112/9789241564823_eng.pdf?sequence=1. Accessed on 20th Dec 2019
https://data.unicef.org/topic/child-health/diarrhoeal-disease. Accessed on 27th Dec 2019
https://www.waterpathogens.org/. Accessed on 20th Dec 2019
https://www.who.int/mediacentre/news/releases/2012/drinking_water_20120306/en/. Accessed on 27th Dec 2019
https://www.who.int/news-room/fact-sheets/detail/malaria 2017. Accessed on 23th Dec 2019
https://www.who.int/water_sanitation_health/publications/qmra/en/. Accessed on 27th Dec 2019
Khan FA, Söderquist B, Jass J (2019) Prevalence and diversity of antibiotic resistance genes in Swedish aquatic environments impacted by household and hospital wastewater. Front Microbiol 10:688. https://doi.org/10.3389/fmicb.2019.00688
Levallois P, Villanueva CM (2019) Drinking water quality and human health: an editorial. Int J Environ Res Public Health 16(4):631. https://doi.org/10.3390/ijerph16040631
Li D, Gu AZ (2019) Antimicrobial resistance: a new threat from disinfection byproducts and disinfection of drinking water? Curr Opin Environ Sci Health 7(1):83–91
Lodder WJ, van den Berg HH, Rutjes SA, de RodaHusman AM (2010) Presence of enteric viruses in source waters for drinking water production in The Netherlands. Appl Environ Microbiol 76 (17):5965–5971
Luby SP, Halder AK, Huda TM, Unicomb L, Islam MS, Arnold BF, Johnston RB (2015) Microbiological contamination of drinking water associated with subsequent child diarrhea. Am J Trop Med Hyg 93(5):904–911
Maes S, Vackier T, Nguyen Huu S, Heyndrickx M, Steenackers H, Sampers I et al (2019) Occurrence and characterization of biofilms in drinking water systems of broiler houses. BMC Microbiol 19(1):77. https://doi.org/10.1186/s12866-019-1451-5
Malik A, Yasar A, Tabinda A, Abubakar M (2012) Water-borne diseases, cost of illness and willingness to pay for diseases interventions in rural communities of developing countries. Iran J Public Health 41(6):39–49
Mari L, Casagrandi R, Bertuzzo E, Rinaldo A, Gatto M (2019) Conditions for transient epidemics of waterborne disease in spatially explicit systems. R Soc Open Sci 6(5):181517. https://doi.org/10.1098/rsos.181517
Mu H, Bai H, Sun F, Liu Y, Lu C et al (2019) Pathogen-targeting glycovesicles as a therapy for salmonellosis. Nat Commun 10(1):4039. https://doi.org/10.1038/s41467-019-12066-z
Omarova A, Tussupova K, Berndtsson R, Kalishev M, Sharapatova K (2018) Protozoan parasites in drinking water: a system approach for improved water, sanitation and hygiene in developing countries. Int J Environ Res Public Health 15(3):495. https://doi.org/10.3390/ijerph15030495
Pandey PK, Kass PH, Soupir ML, Biswas S, Singh VP (2014) Contamination of water resources by pathogenic bacteria. AMB Express 4:51. https://doi.org/10.1186/s13568-014-0051-x
Pärnänen K, Narciso-da-Rocha C, Kneis D, Berendonk TU, Cacace D et al (2019) Antibiotic resistance in European wastewater treatment plants mirrors the pattern of clinical antibiotic resistance prevalence. Sci Adv 5(3):eaau9124. https://doi.org/10.1126/sciadv.aau9124
Pires SM, Majowicz S, Gill A, Devleesschauwer B (2019) Global and regional source attribution of Shiga toxin-producing *Escherichia coli* infections using analysis of outbreak surveillance data. Epidemiol Infect 147:e236. https://doi.org/10.1017/S095026881900116X
Pons W, Young I, Truong J, Jones-Bitton A, McEwen S, Pintar K, Papadopoulos A (2015) A systematic review of waterborne disease outbreaks associated with small non-community drinking water systems in Canada and the United States. PLoS One 10(10):e0141646. https://doi.org/10.1371/journal.pone.0141646
Ramírez-Castillo FY, Loera-Muro A, Jacques M, Garneau P, Avelar-GonzÃlez FJ, Harel J, Guerrero-Barrera AL (2015 May 21) Waterborne pathogens: detection methods and challenges. Pathogens 4(2):307–334.

Rath S, Patra BK (2018) Dispersal of antibiotic resistant Bacteria into aquatic environment – an overview. J Water Pollut Control 1(2):1–3

Ribas A, Jollivet C, Morand S, Thongmalayvong B, Somphavong S, Siew C et al (2017) Intestinal parasitic infections and environmental water contamination in a Rural Village of northern Lao PDR. Korean J Parasitol 55(5):523–532. https://doi.org/10.3347/kjp.2017.55.5.523

Rijal N, Acharya J, Adhikari S, Upadhaya BP, Shakya G, Kansakar P, Rajbhandari P (2019) Changing epidemiology and antimicrobial resistance in *Vibrio cholerae*: AMR surveillance findings (2006-2016) from Nepal. BMC Infect Dis 19(1):801. https://doi.org/10.1186/s12879-019-4432-4432

Rooney RM, Bartram JK, Cramer EH, Mantha S, Nichols G, Suraj R, Todd EC (2004) A review of outbreaks of waterborne disease associated with ships: evidence for risk management. Public Health Rep (Washington, DC: 1974) 119(4):435–442

Schijven J, Derx J, de RodaHusman AM, Blaschke AP, Farnleitner AH (2015) QMRA catch: microbial quality simulation of water resources including infection risk assessment. J Environ Qual 44(5):1491–1502. https://doi.org/10.2134/jeq2015.01.0048

Shanan S, Abd H, Bayoumi M, Saeed A, Sandström G (2015) Prevalence of protozoa species in drinking and environmental water sources in Sudan. Biomed Res Int 2015:345619. https://doi.org/10.1155/2015/345619

Strunz EC, Addiss DG, Stocks ME, Ogden S, Utzinger J, Freeman MC (2014) Water, sanitation, hygiene, and soil-transmitted helminth infection: a systematic review and meta-analysis. PLoS Med 11(3):e1001620. https://doi.org/10.1371/journal.pmed.1001620

Strydom H, Wang J, Paine S, Dyet K, Cullen K, Wright J (2019) Evaluating sub-typing methods for pathogenic *Yersinia enterocolitica* to support outbreak investigations in New Zealand. Epidemiol Infect 147:e186. https://doi.org/10.1017/S0950268819000773

UNICEF (2019) https://data.unicef.org/topic/child-health/diarrhoeal-disease/

WHO (2012) Consultation on the development of a strategy on water quality and health. UNICEF/WHO, GENEVA/NEW YORK. https://www.who.int/mediacentre/news/releases/2012/drinking_water_20120306/en/

WHO (2017) https://www.who.int/gho/malaria/en/

Williams P, Berkley JA (2018) Guidelines for the treatment of dysentery (shigellosis): a systematic review of the evidence. Paediatr Int Child Health 38(sup1):S50–S65. https://doi.org/10.1080/20469047.2017.1409454

Xi C, Zhang Y, Marrs CF, Ye W, Simon C, Foxman B, Nriagu J (2009) Prevalence of antibiotic resistance in drinking water treatment and distribution systems. Appl Environ Microbiol 75(17):5714–5718

Younis G, Mady M, Awad A (2019) *Yersinia enterocolitica*: prevalence, virulence, and antimicrobial resistance from retail and processed meat in Egypt. Vet World 12(7):1078–1084

Zha L, Garrett S, Sun J (2019) *Salmonella* infection in chronic inflammation and gastrointestinal cancer. Diseases (Basel, Switz) 7(1):28. https://doi.org/10.3390/diseases7010028

Chapter 2
Landscape Perspective to River Pollution: A Case Study of Bentong River, Malaysia

Nor Rohaizah Jamil and Zakariya Nafi Shehab

2.1 Introduction

Up to March 2012, the global population surpassed 7 billion people, a double of what it was less than 50 years ago (United States Census Bureau 2012). United Nations (UN 2018) estimated that 55% of the people live in urbanized cities, and it was predicted that by 2030, 68% of people will live in urbanized zones, and even reach 75% by 2050. If this trend of population growth and urban expansion continues at a rapid pace, the risks and dangers posed to the availability of natural resources and the quality of life will greatly escalate (McGrane 2016).

Water perhaps is the most threatened and vulnerable medium by this uncontrolled growth. With the expansion of population growth, now water is becoming a scarce commodity. By 2030, according to a recent United Nations report (UN 2018), more than 700 million people may be forced to migrate to urban areas, due to intense water shortage. The quality of water plays an important role, as it affects the health of soil, crops, animals, human beings, and all surrounding environment (McGrane 2016).

Surface water quality is dependent upon natural processes as well as anthropogenic activities, either collectively or separately. Particularly, the human interference has deteriorated the water quality through point and non-point pollution sources in both urban and rural areas (Khatri and Tyagi 2015). For their benefits, they have utilized the natural resources so badly that affect the setting of landscape and other environmental components. With the landscape changes, the sediment load, nitrogen, phosphorus, and organic matter of streams and rivers get distributed (Rosso and Cirelli 2013). The human beings have utilized the natural resources including landscape for the need of their houses, food, and for the development of other

N. R. Jamil (✉) · Z. N. Shehab
Department of Environment, Faculty of Forestry and Environment, Universiti Putra Malaysia, Serdang, Malaysia
e-mail: norrohaizah@upm.edu.my

A. Singh et al. (eds.), *Water Pollution and Management Practices*,
https://doi.org/10.1007/978-981-15-8358-2_2

infrastructures. These practices increase the impervious surface such as roads, sidewalks, rooftops and parking lots and that lead to more transportation of the pollutants from landscape into the water bodies (Wilson and Weng 2010). Along with this, agricultural runoff also discharges extra amount of nutrients, pesticides, and herbicides into thc nearest water bodies. These changes lead to eutrophication of water bodies and create suitable conditions for algal bloom and phytoplankton biomass, that consequently impair odor and taste of water (Tsegaye et al. 2006; Zhang et al. 2013). The higher content of suspended solids increases the turbidity of water, which lessen penetration of light and affects the growth of fish and other aquatic animals (Giri 2013).

Land use patterns in certain areas influence river water systems through non-point pollutants, which pose a significant risk to water quality, particularly in residential areas (Jia et al. 2013). Furthermore, natural landscapes have been fragmented and transformed into impervious areas, which often lead to a surge in surface runoff. These continuous transformations alter the hydrological processes and increase the load of pollutants to the river bodies (Barbosa et al. 2012). Water quality variables have been strongly related to the configuration and proportion of land uses inside the multiple areas, due to variations in the types of pollutants released (Schoonover and Lockaby 2006).

Similarly, the relation between landscape patterns and water quality deterioration is considered a pattern-process relationship, in which energy, nutrients, and material in a landscape are influenced by permeability, connectivity, aggregation, and configuration (Mitchell et al. 2013). Several studies have highlighted the impact of landscape patterns on water chemistry and quality variation, and the significance of landscape features on stream health (Beckert et al. 2011; Zhang et al. 2013; Griffith 2002; Rothwell et al. 2010). Griffith (2002) even mentioned that the association between landscapes and water quality is area-specific and non-stationary and above all complex; still, there has been a notable correlation among water quality variables and landscape metrics within watersheds (Buck et al. 2004). The spatial structure of landscapes in particular is essential for assessing the connection between landscape and water quality at different scales (Alberti et al. 2007). Furthermore, landscape composition and configuration might be a key factor that has some impacts on hydrological processes, chemical cycles, energy flows, and natural habitats (Mitchell et al. 2013).

2.2 Impact of Urban and Agricultural Runoff on River Water

Urban areas occupy only a small percentage of the earth's surface. Still, they are home to well over half of the world's population (UN 2018). The surge in urban growth prompted scientists to study the impacts of landscape change on hydrological dynamics. Researchers focused on evaluating catchment response to urban development in an attempt to identify and quantify the impacts of this development on surrounding water or downstream water (Rodriguez et al. 2013). However, the dynamics of urban development strategies vary throughout the world. In developing countries, urban growth takes place along large-scale areas which include entire cities built in very short time periods such as new projects to develop cities in China. Contrarily, urban growth in developed countries is often locally scaled, with construction of single house estates or individual buildings in a spatial vicinity with the help of modern monitoring technologies that offer perspective regarding the surrounding environment (Blocken et al. 2013). In spite of that, having a comprehensive metrics for evaluating urban expansion is rather tricky; however, several studies tackling this issue have considered several approaches such as population density, growth rate, spatial geometry, total impervious areas, accessibility, and aesthetics (Bowyer 2015).

Urban landscape has a noticeable influence on hydrological and meteorological processes. Urban expansion increases artificial drainage systems that alter significantly the amount and route of runoff generated from certain vicinity (Dams et al. 2013). The increase in particulate matter and artificial thermal characteristics from urban areas affects the rainfall pattern and may lead to convective summer thunderstorms and downwind precipitation (Jin and Shepherd 2005). Moreover, sewage treatment facilities and sewerage conveyance system discharge huge quantities of pollutants into the water bodies that alter the natural composition of water and change the dynamic of the aquatic ecosystem (Leung and Jiao 2006).

The implications of urban expansion on water quality have been the focus of several researches, in order to find out some mitigation strategies to reduce the risk of water quality of rivers and streams. Effluents from point sources and runoff from non-point sources increase the pollution load in rivers with diverse contaminants including heavy metals, major nutrients, and organic matter. Recently, efforts have shifted to address more urgent and crucial issues such as synthetic chemicals, nano pollutants in rivers, and the fate and transport of pollutants in general (McGrane 2016). In developed countries, steps are being taken to treat storm water and runoff as a renewable resource. Sustainable management strategies are being implemented to make use of this resource and at the same time reduce its dire effects on river water, which would help to restore and enhance river water quality (Blocken et al. 2013). In Canada, urban ponds are being used increasingly to retain storm water runoff to protect water systems and reduce downstream flooding. This has become a prominent feature in many parts of Canada including Ontario and Toronto, which has about 500 ponds within its realm. Moreover, samples taken from some of these ponds indicated that they could improve water quality downstream by retaining pollutants; also the in-pond vegetation can help in reducing nutrient content (Drake et al. 2016).

Agricultural practices like excessive use of fertilizers to obtain high product yield, use of herbicides and pesticides, and impractical irrigation practices also lead to surface water contamination. These applications increase the flux of sediment, nutrients, pathogens, and various chemical compounds, which alter the composition of river water and affect the aquatic ecosystem.

Overall, the pressure of agriculture on river water quality has slightly eased since early 1990s due to the reduction in fertilizers and pesticides use in light of the environmental movement. However, despite this minor improvement, pollutions from agricultural practices remain significant and substantial. In comparison to the point source the agricultural runoff (non-point pollution) contributes more in the degradation of water quality and that can be controlled only up to certain stage (Giri and Qiu 2016). Fertilizers having high concentration of nitrogen, phosphorous, and other compounds are constantly added to the soil without following any recommended dose. Therefore, the surplus amount of these compounds often remains in the soil get leached into the ground as well washed off to the water bodies. The enhanced concentrations of nutrients in river water decrease the oxygen level and lead to toxic conditions by producing more algal boom (Evans et al. 2019).

Furthermore, in most developing countries, the decline in forest and woodlands is attributed largely to land transformation, especially the expansion of agricultural fields for crop production. Various studies also alluded that these structural changes are influenced by political and economic policies (FAO 2018).

Unfortunately, these perpetual transformations in natural landscapes occasionally invoke unexpected problems that sometimes extent to the other regions also. For example, the algae bloom phenomenon that is attributed to *Sargassum* seaweed on the Caribbean coast of Mexico can spread to the entire tropical Atlantic Ocean. This area is named as Great Atlantic *Sargassum* Belt (GASB) known for largest macroalgal bloom in the world. The major reasons for the propagation of algal bloom are deforestation and use of fertilizers in the Amazon forest that steadily increased nutrient runoff. The aggregations of macroalgae in coastal waters deplete oxygen and affect the other aquatic organism. At excess level, they may produce rotten egg smells that consequently affect the number of tourists as well as to the local fisheries. So, the landscape alteration can affect the whole water systems (Wang et al. 2019).

2.3 Landscape Influence on Water Quality: A Case Study in Bentong River

Landscape patterns and ecological processes interact with one another. By establishing relationships between the two, we can take further steps to comprehend the influence of landscape on water quality and assess more information that strengthen our grasp of landscape ecology. Hence, it has become very imperative to identify and describe the relationships among water parameters and land use patterns and landscape metrics in order to implement sustainable water management strategies (Crosa et al. 2006). Lately, studies have started to focus on spatial configuration of land uses and on different landscape metrics in an attempt to further understand the association and interconnection between land use patterns and water characteristics of watersheds. Incorporating advanced statistical analysis and spatial analysis techniques have made a notable progress in these studies (Ierodiaconou et al. 2005).

Therefore, a case study of Bentong River basin in Pahang state, Malaysia has been analyzed to illustrate the effects and significance of spatial land uses and landscape patterns on water quality. Normally, it is somewhat difficult to assert and clarify these relationships in rivers, due to the continuous inputs from upstream. Nevertheless, this study utilized the basic concept that surface settings and spatial composition will categorically change water quality in adjacent water systems.

Bentong River is constituted of multiple tributaries with Kelau River, Semantan River, Benus River, and Telemung River representing the major ones. Bentong catchment is located in the western part of Pahang State, Malaysia with an estimated area of 3401 km^2. The average annual total rainfall received in Bentong catchment area is approximately 1855 mm, the mean rainfall distribution throughout the year, according to the previous rainfall records of the past 35 years (Malaysian Meteorological Department 2019). The river basin contains various agricultural and industrial areas with a vast dispersion of forests. Wastewaters are discharged into the middle and lower reaches of the river. Large-scale development projects in the entire state, including Bentong, have resulted in the clearing of hundreds of square miles of land for oil palm and rubber plantations, and the resettling of several hundred thousand people in new villages under the federal agencies and institutions like the Federal Land Development Authority (FELDA), Federal Land Consolidation and Rehabilitation Authority (FELCRA), and Rubber Industry Smallholders Development Authority (RISDA) has also affected the quality of river basin.

With the help of Fig. 2.1, a Digital elevation model (12.5 resolution) data and maps interpreted from 2011 ASFs (Vertex) images used to delineate and extract the stream network within the Bentong watershed and land use and cover composition map are observed by using ArcGIS 10.2 techniques (Shehab et al. 2020). Depending upon the types of land uses, seven different zones are categorized and further reclassified into four major categories based upon their significance level. These major categories are: (1) forests; (2) agricultural land; (3) residential areas; and (4) facilities and industrial areas (Fig. 2.1B). Shehab et al. (2020) have divided the study area into seven zones to better understand and quantify the impacts of certain land types or landscape patterns on water quality and to compare their relationships with water characteristics. Composition percentage of land use types for seven different zones is illustrated in (Fig. 2.2. The classification is largely based on stream network and sampling sites.

Furthermore, several landscape metrics are selected to determine the impact of landscape structure on water quality. FRAGSTATS 4.1 software, which is a spatial pattern analysis program for quantifying the structure of landscapes, is used to calculate these metrics using the land use data in the study area. The applied metrics are patch density (PD), edge density (ED), Shannon's diversity index (SHDI), Contagion (CONTAG), largest patch index (LPI), cohesion index (COHE), shape index (SHMN), aggregation index (AI), and mean Euclidean nearest neighbor index (ENNMN). These metrics have been established based on landscape, class, and patch scale, all of which differ in definition and measurement. Here, the indices will

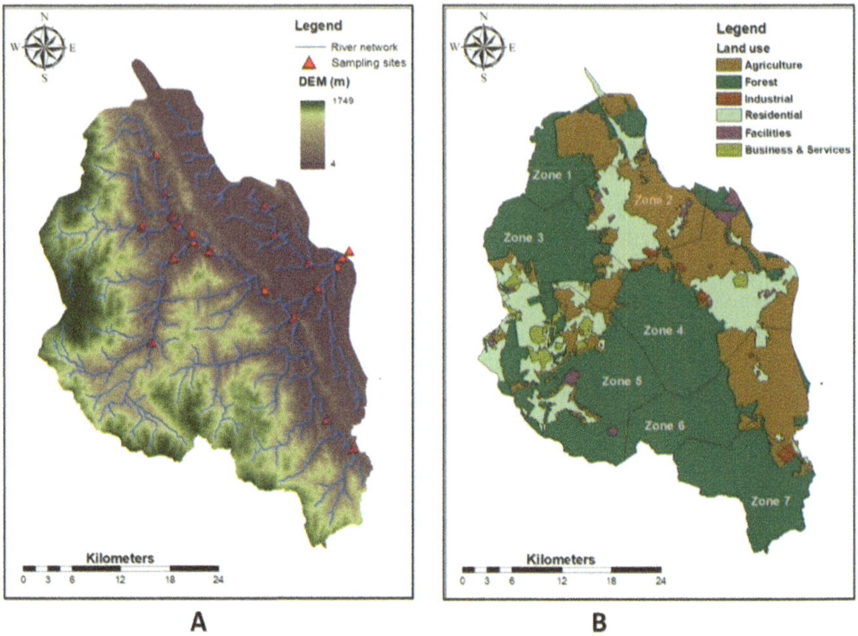

Fig. 2.1 (**a**) Water sampling sites in Bentong basin, the digital elevation model (DEM) and the drainage system, and (**b**) land use patterns and basin zones

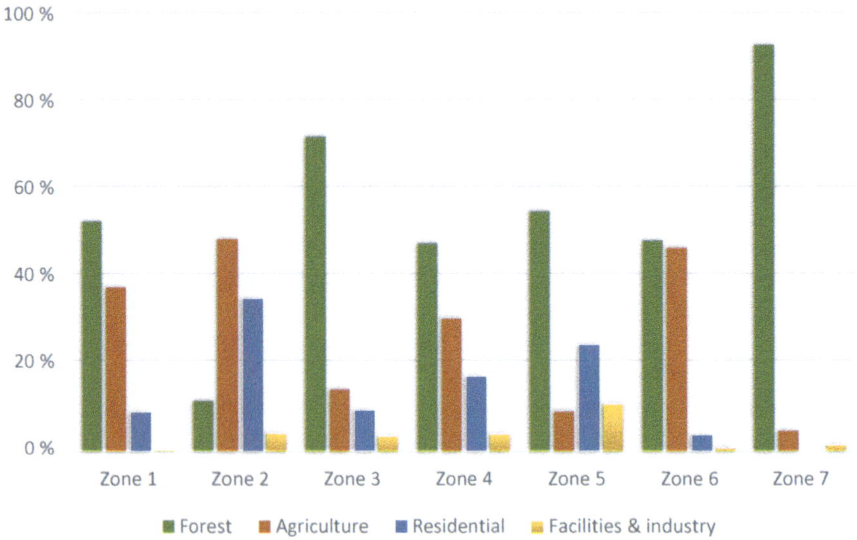

Fig. 2.2 Composition percentage of land use types for seven different zones in Bentong basin

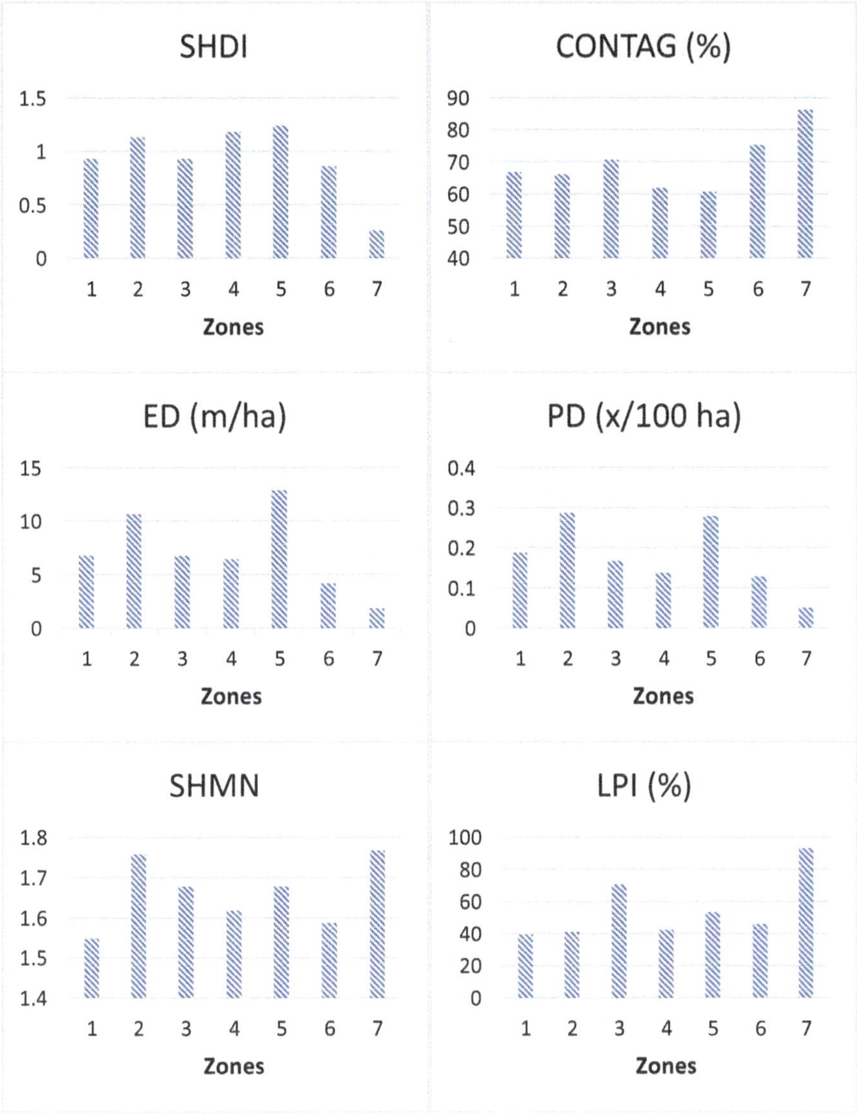

Fig. 2.3 Landscape pattern metrics in different zones in Bentong basin

separately quantify the spatial distribution and pattern of a single patch or land cover type in the landscape, which is more suitable to characterize each zone independently (Shen et al. 2014). It is discussed in detail later on (Fig. 2.3).

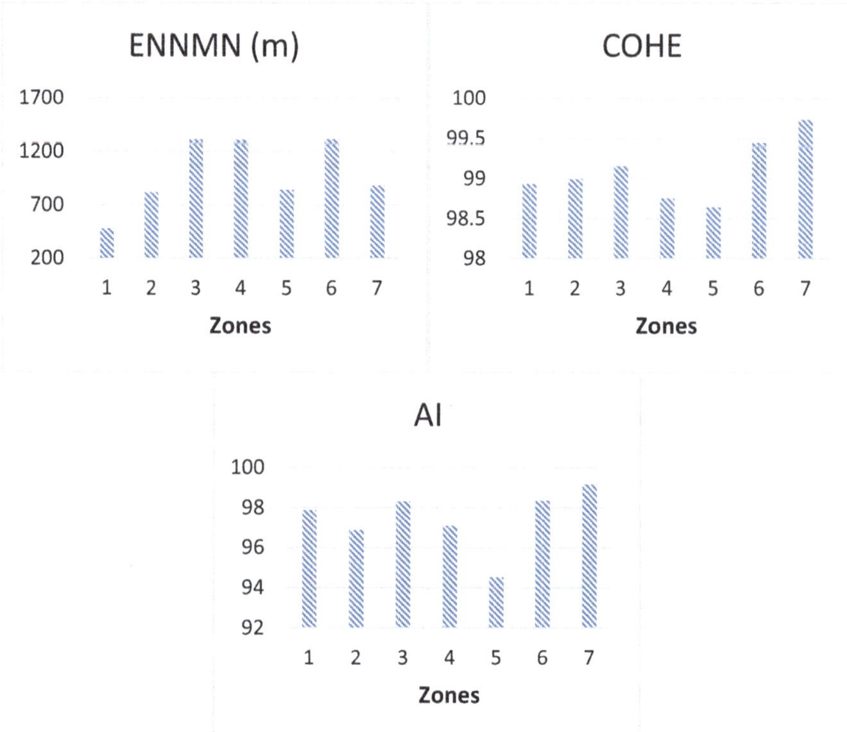

Fig. 2.3 (continued)

2.4 Understanding the Relationship Between Landscape Patterns and Water Characteristics

Shehab et al. (2020) have analyzed the Pearson correlation that showed significant relationships between water parameters and land use types in both the normal and wet season (Table 2.1). During normal season, forest areas of various zones classified in this study area are found to be positively correlated with DO and pH and negatively correlated with *E. coli*, sulfate, E.C, phosphate, and ammoniacal nitrogen. Agricultural lands are positively related to nutrients, COD, E.C, and temperature, and negatively related to DO and pH values. Residential lands showed positive correlation with BOD, TSS, E.C, *E. coli*, and nutrient variables, and are negatively correlated with pH and dissolved oxygen. Facilities and industrial lands showed negative relation with DO, but positive correlation with phosphate, E.C, COD, TSS, TDS, and ammoniacal nitrogen (Shehab et al. 2020).

For the wet season, forests are also positively correlated with DO and pH, but showed a negative correlation with BOD, COD, E.C, TDS, *E. coli*, and the nutrient variables. In contrast, agricultural lands are positively correlated with most of the parameters except with DO and pH values. Residential lands showed positive

Table 2.1 Pearson correlation coefficients between land uses and water quality variables in both seasons

	Forest	Agriculture	Residential	Facilities & industrial
Normal season				
pH	0.147*	−0.498*	−0.193*	−0.207
Temperature	0.197	0.602*	0.359*	0.172
DO	0.453*	−0.506*	−0.196*	−0.189*
BOD	0.145	−0.004	0.390*	0.302*
COD	0.030	0.160*	0.047	0.096*
TSS	0.232	0.315	0.423*	0.391*
TDS	−0.201	0.173	0.235	0.354
E.C	−0.087*	0.359*	0.482*	0.321
Turbidity	0.407	0.402	0.235	0.237*
E. coli	−0.337*	0.90	0.342*	0.183
Ammoniacal nitrogen	−0.385*	0.333*	0.145*	0.223*
Nitrate	−0.219	0.200	0.082	0.091
Phosphate	−0.324*	0.130*	0.130*	0.276*
Sulfate	−0.208*	0.411*	0.271*	0.091
Wet season				
pH	0.115*	−0.417*	−0.134	0.195
Temperature	−0.125	0.304*	−0.203	−0.463
DO	0.057*	−0.228*	−0.044*	−0.217*
BOD	−0.084*	0.154	0.126*	0.043
COD	−0.61*	0.088*	0.206*	0.305*
TSS	−0.008	0.347	0.234*	0.305
TDS	0.312*	0.345*	0.176	0.150*
E.C	−0.113*	0.315*	0.197*	0.068*
Turbidity	0.108	0.442*	0.321*	0.097
E. coli	−0.512*	0.179	0.294*	−0.178
Ammoniacal nitrogen	−0.427*	0.311*	0.384	0.083*
Nitrate	−0.154	0.254*	−0.238	0.068
Phosphate	−0.154*	0.395*	0.294*	0.053*
Sulfate	0.120	0.427*	0.235	0.022*

*$p < 0.05$; **$p < 0.01$ (*DO* dissolved oxygen, *BOD* biological oxygen demand, *COD* chemical oxygen demand, *TSS* total suspended solid, *TDS* total dissolved solid, *EC* electrical conductivity)

correlation with turbidity, BOD, COD, TSS, E.C, *E. coli*, and phosphate, while it has negative correlation with DO. Facilities and industrial lands are also negatively related to DO only, but positively correlated with sulfate, E.C, COD, TDS, and ammoniacal nitrogen. There is a noticeable pattern observed by Pearson correlation analysis, that forest areas have a positive effect on water quality in both seasons, whereas the other land uses have an adverse impact on the river quality throughout the area (Neary et al. 2009).

Along with this, calculated landscape metric values have demonstrated large variations among different study areas (Fig. 2.3). In landscape ecology, the principal approach to illustrate landscape patterns and measure spatial heterogeneity is via landscape metric analysis (Yu et al. 2019). These metrics function as indicators of structural composition and spatial configuration of landscapes. The highest values of CONTAG (86.34%), AI (99.2%), COHE (99.74), SHMN (1.77), and LPI (93.75%) and the lowest values of SHDI (0.27), PD (0.052/100 ha), and ED (1.92 m/ha) were all recorded in zone 7. While the highest values of SHDI (1.25), ED (12.95 m/ha), and PD (0.28/100 ha) were recorded in zone 5. Whereas, zone 1 had displayed the lowest values of ENNMN (488.4 m), SHMN (1.55), and LPI (40.1%). There were clear differences with correlations between these metrics and water quality parameters in both seasons, which highlight the sensitivity toward seasonality and hydrological changes.

By using cluster analysis, these zones are further categorized into 22 sampling sites based upon pollution load, i.e., signified slight pollution, moderate pollution, and high pollution load to analyze the seasonal variations (Fig. 2.4). During the normal season, cluster 1 included sites 4, 13, and 14. These sites are located at the brinks or on the perimeter of Bentong River and are mainly surrounded with forest cover and limited population. Naturally, they should be much less polluted than the other sites. Cluster 2 comprised of sites 3, 6, 7, 9, 11, and 22; and cluster 3 covered the rest of the sites. Clusters 2 and 3 are somewhat similar but vary in pollution severity; both are represented by mostly residential and agricultural areas. Still, the sites in cluster 3 are located in areas that contain more industrial activities, infrastructure, and business practices besides the residential and agricultural parts. Therefore, these sites are subjected to more wastewater discharge and much more contaminated runoff.

During the wet season, cluster 1 included sites 1, 2, 3, 5, 6, 7, 8, 9, 10, 12, 14, 15, and 21. While cluster 2 contained sites 4, 13, and 22. Interestingly, most of the sampling sites in cluster 1 from the normal season shifted to cluster 2 in the wet season, which is significantly different. The reason could be attributed to runoff resulting from rainfalls, which carries various pollutants and compounds from nearby and surrounding areas, especially agricultural lands, and deposit them in waterways. Cluster 3 comprised sites 11, 16, 17, 18, 19, and 20, which are primarily positioned in zone 4. These sites, particularly sites 16 and 17, are located downstream of the study area, thus the generated runoff passing through various areas would ultimately worsen the water quality downstream as cited by (Absalon and Matysik (2007)). Additionally, Ravichandran (2003)) stated that the confluence of rivers sometimes leads to water quality deterioration due to different water chemical composition and the alterations in hydrology.

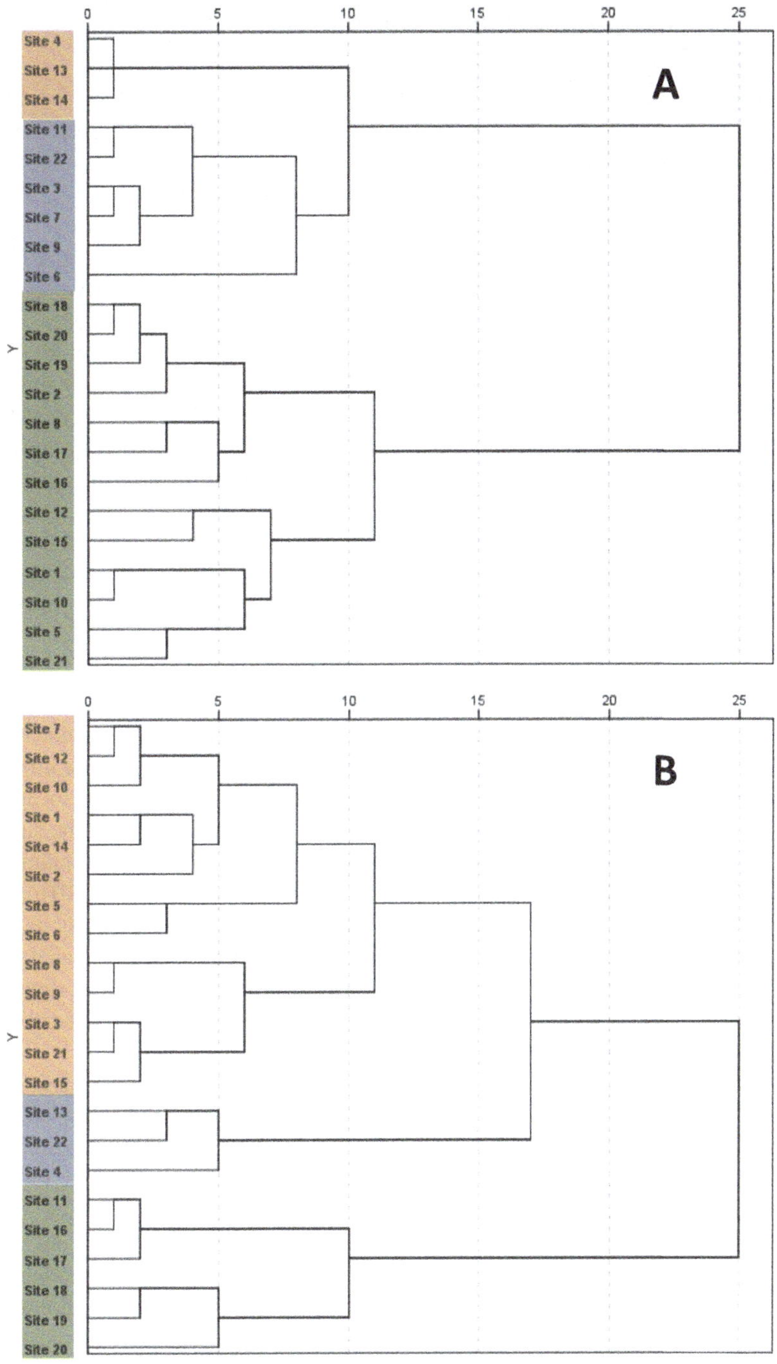

Fig. 2.4 Dendogram of the cluster analysis, (**a**) in the normal season and (**b**) in the wet season according to the 22 sampling sites in Bentong basin

2.5 Landscape Impact on Water Characteristics

Generally, tropical areas are distinguished by having substantial seasonality in climate with distinct dry and wet seasons (Ferreira et al. 2017). Furthermore, in humid tropics, hydrological processes have higher energy inputs and are subjected to quick changes, basically related to human intervention (Wohl et al. 2012). Surface runoff that affects water quality is itself affected by climate and land uses in the area (Singh et al. 2004). Several researchers have reported that forest and grass areas normally have a positive impact on water quality (Lee et al. 2009; Walker et al. 2009), while residential and agricultural areas have a negative influence on the quality of water (Bu et al. 2014; Namugize Jean et al. 2018). Shehab et al. (2020) have also found that forest cover is normally related to good water quality and plays a considerable role in enhancing water quality in different watersheds worldwide (Oliveira et al. 2016). Forest areas are found to be negatively related to numerous water quality characteristics in each season. This result reinforces the idea that forest and grass areas have a fixation and absorption impact on pollutants in river systems (Bu et al. 2018). In addition, forest cover functions as a filter or a sieve that screens and reduces the contaminants and sediments conveyed with the runoff (Ding et al. 2013). Furthermore, grass acts as a retaining agent that lessens the amount of surface runoff (Ouyang et al. 2010).

Several studies have illustrated that mostly, urban and agricultural lands have adverse effects on water quality (Ou et al. 2016; Huang et al. 2016). Excessive use of fertilizers and soil erosion can result in higher deposition of nutrients, organic matter, and sediments in water (Poudel 2016). Additionally, the level of negative impact of agricultural lands on water quality is highly influenced by the geographic location as well as farming techniques and habits (Jeon et al. 2004). Comparably, the surge in wastewater discharge and effluent from residential areas as well as contaminants buildup on impervious tops leads to decline in water quality (Ding et al. 2016). It is a well-known fact that residential areas contribute primarily to water quality degradation (White and Greer 2006), even with relatively small percentage of urban domain, the effects that it exerts on water quality are exceptionally strong and significant (Zhou et al. 2012). Residential areas are found to be positively related to most variables in both seasons (Shehab et al. 2020).

In addition to the spatial effect on water quality, there is an obvious temporal impact related to seasonal changes. Huang et al. (2016) stated that seasonal variations in hydrological processes like precipitation, runoff, interception, and abstraction have a major influence on the river flow, which lead to different levels of sediments and nutrients from one season to another. However, Woli et al. (2004) mentioned that during the normal season, agricultural applications should not be used to determine the nutrients variability in spite of the fertilizer usage. Therefore, river quality deterioration would have to be attributed to other sources. In this case, the residential and industrial areas, which are also heavily correlated with most water quality parameters in the normal season, become the main suspect. The sewerage points that are normally scattered in the populated areas as well as the industrial effluents affect water quality immensely, particularly where the population density is substantial.

Conversely, the strong correlation of water variables with agricultural lands in the wet season cannot be overlooked. Runoff resulting from rainfall carries fertilizer ingredients and eroded soil into waterways, which worsen the water quality in the river. Still, the impact of residential areas, which are also strongly correlated with most water parameters, should be counted since the input or effluent from residential areas is relatively consistent regardless of the season albeit with different quantities (Namugize Jean et al. 2018). Moreover, residential areas tend to increase inorganic nitrogen concentrations in river water, which would also explain the high correlation with nutrients especially NH_3-N (Foley et al. 2005). Although Oliveira et al. (2016) mentioned that in areas with distinctive rainy seasons (like Malaysia), non-point pollution sources typically have a reduced impact on water characteristics due to the drop in runoff rate in the dry and normal seasons. Thus, attentions to hydrological and temporal data are necessary to illustrate precisely the effects of land uses on water quality.

Moreover, Yu et al. (2013)) stated that converting natural lands and forests into residential areas alters the soil surface conditions, and elevates the imperviousness levels in the area and ultimately runoff with rainstorms (White and Greer 2006). Thus, pollution and changes in water characteristics tend to increase in these areas. Other researchers also observed that water quality deterioration is directly linked with increasing residential areas, along with agricultural and cultivated lands (Zhao et al. 2015; Wan et al. 2014; Du Plessis et al. 2014). Similarly, transforming forest areas into cultivated lands might change and raise water and sediment connectivity, consequently affecting water quality (Masselink et al. 2017). From the correlation results, it is evident that most water quality variables are significantly related more to agricultural lands than to forests. Thus, land use types had a significant influence not merely on the correlation coefficients with water parameters but also on the extent of effect that land use itself had on every water quality parameter (Yu et al. 2016). Still, using Pearson correlation analysis to support the presumption that land uses are in fact the main driver of water quality changes must be taken cautiously (Ferreira et al. 2017).

2.6 Landscape Metrics Relationship with Water Quality

Landscape configuration signifies the physical distribution or spatial makeup of patches within a class or landscape (Mcgarigal et al. 2002). Landscape metrics have been established to measure and quantify land use patterns and to better comprehend the landscape configuration and spatial diverseness (Griffith 2002). Understanding the influence of landscapes and human interference with land composition on water quality is required, in order to implement protection and rehabilitation practices on water bodies (Tudesque et al. 2014). It has been found that landscape metrics might be important parameters in predicting water quality at different sites exposed to mine wastes (Xiao and Ji 2007). In fact, several studies have demonstrated that landscape metrics are notable factors that illustrate the association between land uses and water quality and its effect on the hydrological processes (Shen et al. 2014).

Landscape impact on water characteristics varies over seasons (Pratt and Chang 2012), and is mainly reliant on scale (Zhou et al. 2012). Correlations of landscape metrics with water quality parameters are shown in Table 2.2. SHMN is considered as an indicator of the complexity of land use shapes. This index is found to be positively correlated with DO ($p < 0.01$) during the normal season and showed a significant negative correlation with BOD, turbidity, and ammoniacal nitrogen, which implies that more complexity in land use shapes that may result in a better water quality (Lee et al. 2009). SHDI reflects the number of land uses in the area and the proportion of change in land shapes. The presence of numerous different land use types in a certain area leads to higher SHDI values (McGarigal et al. 2002). This metric is positively correlated with almost all the variables and negatively with DO, which signifies that water quality tends to decline with high intersperse of different land use types as well as high spatial occupancy (Lee et al. 2009). Furthermore, this consequently alters water purification functions due to the constant changes in the natural environment (Xia et al. 2012). CONTAG percentage echoes the aggregated proportion or dispersion of land use patterns that are present in a certain area, which tends to be high when there is a low level of land use dispersion (Bu et al. 2014). During both seasons, CONTAG correlated negatively with almost all of the variables (significantly with E.C, turbidity, pH, temperature, DO, and TSS), this would indicate that water deterioration typically happens in landscapes that have a high level of dispersion and are greatly fragmented and uneven (Uuemaa et al. 2005). Likewise, ED and PD metrics are considered as pointers to the unevenness and fragmentation of a landscape (Fichera et al. 2012). McGarigal and Marks (1995) observed that PD and ED values would rise, while LPI value would drop when there are numerous small spots and patches of land covers in the area. These metrics are positively correlated with most water quality parameters in the normal season. This could mean that large populace concentration in areas near water bodies would relatively increase pollution levels (Shen et al. 2015). In addition, PD & ED are positively correlated with nutrients in the wet season, which indicates problems with runoff and soil erosion (Griffith 2002).

Since it has been reported that forests and natural areas contribute to better water characteristics (Tong and Chen 2002), it is more probable that point source pollution in the area is degrading the water quality more than non-point sources. LPI gives an indication to the largest cover or use inside a certain area (Ding et al. 2016). Forests are the dominant landscape in this case study area and most of the water quality variables are negatively correlated with this metric. This result is compatible with the findings of Lee et al. (2009)), and indicates that extensive forests cover could enhance water quality to some degree (Shen et al. 2014). Moreover, this reinforces the previous suggestion that non-point sources had a lesser effect on the variables than point source pollution. AI &COHE echo the physical and structural linkage and connectedness of a particular land cover or use inside an area, and consequently,

Table 2.2 Pearson correlation coefficients between landscape metrics and water quality variables in both seasons

	PD	LPI	ED	SHMN	ENNMN	COHE	AI	SHDI	CONTAG
Dry season									
pH	−0.080	−0.637*	−0.692	−0.331	−0.072	−0.513*	−0.045	0.456*	−0.530*
Temperature	0.103	−0.619*	−0.423	−0.479*	0.066	−0.608*	−0.261	0.474*	−0.592*
DO	−0.396*	0.385	−0.297*	0.624**	0.254	0.521*	0.076	−0.302*	0.471*
BOD	0.360	−0.007	−0.418	−0.461*	−0.047	0.054	0.382	−0.223	0.098
COD	0.063	−0.295	0.079	−0.274	0.095	−0.380	0.010	0.363	−0.395
TSS	0.014	−0.355	0.140*	−0.240	0.076	−0.536*	−0.401	0.426*	−0.510*
TDS	0.226	−0.199	0.225	0.203	−0.042	−0.113	−0.159	0.205	−0.149
E.C	0.154*	−0.490*	0.264	−0.198	−0.028	−0.624*	−0.447*	0.533*	−0.610*
Turbidity	0.307*	−0.288	0.199*	−0.466*	0.260	−0.479*	−0.121	0.299	−0.436*
E. coli	0.548*	−0.170	0.574*	0.381	−0.293	−0.171	−0.431*	0.281	−0.202
Ammoniacal nitrogen	0.282*	−0.320	0.192*	−0.562*	0.107	−0.445*	−0.135	0.238*	−0.393
Nitrate	0.218	−0.031	0.176	−0.137	0.229	−0.100	0.018	0.053	−0.088
Phosphate	0.203	−0.074	0.120*	0.106	−0.378	0.091	0.079	−0.096	0.089
Sulfate	0.215	−0.226	0.151	−0.154	0.216	−0.324	−0.051	0.215	−0.300
Wet season									
pH	−0.303	−0.355	−0.352	−0.497	−0.030	−0.063	0.163	−0.030	−0.056
Temperature	−0.164	−0.349	0.259	−0.359	−0.142	−0.093	0.315	−0.001	−0.089
DO	−0.158	0.090	−0.198*	0.680*	0.105	−0.022	−0.155	−0.627*	0.531*
BOD	0.081	−0.095	0.086	0.099	0.181	−0.097	−0.005	0.197*	−0.137
COD	0.138	0.055	0.195*	−0.065	−0.191	0.049	0.279	0.201*	0.096
TSS	0.008	−0.283	0.022*	−0.071	0.135	−0.285	−0.074	0.282*	−0.299*
TDS	0.254*	−0.254	0.174*	−0.391	0.213	−0.395	−0.067	0.249	−0.362*
E.C	0.641*	−0.413	0.414*	−0.376	−0.003	−0.491*	−0.141	0.357*	−0.468*
Turbidity	0.465*	−0.333	0.683*	−0.138	0.221	−0.388	−0.131	0.360*	−0.395*

(continued)

Table 2.2 (continued)

	PD	LPI	ED	SHMN	ENNMN	COHE	AI	SHDI	CONTAG
E. coli	0.507*	−0.430*	0.466*	0.282	−0.325	−0.262	−0.215	0.361*	−0.310
Ammoniacal nitrogen	0.425*	−0.363	0.400*	0.393	−0.096	−0.162	−0.231	0.346*	−0.234*
Nitrate	0.169*	0.307	0.159*	−0.087	0.254	0.066	0.255	−0.087	0.078
Phosphate	0.150*	−0.425*	0.158*	−0.012	0.003	−0.334	−0.142	0.367*	−0.365
Sulfate	0.114	−0.386	0.066	−0.285	0.121	−0.434*	−0.072	0.332	−0.421*

$^*p < 0.05$; $^{**}p < 0.01$

reflects the impacts of landscape dissemination or dispersal on variables (Ding et al. 2016). The connectivity theme of landscapes is clearly noticeable in Bentong basin with some separations in few areas. Most of the variables are negatively relatedto these metrics, which may indicate that a more separated landscape can lead to a worse water quality. ENNMN is normally applied to measure patch separation and seclusion in an area. The high values of ENNMN in the different zones are probably attributed to large proportion of one land use or cover with a small number of different patches (Zhou et al. 2012).

Occasionally, landscape metrics can be used to characterize water quality in certain water bodies. Land use or cover types have a measurable and definitive impact on water quality by affecting the runoff composition and process. Following rainfalls, the runoff conveys substances and pollutants from different lands into a water system. These spatial patterns of the area exemplify the influence of land uses on the water quality (Galbraith and Burns 2007).

2.7 Conclusion

Fresh water is a precious resource as it constitutes only 0.3% of total water resources across the globe. Availability of water is subjected to natural influences and anthropogenic activities. However, anthropogenic influences on water quality have the major impact on life. Undoubtedly changes in land use and land cover pattern lead to a decline in river water quality. In any given area, selected landscape pattern metrics that are correlated with river water characteristics may be used as indicators to assist with the monitoring and management of river systems.

Furthermore, river monitoring and restoration practices must consider land use and land cover patterns on multiple scales, ranging from buffer zones, catchments, sub-watersheds to the entire watershed. Focusing only on smaller regions often leads to a failure of river management. Future management plans must consider multiple scales and landscape planning at the sub-watershed and watershed scales seems vital in particular. The case study yielded relationships that enabled us to interpret and describe the effects of multiple factors such as spatial and temporal changes and landscape patterns on water variables. Understanding these relationships provides information and valuable data regarding water quality conditions and helps to outline suitable water management practices.

Moreover, in order to curtail or limit landscape effects on river water, urbaniza tion must be restricted in and around the riparian areas of rivers. As this case study demonstrated, urbanization is a key factor affecting the river water quality. Future river restoration or management efforts should be focused to plan more sustainable urban landscapes around rivers that control expansion of urban land, and minimize pollution load. Additionally, river restoration entails interdisciplinary collaborations and engagement with policy makers. In spite of the importance of natural environmental settings, anthropogenic factors play a dominant role in influencing the flow and water quality of rivers everywhere. Consequently, this requires natural and social scientists to work together in order to fully comprehend and successfully restore degraded river systems.

References

Absalon D, Matysik M (2007) Changes in water quality and runoff in the Upper Oder River Basin. Geomorphology 92:106–118

Alberti M, Booth D, Hill K, Coburn B, Avolio C, Coe S, Spirandelli D (2007) The impact of urban patterns on aquatic ecosystems: an empirical analysis on Puget lowland sub-basins. Landscape Urban Plan 80(4):345–361

Barbosa AE, Fernandes JN, David LM (2012) Key issues for sustainable urban storm water management. Water Res 46:6787–6798

Beckert K, Fisher T, Neil O, Jesien JR (2011) Characterization and comparison of stream nutrients, land use, and loading patterns in Maryland coastal bay watersheds. Water Air Soil Pollut 221 (1–4):255–273

Blocken B, Derome D, Carmeliet J (2013) Rainwater runoff from building facades: a review. Build Environ 60:339–361

Bowyer D (2015) Measuring urban growth, urban form and accessibility as indicators of urban sprawl in Hamilton, New Zealand. Master Thesis in Geographical Information Science

Bu H, Meng W, Zhang Y, Wan J (2014) Relationships between land use patterns and water quality in the Taizi River basin, China. Ecol Indic 41:187–197. https://doi.org/10.1016/j.ecolind.2014.02.003

Bu H, Song X, Xia J (2018) The hydrogen and oxygen isotopic compositions of precipitation in a forested watershed of the South Qinling Mts., China. Environ Sci Pollut Res 25:6720–6728

Buck O, Niyogi DK, Townsend CR (2004) Scale-dependence of land use effects on water quality of streams in agricultural catchments. Environ Pollut 130:287–299

Crosa G, Froebrich J, Nikolayenko V, Stefani F, Gallid P, Calamari D (2006) Spatial and seasonal variations in the water quality of the Amu Darya River (Central Asia). Water Res 40:2237–2245

Dams J et al (2013) Mapping impervious surface change from remote sensing for hydrological modeling. J Hydrol 485:84–95

Ding S, Zhang Y, Liu B, Kong W, Meng W (2013) Effects of riparian land use on water quality and fish communities in the headwater stream of the Taizi River in China. Front Environ Sci 7 (5):699–708

Ding J, Jiang Y, Liu Q, Hou Z, Liao J, Fu L, Peng Q (2016) Influences of the land use pattern on water quality in low-order streams of the Dongjiang River basin, China: a multi-scale analysis. Sci Total Environ 205:551–552

Drake J, Young D, McIntosh N (2016) Performance of an underground Stormwater detention chamber and comparison with Stormwater management ponds. Water 8:211. https://doi.org/10.3390/w8050211

du Plessis A, Harmse T, Ahmed F (2014) Quantifying and predicting the water quality associated with land cover change: a case study of the Blesbok Spruit Catchment, South Africa. Water 6 (10):2946–2968

Evans A, Mateo-Sagasta J, Qadir M, Boelee E, Ippolito A (2019) Agricultural water pollution: key knowledge gaps and research needs. Curr Opin Environ Sustain 36:20–27

FAO (2018) Sustainable food systems - concept and framework

Ferreira MC, Lucas-Borja ME, Rodrigues V, Zema D (2017) Effects of land use and sampling distance on water quality in tropical headwater springs (Pimenta creek, São Paulo State, Brazil). Sci Total Environ:690–701

Fichera CR, Modica G, Pollino M (2012) Land cover classification and change-detection analysis using multi-temporal remote sensed imagery and landscape metrics. Eur J Remote Sens 45:1–18

Foley JA, DeFries R, Asner GP, Barford C, Bonan G, Carpenter SR et al (2005) Global consequences of land use. Science 309:570–574
Galbraith L, Burns C (2007) Linking land-use, water body type and water quality in Southern New Zealand. Landsc Ecol 22(2):231–241
Giri S (2013) Assessing best management practices and implementation strategies to improve water quality. A Doctoral Dissertation. Michigan State University, East Lansing, Michigan
Giri S, Qiu Z (2016) Understanding the relationship of land uses and water quality in twenty first century: a review. J Environ Manag 173:41–48
Griffith JA (2002) Geographic techniques and recent applications of remote sensing to landscape-water quality studies. Water Air Soil Pollut 138:181–197
Huang Z, Han L, Zeng L, Xiao W, Tian Y (2016) Effects of land use patterns on stream water quality: a case study of a small-scale watershed in the three gorges reservoir area. China Environ Sci Pollut R 23:3943–3955
Ierodiaconou D, Laurenson L, Leblanc M, Stagnitti F, Duff G, Salzman S, Versace V (2005) The consequences of land use change on nutrient exports: a regional scale assessment in south-west Victoria. Australia J Environ Manage 74:305–316
Jeon JH, Yoon CG, Ham JH, Jung KW (2004) Model development for nutrient estimates from paddy rice fields in Korea. J Environ Sci Health B 39:845–860
Jia H, Yao H, Yu SL (2013) Advances in LID BMPs research and practice for urban runoff control in China. Front Environ Sci Eng 7:709–720
Jin M, Shepherd JM (2005) Inclusion of urban landscape in a climate model: how can satellite data help? Bull Ame Meteorol Soc 86(5):681–689
Khatri N, Tyagi S (2015) Influences of natural and anthropogenic factors on surface and groundwater quality in rural and urban areas. Front Life Sci 8(1):23–39
Lee SW, Hwang SJ, Lee SB, Hwang HS, Sung HC (2009) Landscape ecological approach to the relationships of land use patterns in watersheds to water quality characteristics. Landscape Urban Plan 92:80–89
Leung C, Jiao JJ (2006) Use of strontium isotopes to identify buried water main leakage into groundwater in a highly urbanized coastal area. Environ Sci Technol 40:6575–6579. https://doi.org/10.1021/es0611487
Malaysian Meteorological Department (MMD) 2019 Website: climate of Malaysia. http://www.met.gov.my
Masselink RJ, Heckmann T, Temme AJ, Anders NS, Gooren H, Keesstra SD (2017) A network theory approach for a better understanding of overland flow connectivity. Hydrol Process 31 (1):207–220
Mcgarigal K, Cushman SA, Neel MC, Ene E (2002) FRAGSTATS: spatial pattern analysis program for categorical maps. Computer software program produced by the authors at the University of Massachusetts, Amherst
McGarigal K, Marks BJ (1995) Spatial pattern analysis program for quantifying landscape structure. Gen. Tech. Rep. PNW-GTR-351. US department of agriculture. Pacific Northwest research station, Forest service, pp 1–122
McGrane SJ (2016) Impacts of urbanization on hydrological and water quality dynamics, and urban water management: a review. Hydrol Sci J 61(13):2295–2311. https://doi.org/10.1080/02626667.2015.1128084
Mitchell ME, Bennett E, Gonzalez A (2013) Linking landscape connectivity and ecosystem service provision: current knowledge and research gaps. Ecosystems 16(5):894–908
Namugize Jean N, Graham J, Mark G (2018) Effects of land use and land cover changes on water quality in the Umngeni river catchment, South Africa. Phys Chem Earth Parts A/B/C. https://doi.org/10.1016/j.pce.2018.03.013
Neary DG, Ice GG, Jackson CR (2009) Linkages between forest soils and water quantity and quality. Forest Ecol Manag 258:2269–2281

Oliveira LM, Mailard P, Pinto EJA (2016) Modeling the effect of land use/land cover on nitrogen, phosphorus and dissolved oxygen loads in the Velhas River using the concept of exclusive contribution area. Environ Monit Assess 188(333):1–19

Ou Y, Wang X, Wang L, Rousseau AN (2016) Landscape influences on water quality in a riparian buffer zone of drinking water source area, Northern China. Environ Earth Sci 75(114):1–13

Ouyang W, Skidmore AK, Toxopeus AG, Hao F (2010) Long-term vegetation landscape pattern with non-point source nutrient pollution in upper stream of Yellow River basin. J Hydrol 389 (3–4):373–380

Poudel DD (2016) Surface water quality monitoring of an agricultural watershed for non-point source pollution control. J Soil Water Conserv 71(4):310–326

Pratt B, Chang H (2012) Effects of land cover, topography, and built structure on seasonal water quality at multiple spatial scales. J Hazard Mater 209–210:48–45

Ravichandran S (2003) Hydrological influences on the water quality trends in Tamiraparani Basin, South India. Environ Monit Assess 87:293–309

Rodriguez F, Bocher E, Chancibault K (2013) Terrain representation impact on periurban catchment morphological properties. J Hydrol 485:54–67. https://doi.org/10.1016/j.jhydrol.2012.11.023

Rosso JJ, Fernández CA (2013) Effects of land use on environmental conditions and macrophytes in prairie lotic ecosystems. Limnologica 43(1):18–26. ISSN 0075-9511

Rothwell JJ, Dise NB, Taylor KG, Allott TEH, Scholefield P, Davies H, Neal C (2010) A spatial and seasonal assessment of river water chemistry across North West England. Sci Total Environ 408:841–855

Schoonover JE, Lockaby BG (2006) Land cover impacts on stream nutrients and fecal coliform in the lower Piedmont of West Georgia. J Hydrol 331:371–382

Shehab, ZN, Jamil, NR, Aris, AZ, Shafie, NS (2020) Impact of landscape patterns and land use types on water quality in Bentong River Basin, Malaysia. (Submitted paper)

Shen ZY, Hou XS, Li W, Aini G (2014) Relating landscape characteristics to nonpoint source pollution in a typical urbanized watershed in the municipality of Beijing. Landsc Urban Plan 123:96–107

Shen ZY, Hou XS, Li W, Aini GZ, Chen L, Gong YW (2015) Impact of landscape pattern at multiple spatial scales on water quality: a case study in a typical urbanized watershed in China. Ecol Indic 48:417–427

Singh P, Haritashya UK, Kumar N (2004) Seasonal changes in meltwater storage and drainage characteristics of the Dokriani Glacier, Garhwal Himalayas (India). Hydrol Res 35(1):15–29

Tong STY, Chen WL (2002) Modeling the relationship between land use and surface water quality. J Environ Manag 66:377–393

Tsegaye T, Sheppardi D, Islam KR, Johnson A, Tadessei W, Atalay A, Marzen L (2006) Development of chemical index as a measure of in-stream water quality in response to land use and land cover changes. Water Air Soil Pollut 174:161–179

Tudesque L, Tisseuil C, Lek S (2014) Scale-dependent effects of land cover on water physicochemistry and diatom-based metrics in a major river system, the Adour-Garonne basin (South Western France). Sci Total Environ 466:47–55

UN (United Nations) (2018) World urbanization prospects: the 2018 revision. United Nations Department of Economic and Social Affairs (Population Division), New York

United States Census Bureau (2012) U.S. and world population clock. Available from: www.census.gov/popclock. Accessed 3 Nov 2013

Uuemaa E, Roosaare J, Mander Ü (2005) Scale dependence of landscape metrics and their indicatory value for nutrient and organic matter losses from catchments. Ecol Indic 5:350–369

Walker TR, Crittenden PD, Dauvalter VA, Jones V, Kuhry P, Loskutova O, Mikkola K, Nikula A, Patova E, Ponomarev VI, Pystina T, Rätti O, Solovieva N, Stenina A, Virtanen T, Young SD (2009) Multiple indicators of human impacts on the environment in the Pechora Basin, north-eastern European Russia. Ecol Indic 9:765–779

Wan R, Cai S, Li H, Yang G, Li Z, Nie X (2014) Inferring land use and land cover impact on stream water quality using a Bayesian hierarchical modeling approach in the Xitiaoxi River Watershed, China. J Environ Manag 133(0):1–11

Wang M, Hu C, Barnes BB, Mitchum G, Lapointe B, Montoya Joseph P (2019) The great Atlantic Sargassum belt. Science 365(6448):83–87. https://doi.org/10.1126/science.aaw7912

White MD, Greer KA (2006) The effects of watershed urbanization on the stream hydrology and riparian vegetation of Los Peñasquitos Creek. California Land Scape Urban Plan 74:125–138

Wilson C, Weng Q (2010) Assessing surface water quality and its relation with urban land cover changes in the Lake Calumet area, Greater Chicago. Environ Manag 45:1096–1111

Wohl E et al (2012) The hydrology of the humid tropics. Nat Clim Chang 2(9):655–662. https://doi.org/10.1038/nclimate1556

Woli KP, Nagumo T, Kuramochi K, Hatano R (2004) Evaluating river water quality through land use analysis and N budget approaches in livestock farming areas. Sci Total Environ 329:61–74

Xia LL, Liu RZ, Zao YW (2012) Correlation analysis of landscape pattern and water quality in Baiyangdian watershed. Procedia Environ Sci 13:2188–2196

Xiao H, Ji W (2007) Relating landscape characteristics to non-point source pollution in mine waste-located watersheds using geospatial techniques. J Environ Manag 82(1):111–119

Yu D, Shi P, Liu Y, Xun B (2013) Detecting land use water quality relationships from the view point of ecological restoration in an urban area. Ecol Eng 53:205–216

Yu M, Huang Y, Cheng X, Tian J (2019) An ArcMap plug-in for calculating landscape metrics of vector data. Eco Inform 50:207–219

Yu S, Xu Z, Wu W, Zuo D (2016) Effect of land use types on stream water quality under seasonal variation and topographic characteristics in the Wei River basin. China Ecol Indic 60:202–212

Zhang G, Guhathakurta S, Dai G, Wu L, Yan L (2013) The control of land-use patterns for stormwater management at multiple spatial scales. Environ Manag 51(3):555–570

Zhao J, Lin L, Yang K, Liu Q, Qian G (2015) Influences of land use on water quality in a reticular river network area: a case study in Shanghai. China Landsc Urban Plann 137:20–29

Zhou T, Wu J, Peng S (2012) Assessing the effects of landscape pattern on river water quality at multiple scales: a case study of the Dongjiang River watershed. China Ecol Indic 23:166–175

Chapter 3
Existence of Antibiotics in Wastewater as a Pollution Indicator

Halah H. Al-Haideri, Fikrat M. Hassan, and Saja H. Abdul-Ameer

3.1 Introduction

Antibiotics are the most successful and fast innovations of human health sector worldwide, which are used extensively to improve human health by killing or inhibiting the growth of bacteria. Antibiotics are antimicrobial drugs that are also used to treat animals and plant infections and for limiting the spread of potential diseases of multidrug infections, as well as promoting growth in animal farming. In recent decades, due to their wide-spectrum advantages, antibiotics have been widely and effectively used in veterinary sector as well as in aquaculture, agriculture, bee-keeping, and livestock as growth potentials (Smith et al. 2002; Singer et al. 2003; Pruden et al. 2013; Gothwal and Shashidhar 2014).

Literally, antibiotic is a term refers to "antibiosis," which practically means "anti-life," and thus define as an organic compound released by one microorganism and being toxic to other microorganisms or can inhibit the growth of or even kill other microorganisms even at low concentrations (Russell 2004); however, recently, the definition has been changed to include those substances as antibiotics that are genuinely produced partially or completely by microorganisms via specific ways. Based upon the type of microorganism, they can be antibacterial, antifungal, and antiviral (Walsh 2003; Russell 2004; Brooks et al. 2004). Antibacterial drugs can act as bactericidal when completely kill the bacteria or bacteriostatic when only inhibit the bacteria. Bacteriostatic drugs can be changed into bactericidal by changing certain conditions such as poor nutrition, temperature variations, toxic proteolytic compounds, or antibodies (Choquet-Kastylevsky et al. 2002; Slatore and Tilles 2004).

H. H. Al-Haideri (✉) · F. M. Hassan · S. H. Abdul-Ameer
The Department of Biology, College of Science for Women, University of Baghdad, Baghdad, Iraq

A. Singh et al. (eds.), *Water Pollution and Management Practices*,
https://doi.org/10.1007/978-981-15-8358-2_3

Antibiotics as chemotherapeutic agents are released in large quantities in natural ecosystems and exist at low concentrations in the environment due to the stability of antibiotic precursors or their metabolic compounds which readily pass through water treatment systems and leak into the ecosystem (Singer et al. 2002; Ferber 2003; Jjemba 2006; Sarmah et al. 2006; Kümmerer 2009). Antibiotics are administered in the treatment of human-infectious diseases as well as prophylactically to enhance yield in the livestock farming, or non-target organisms, such as degradation of pollutants and nutrient cycling (Flaherty and Dodson 2005). The native organisms can interact with the antibiotics that entered the aquatic ecosystems and can then alter and modify the microbial community genetically and structurally (Singer et al. 2016). The extensive use of chemotherapeutic drugs and the continuous exposure of environment to these drugs can boost the selection of potentially resistant pathogens, in other words, some species that are not directly affected by antibiotics can be affected by antibiotic-indirect cascade effect, when feed on microorganisms (Kümmerer 2004; Cabello 2006). Most notably, the environmental microorganisms produce several antibiotics, whereas antibiotic resistance genes come from pathogenic bacteria via horizontal gene transfer (HGT) from environmental microorganisms as well; however, treatment of infections may generate a strong pressure for antibiotic selectivity (Kümmerer 2004). The prevalence and persistence of bacteria in the presence of antibiotics is due to the selective pressure that modifies their genetic requirements and generates antibiotic-resistant bacteria (ARB), which act as an ultimate threat for public health. ARBe not only have the action for proliferation, but also have the ability to transfer resistance genes to sensitive bacteria (Martinez et al. 2007). Antimicrobial-resistant bacteria possess several mechanisms to get rid of the effect of these drugs such as secretion of enzymes that are able to degrade the antibiotics, efflux pumps which are able to repel the drugs out of the bacterial cells, modify bacteria cell wall to prevent the electrostatic interaction of antibiotics with host cell wall, and downregulation of porins which in turn prevent antibiotics from cross the cell, etc. (Bird et al. 2019).

Generally, antibiotics can be classified according to many different aspects such as the mode of action, chemical structure, and the way of entrance (by injection, oral and/or topical); however, the mechanism of action and activity spectrum are considered to be the most common ways to classify antibiotics to β-lactams, carbapenem, aminoglycoside, macrolides, sulfonamides, quinolones, tetracycline, glycopeptides, and fluoroquinolone (Calderon and Sabundayo 2007; Vila et al. 2007; van Hoek et al. 2011; Frank and Tacconelli 2012). According to WHO classification (2019), antibiotics are further classified based on the importance of their appropriate use, and to be used as a tool to better support antibiotic monitoring and activities management in countries. Table 3.1 represents the WHO classification of antibiotics and their mode of action in some details (Adzitey 2015).

The β-lactam antibiotics activity is restricted in their 3-carbon and 1-nitrogen ring in the chemical structure, by which are able to destroy the bacterial cell wall by interfering with proteins during peptidoglycan synthesis. The modification of beta-lactam drugs side chain increases the ability to evade the degradative capacity of enzymes produced by some bacterial species and make the movement of antibiotics

Table 3.1 WHO AWaRe antibiotic classification for monitoring and evaluation, and their mode of action

No.	Class of antibiotic	Representative drug	Mechanism of action
1	β-lactam	Amoxicillin, Ampicillin, Azlocillin, Bacampicillin, Benzathinebenzylpenicillin, Benzylpenicillin, Clometocillin, Cloxacillin, Dicloxacillin, Flucloxacillin, Nafcillin, Oxacillin, Phenoxymethylpenicillin, Penamecillin, Pheneticillin, Piperacillin, Pivampicillin, Pivmecillinam, Procaine benzylpenicillin, Sulbenicillin	Inhibitors for cell wall synthesis and murine assembly
2	Aminoglycosides	Amikacin, Arbekacin, Dibekacin, Gentamicin, Isepamicin, Kanamycin, Micronomicin, Neomycin, Netilmicin, Netilmicin, Ribostamycin, Sisomicin, Streptomycin, Tobramycin	Inhibitors of protein synthesis (Translation)
3	Macrolides	Azithromycin, Clarithromycin, Dirithromycin, Erythromycin, Josamycin, Lincomycin, Midecamycin, Oleandomycin, Roxithromycin, Spiramycin, Telithromycin	Protein synthesis inhibitors (Translation)
4	First Generation of Cephalosporins	Cefacetrile, Cefadroxil, Cefalexin, Cefalotin, Cefapirin, Cefatrizine, Cefazedone, Cefazolin	Inhibition of the growth of Gram-negative bacteria by binding to penicillin-binding proteins (PBPs)
	Second Generation of Cephalosporins	Cefaclor, Cefamandole, Cefbuperazone, Cefmetazole, Cefminox, Cefonicid, Ceforanide	
	Third Generation of Cephalosporins	Cefcapenepivoxil, Cefdinir, Cefditorenpivoxil, Cefixime, Cefmenoxime, Cefodizime	
5	Monobactams	Aztreonam	Cell wall inhibitors
6	Lincosamides	Clindamycin	Protein synthesis inhibitors
7	Tetracyclines	Tetracyclines, Doxycycline, Eravacycline, Lymecycline, Metacycline, Minocycline (IV), Minocycline (oral), Omadacycline	Protein synthesis inhibitors
8	Amphenicols	Chloramphenicol	Protein synthesis inhibitors
9	Fluoroquinolones	Ciprofloxacin, Enoxacin, Delafloxacin, Fleroxacin, Flumequine, Garenoxacin, Gatifloxacin, Gemifloxacin, Levofloxacin, Lomefloxacin, Moxifloxacin, Norfloxacin,	DNA replication inhibitors

(continued)

Table 3.1 (continued)

No.	Class of antibiotic	Representative drug	Mechanism of action
		Ofloxacin, Pazufloxacin, Pefloxacin, Sitafloxacin, Sparfloxacin, Tosufloxacin	
10	Glycopeptides	Dalbavancin, Oritavancin, Telavancin, Teicoplanin, Vancomycin (IV), Vancomycin (oral)	Inhibitors for murine synthesis and assembly
11	Trimethoprim	Trimethoprim	Beta-lactamase inhibitors
12	Carbapenems	Biapenem, Doripenem, Imipenem/cilastatin, Ertapenem, Meropenem, Meropenem-vaborbactam, Panipenem	Peptidoglycan inhibitors
13	Rifamycins	Rifaximin, Rifamycin, Rifampicin, Rifabutin	Bacterial DNA polymerase inhibitors

Adzitey (2015) and WHO (2019)

across the outer membrane more easier (Poirel et al. 2005). Cephalosporins are similar to beta-lactam with semi-synthetically side chains (3,6-dihydro-2 H-1,3-thiazane) rings, which enable them to be resistant to penicillinase by binding to different penicillin-binding proteins (PBPs) in the cell wall, as well as to invade the blood-brain barrier (Etebul and Arikekpar 2016). Unlike beta-lactams, macrolides characterized with unique macrocyclic lactose ring contains l-cladinose and D-desosamine sugars, which increase their spectrum activity (Moore 2015). Macrolides are able to kill or inhibit bacterial protein synthesis by binding to ribosomes and preventing the addition of amino acids to polypeptide chains (Kapoor et al. 2017). Teteracyclines and doxycycline are firstly semi-synthetic derivatives, which target protein synthesis processes (Yoneyama and Katsumata 2006). Furthermore, other pharmaceuticals target the DNA replication and transcription processes by interfering with DNA gyrase such as quinolones and nalidixic acids (Higgins et al. 2003). The wide-spectrum activity and high efficiency of quinolones are due to their chemical structure with two rings in their side chains, with additional ring in the recent generation (Etebul and Arikekpar 2016). The unusual structure of aminoglycoside with 3-amino glycans joined by glycosidic bonds stands behind the wide-spectrum activity against microorganisms. Protein synthesis inhibition is the main target of aminoglycoside (Peterson 2008). The scope of this review is to outline the presence, dissemination, and impact of macropollutants in different environmental compartments around the world, and the ways by which these drugs are delivered to the surface water. In addition, the review has pointed out the mode action of most detectable antibiotics and the rate of detection in the environment.

3.2 The Source, Release, and Fate of Antibiotics

After the discovery of ampicillin in 1928 by Alexander Fleming (English Bacteriologist), from a soil fungus *Pencillium notatum*, it was widely used in human clinical treatment in 1940 (Schlegel 2003; Russell 2004; Etebul and Arikekpar 2016). The continuous uses of ampicillin in human health care system have propagated to develop and transform the management for bacterial infections disease treatment (Aminov 2010), as well as, for promoting faster growth of livestock (White and Cox 2013). The additional use of antibiotics in intensive fish farming and infected plants leads to the discharge of these compounds either to sewage treatment plants or directly into water and soil. The concentration of drugs used in agriculture is low compared to that used in human, veterinary care, and animal production (Cabello 2006; Dolliver and Gupta 2008).

According to the basic structure, antibiotics are either natural, which are produced by bacteria and fungi, to kill or inhibit other competitive microorganisms with bacteriostatic or bactericidal effect, such as penicillin that is easily degraded, or semisynthetic/synthetic compounds that are chemically modified to increase the efficiency by insertion of additives to their active site (Piorel et al 2005). Thus would be more stable, less degradable, and accumulate to higher concentrations, such as fluoroquinolones and tetracyclines (Kümmerer 2009). After administration and release of these compounds into different environmental compartments, they are not fully metabolized and a large quantity (30-90 %) are exerted as active compounds through urine and feces (McManus et al. 2002; Jjemba 2006; Lienert et al. 2007); therefore, antibiotics from human use can enter to the wastewater treatment plants (WWTPs) (Kemper 2008). The WWTPs are considered to be one of the major pathways for occurrence and existence of antibiotics in different aquacultures, because all WWTPs are not fully designed to remove antibiotic residues, resulting in a direct release to the receiving environment (Louvet et al. 2010). In megacities with well-developed sewage infrastructure, antibiotics enter the aqueous environment via sewage.

Antibiotics are released to the water stream as a natural product with their metabolites in their partially metabolized forms, and thus the concentration becomes lower. However, within treatment plants, microbial communities may expose to higher concentration of selected antibiotics. Around 50–80% of total natural compounds are released to the sludge via urine and partially through feces as a complex of metabolite-binding compounds, if not completely eliminated and accumulate in the sludge and reach the farmlands and receiving surface water (McManus et al. 2002; Jjemba 2006; Singer et al. 2008). In some cases, antibiotic contamination can be delivered in non-routine pathway like breakage in sewer or industrial effluent pipe, or mixing of storm water with un-treated effluent (Zuccato et al. 2010). Antibiotics can also reach the groundwater, surface drainage systems, and soils by application of municipal biosolids on land. The pharmaceutical industries and hospitals allow the direct discharge of discarded or unused or expired antibiotics into sludge or waste bins and later into wastewater treatment plants and landfills

(Peng et al. 2009). Use of sludge in place of fertilizers is another source of antibiotics in soil, and along with water irrigation can reach in the soil (Park et al. 2007), which might cause allergy or toxic reaction, and develop antibiotic resistance in human (Reddersen et al. 2002).

In general, antibiotic exposure routes are different particularly when drugs are used for animal purpose rather than for human, as these drugs are discharged directly to the soils or lands. Nevertheless, the concentration of antibiotics in manure or dropping-amended soil is significantly higher than those in aquatic ecosystem, although, some of these drugs are removed or eliminated rapidly over time by microbial activity in the soil, while others might persist for many years (Sarmah et al. 2006). Similarly, in sewage sludge, antibiotics exhibit robust interaction with particles, and merely a small portion of the total amount is being active, but is still a major issue (Kumar et al. 2005). In another way, antibiotics can be removed or eliminated from water that comes from soil or sediment, due to ability of these antibiotics to bind with soil particles, although this binding might delay their degradation (Boxall et al. 2012).

3.3 Antibiotics Existing in the Environment

It is known that the increasing consumption of antibiotics in the world resulted in their presence in detectable concentrations in many environmental areas, such as surface water, groundwater, drinking water, domestic sewage, sediment, soil, agricultural ecosystem, etc. (Christian et al. 2003; Pei et al. 2006; Roberts and Thomas 2006; Baquero et al. 2008; Standley et al. 2008; Fick et al. 2009; Lindberg et al. 2010; Teijon et al. 2010; Zhang et al. 2011; López-Serna et al. 2013; Li et al. 2014). The concentration of detected antibiotics in natural environment is variable and ranges from few nanograms to hundreds of nanograms based on their behavior in the nature like adsorption, biodegradation, and photodegradation, as well as on their usage and environmental parameters (Table 3.1) (Göbel et al. 2005; Verlicchi et al. 2015). For example, ciprofloxacin was detected in high quantities (50–80 %), followed by tetracyclin (80–90 %), clarithromycin (25%), and sulfamethazole (15–30 %), whereas, erythromycin was detected in very low amount (5–10 %) (Mompelat et al. 2009; Al Aukidy et al. 2012; Orya et al. 2016). However, the most detectable antibiotics in water are trimethoprim, quinolones, and sulfonamides (Blackwell et al. 2005). These were detected from sewage treatment plant and flow through of receiving river in China, while in the secondary effluents in the rivers also contained high concentrations of these antibiotics ranged 195, 2001, and 3866 ng L^{-1}, respectively (Carvalho and Santos 2016). In Shanghai, animal manures and agricultural soils appeared to have chloramphenicol, sulfonamides, and tetracycline with concentrations of 3.27–17.85, 5.85–33.37, and 4.54–24.66 mg kg^{-1}, respectively (Xu et al. 2015). On the other hand, fluoroquinolone has predominantly appeared in sewage sludge, whereas sulfonamides are selectively present in the wastewater (Giger et al. 2003; Ji et al. 2012).

In another study, it was found that ofloxacin and ciprofloxacin were present in hospital effluent close to the Terriver in Spain (Heise et al. 2006). In addition, erythromycin and ciprofloxacin were also noted in both receiving sewage water and treated effluent in Sweden with concentrations 0.47 μg L^{-1} and 1.41 μg L^{-1} and 0.35 μg $L^{-1}$and 0.06 μg L^{-1}, respectively (Rodriguez-Mozaz et al. 2015; Östman et al. 2017).

3.4 Existence in Water

3.4.1 Occurrence in Drinking Water

Natural and fresh aquatic environment is threatened by the presence of pharmaceutical compounds and raises the attention for risk assessment and distribution of antibiotic-resistance microorganisms (Eguchi et al. 2004; Xu et al. 2007). The high-quality drinking water and tap water became a big issue and far from trust from contamination by antibiotic residues. For instance, macrolides, erythromycin, and clarithromycin were detected in tap water in Madrid, and norfloxacin, ciprofloxacin, lomefloxacin, and enrofloxacin, belonging to fluoroquinolone, were reported in tap water with concentration ranged from 1.0–679.7 ng L^{-1} to 2.0–37 ng L^{-1}, respectively, in China (Yiruhan et al. 2010; Valcarcel et al. 2011). Drinking water contamination by antibiotics is not restricted due to the improper discarded drugs in the water supply worldwide.

Many countries have been found to have the presence of several classes of antibiotics like lincomycin, erythromycin, and roxithromycin in different drinking water sources (Stackelberg et al. 2007; Benotti et al. 2009; Huerta-Fontela et al. 2011), such as USA, Canada, and many European countries (Focazio et al. 2008; Benotti et al. 2009; Mompelat et al. 2011; DeJongh et al. 2012; Zhang et al. 2015). Mahmood et al. (2019) reported that fluoroquinolones and β-lactams were the most abundant drugs among other antibiotics in drinking water in Baghdad, Iraq; however, their concentrations varied according to water treatment management. In raw and finished water, ciprofloxacin was detected with relatively high concentration of 1.344 and 1.312 μg L^{-1}, respectively, whereas both levofloxacin and amoxicillin were detected in raw water at 0.414 μg L^{-1} and 1.50 μg L^{-1}, respectively. The presence of high concentration of ciprofloxacin and erythromycin in drinking water is mainly attributed to the human consumption and the influence of wastewater discharge, thus the possibility and extent of pollution from animal sources should not be underestimated, where enrofloxacin is usually used for animal treatment only, but could be metabolized to ciprofloxacin under certain conditions (USP 2003; Zhengqi et al. 2004). Another study showed that macrolid antibiotic like tylosin was found in drinking water in UK, Germany, and Italy, at concentrations of 10, 6, and 1.7 ng L^{-1}, respectively (Waggot 1981). In China, several antibiotics like sulfapyridine, sulfamethoxazole, ciprofloxacine, enrofloxacin, levofloxacin chloramphenicol, and doxycycline were detected in drinking water with concentrations ranging from 0.5 to 21.4 ng L^{-1} (Hannaa et al. 2018).

3.4.2 Surface Water, Streams, and Lakes

Rivers and lakes are another place where antibiotics could be directly discharged from human consumption or veterinary use. In Spain, five rivers were claimed to be contaminated by antibiotics, ciprofloxacin, clarithromycin, erythromycin, metronidazole, norfloxacin, ofloxacin, sulfamethoxazole, tetracyclin, and trimethoprim with mean concentration of 3, 235, 320.5, 1195.5, 10, 179, 179, 326, 23, and 424 ng L^{-1}, respectively (Valcarcel et al. 2011). Many studies from South Korea, Italy, USA, France, Taiwan, and Sweden also reported the presence of similar types of antibiotics in the rivers. Furthermore, a cluster of antibiotics, sulfapyridine, sulfamethoxazole, ciprofloxacin, enrofloxacin, levofloxacin chloramphenicol, and doxycycline were investigated in Shandong river in China, with the relative concentration ranging from 0.3 to 3.9 ng L^{-1} (Hannaa et al. 2018). In another study in China, four rivers were shown to be contaminated by norfloxacin, ofloxacin, ciprofloxacin, and oxytetracycline with concentrations of 5770, 1290, 653, and 652 ng g^{-1}, respectively (Peng et al. 2008, 2011).

The distribution of antibiotics residues in surface water in North America is due to not only processing, but also due to discarding of treated water, which resulted in a good media for a complex of bacteria to be exposed to several antibiotics and dissemination of antibiotics rapidly (Rodriguez-Rojas et al. 2013). Using liquid chromatography-mass spectrometry (LC-MS) to analyze and measure the concentration of antibiotics in surface water, many antibiotics appeared to be over the limits of quantification, for example, the concentration of sulfapyridine and sulfamethoxazole were above 10 μg L^{-1} in Europe (Díaz-Cruz et al. 2008; Ginebreda et al. 2010). In the USA, the concentration of some detected antibiotics was less than 2 μg L^{-1}, whereas sulfadimethoxine was 15 μg L^{-1}(Lindsey et al. 2001). However, ciprofloxacin and nalidixic acid were present in such high concentration up to 14 μg L^{-1} and 23 μg L^{-1}, respectively, in rivers and streams in South Africa (Agunbiade and Moodley 2016). Lack of good sanitation facilities and well-designed wastewater treatment approaches in low-income countries in particular is another cause of increase in the concentration of antibiotic residues in fresh and surface water. It was found that the mean concentration of antibiotics in contaminated fresh water in Asia-Pacific and Africa ranged between 17.7 μg L^{-1} to 11.3 μg L^{-1}, respectively, and 0.9 μg L^{-1} to 0.4 μg L^{-1} in America and Europe, respectively, unlike other high-income countries, where the consumption rate of antibiotics is high (Segura et al. 2015; Klein et al. 2018). However, a very small amount of several antibiotics was also detected in surface water in these countries such as Saudi Arabia, with less than 1 μg L^{-1}(Alidina et al. 2014).

3.4.3 Wastewater Treatment Plant

Antibiotics have a main role in the environmental pollution through wastewater treatment processes. Some of these pharmaceuticals may be eliminated from wastewater by adsorption onto solids; however, these can be released into the groundwater by sludge application to land, landfill, or soil erosion, thus the removal of these compounds are variable and depend mainly on the complexity of the environmental matrices and the properties of individual compound, in terms of the ability of this compound to be accumulated or degraded, and of the treatment unit in sludge treatment plants (STPs), leading to removal of the individual drugs at any step of treatment stages (Meakins et al. 1994; Diaz-Cruz et al. 2003). Several medicinal compounds are degraded by biodegradation at both aerobic and anaerobic pathway, in the presence of intra or/and extracellular enzymes excreted by bacterial cells (Meakins et al. 1994). However, this process completely depends on the chemical properties of these drugs. When a compound is composed of molecules with long, highly branched side chain then biodegradation is not easy compared to those with unbranched, short side chain. In addition, the biodegradation of aromatic and saturated aliphatic compounds are less accessible than that of unsaturated and simple aromatic compounds (Voulvoulis et al. 2016). For instance, in Germany, four iodinated compounds (diatrizoate, iopamidol, iopromide, and iomeprol) were detected in raw sewage as parent compounds, in aqueous phase, as these will not be degraded or absorbed through the sewage treatment processes. The concentration of these compounds is usually more than 1 μg L^{-1} in raw sewage and final effluents, except for iopamidol with concentration of 15 μg L^{-1} (Voulvoulis et al. 2016). In STP, the microbial community may adapt gradually to particular compounds like clofibric acid, ibuprofen, and diclofenac and may degrade them efficiently over time; however, clofibric acid shows more persistence and is highly mobile in sewage treatment plant (Zwiener and Frimmel 2003). In conventional STPs, many factors affect the removal of these compounds, i.e., sludge retention time (SRT), hydrolic retention time (HRT), and temperature (Suarez et al. 2008).

In conventional wastewater treatment, after primary sedimentation, biological treatment by activated sludge and trickling filters are used to eliminate drugs, similar to that used to remove organic micropollutants by adsorption and biodegradation or chemical degradation, and biotransformation. However, it was found that removal of drugs by activated sludge is much greater than that of percolating filter, which is due to the high efficient activity of bacteria in activated sludge. Nevertheless, WWTPs are not fully designed to eliminate a very small amount of organic compounds; therefore, activated sludge should have more attention for removal of biological nutrients and pharmaceuticals.

Generally, the influent of contaminated water that enters the conventional municipal STPs undergoes three main treatment stages: preliminary, primary, and secondary steps; which then come out with final effluent that passes to the surface water body to be reused indirectly for agricultural purposes; although some of chemical or drug compounds usually exist at low concentrations ranging from 10^{-3} to

10^{-2} μg L^{-1} or higher in raw influent, not all of them are completely removed by STPs. The total removal varied for different compounds, which depends mainly on the chemical and physical properties of these substances and the environmental conditions where the treatment occurs, such as pH, temperature, oxygen availability, and type of reactors. As a consequence, final effluent from secondary treatment step may contain trace amounts of pharmaceuticals that would probably be discharged to the surface water and creates a high-acute risk medium for aquatic environment (Voulvoulis et al. 2016). Similarly, all other compounds that show low environmental risk would probably exert a negative impact on aquatic biota, due to their discharge at massive daily mass loads and their chronic toxicity in the long term (Verlicchi et al. 2012). In the UK, several analytical approaches were used to detect the existence of some pharmaceuticals in the influent of six sewage treatment plants. It was found that ibuprofen was present in both influent and effluent samples, which suggested that the ibuprofen was removed by 80–100 % by different STPs, whereas the efficiency of STPs to remove ibuprofen was low up 14.4–44 % (Kanda et al. 2003). Another study found the occurrence of several pharmaceuticals (ibuprofen, paracetamol, salbutamol, propranolol HCl, and mefenamic acid) in the STPs at a concentration of ng L^{-1} in the UK (Jones et al. 2002). Most notably some pharmaceuticals are highly eliminated like ibuprofen, oxybenzone, chloroxylenol, whereas galaxolide and others are removed by biological process at low level, which means that some hydrophobic and recalcitrant compounds are eliminated from the aqueous phase but would be accumulated in the biosolids.

Some compounds are not totally removed via oxidative and biological process and resulted in transformation products (Khan and Ongerth 2002). Based on some researchers (Ternes and Joss 2006; Yasojima et al. 2006; Watkinson et al. 2007), the efficient removal of pharmaceuticals via preliminary and primary treatment processes is poor and in some cases, the presence of human metabolites lead to release these compounds in raw influents during the process (Carballa et al. 2004; Göbel et al. 2007). For instance, ibuprofen and naproxen were not remarkably reduced through pre-treatment and sedimentation, which indicated their poor sorption onto sludge probably because of their acidic structure with very low solid-phase partition coefficient K_d value (K_d ⍰ 500 l Kg^{-1}or log K_d ⍰ 2.7) (Yasojima et al. 2006).

Furthermore, different behavior of all pharmaceuticals found in wastewater during treatment processes makes their removal not possible during initial treatment stage and may also be related to conditions varying in different seasons (Jones et al. 2002; Lindberg et al. 2006. In conventional STPs, both hospital and domestic wastes are treated together and then discharged to the environment; however, the removal efficiency of some pharmaceuticals from both sources ranged from 10 to 90% due to their resistance to conventional treatments (Pauwels and Verstraete 2006). In Spain, a significant concentration (13 μg L^{-1}) of ofloxacin and ciprofloxacin has been detected in the hospital effluents near to the Ter river (Rodriguez-Mozaz et al. 2015), whereas erythromycin and ciprofloxacin were both detected in influent sewage water (0.47 and 1.14 μg L^{-1}) and in the final effluent (0.35 and 0.06 μg L^{-1}), respectively, in Sweden (Östman et al. 2017). In Germany, the persistence of triclosan in the influent and sludge and effluent of wastewater treatment plant was established

(Bester 2003). It was noted that the concentration of triclosan (100 ng L^{-1}) in the influent was reduced by 30% and only about 5% of all discharged influent was found in the effluent. This reflects the weak adsorption of triclosan onto sludge and also impose that these compounds will not recover as a parent form and may transform with other metabolic substances. Similarly, the fate and behavior of triclosan in wastewater treatment plant was also reported (Singer et al. 2002), where about 79% of overall concentration of triclosan was removed by biological degradation; 15% by adsorption on to sludge and 6% released to the surface water, and the concentration was estimated in the range of 42–213 ng L^{-1} in wastewater effluent and 11–98 ng L^{-1} in the receiving surface water. In a study, Cha et al. (2006) reported that the concentration of antibiotics oxacillin and ampicillin in the effluents of WWTPs toward cache la Poudre river is 95 μg $L^{-1}$and 86 μg L^{-1}, respectively, in Colorado, whereas their concentration from WWTPs effluent to Msunduzi river in South Africa was about 27 μg L^{-1} and 9 μg L^{-1}, respectively (Matongo et al. 2015).

The tendency of some pharmaceuticals in the secondary effluent of STPs varies. The most detectable antibiotics in Italy is tramadol, with a concentration of 57 μg L^{-1}, followed by ibuprofen (48 μg L^{-1}), diclofenac (11 μg L^{-1}), trimethoprim (6.7 μg L^{-1}), erythromycin (6.3 μg L^{-1}), ciprofloxacin (5.7 μg L^{-1}), sulfamethoxazole and roxithromycin (5 μg L^{-1}) (Verlicchi et al. 2013). The effluent of WWTPs from industrial area act as a massive mass of accumulated antibiotics that would be much more alarming to threat the aquatic environment, even at low concentration. For instance, ciprofloxacin and ofloxacin were detected at high level (31,000 μg $L^{-}1$ and 160 μg L^{-1}, respectively) in the effluent WWTPs of well-known industrial city in India (Larsson et al. 2007; Larsson 2014). The dissemination of pharmaceuticals is not limited in the effluent from WWTPs only, but may reach to far areas found in the study of Li et al. (2008), where oxytetracycline was detected 20 km downstream from WWTPs effluent in Xiao River in Asia, with a relative high concentration of 484 μg L^{-1}.

Several studies have shown the presence of pharmaceutical compounds in the effluent from WWTPs universally. Sulfamethazine was detected in wastewater treatment plant in Vietnam at a concentration of 19 μg L^{-1} (Managaki et al. 2007), whereas sulfamethoxazole; another form of sulfamethazine; and ofloxacin were found in Latin America with concentrations of 14.3 μg L^{-1} and 17.7 μg L^{-1}, respectively (aus der Beek et al. 2016). The concentration of sulfamethoxazole was significantly high in Kenya and Mozambique, with estimated concentration of 38.9 μg L^{-1} and 53.8 μg L^{-1}, respectively (Lindsey et al. 2001; Madikizela et al. 2017. Quinolones, sulfonamide, and macrolides were also administered in municipal wastewater plant in Beijing, with the mean concentration of 4916, 2916, and 356 ng L^{-1}, respectively (Li et al. 2013).

The fate and presence of quinolones and fluoroquinolones in municipal wastewater treatment plants were studied during anaerobic, anoxic, and aerobic treatment processes. All antibiotics showed to bind with biosolids and being in contact with microbial mass, which then create a medium to transfer genetic material through horizontal transfer, therefore WWTPs are considered as a prominent place for antibiotic-resistance bacteria (Jia et al. 2012; Michael et al. 2013; Rizzo et al. 2013). In South China, the removal efficiency of eleven classes of pharmaceutical compounds in WWTPs in Guangdong varied and ranged from 21 to 100% (Zhou et al. 2013). Furthermore, sulfonamide has gained more attention due to its distribution pattern between water and sediments. Although it was detected significantly in sludge and wastewater, many studies refer to its occurrence in the outlet of sediment and agriculture. This is due to low water distribution coefficient (K_d), which poses the low sorption affinity to water and sediment particles (Thiele-Bruhn 2003). In China, the most common antibiotic used for human and animal body is fluoroquinolones, which can enter the aquatic system via urine and ultimately the environment though wastewater (Renew and Huang 2004). The massive consumption of fluoroquinolones resulted in development of fluoroquinolone-resistance bacteria initiated from livestock to wastewater and from municipal STPs, hospitals, and rivers to effluent and sewage sludge (Reinthaler et al. 2003; Polk et al. 2004; Hu et al. 2008; Taylor et al. 2008).

In WWTPs, the combined sewage from hospitals and other places is subjected to several biological and physicochemical processes for biodegradation, where the number of pathogens is reduced, and nitrogen and phosphorous are removed prior to discharge in surface water; however, sewage acting as a special nutrient-rich environment, with high bacterial density, antibiotics, boosts the antibiotic resistance in bacteria, and antibiotic-resistance genes (Szczepanowski et al. 2009; Ju et al. 2019). As a consequence, sewage is considered as an initiator for development, recombination, and distribution of antibiotic resistance (Rizzo et al. 2013). Furthermore, WWTPs are not particularly designed to eliminate antibiotic-resistance genes, although the biomass of pathogens is decreased through WWTPs, and strong selection for antibiotics may occur during treatment processes, which may eventually increase the fraction of resistant pathogens (Hocquet et al. 2016; Bürgmann et al. 2018). The richness of antibiotic-resistance bacteria and antibiotic-resistance genes is highly reduced in the water portion of sewage, whereas their abundance is more in biosolids used as a fertilizer for crops (Munir et al. 2011; Chen and Zhang 2013; Yang et al. 2014). Most notably the different kinds of sewage treatments employed in WWTPs, such as anaerobic digestion, drying, and application to agricultural soils, may limit the abundance of antibiotic-resistance genes in both effluent and biosolids to some extent (Diehl and Lapara 2010).

3.4.4 Occurrence in Groundwater

The ground water bodies are exposed to antibiotic contamination by artificial man-made activities. Soil acts as a natural barrier to curb the contaminant into the subsurface water; however, if the contamination occurs, it is difficult to suppress its impact. Pollutants could be administered to the groundwater by many sources, such as infiltration of wastewater, natural bank filtration, water supply pipes, and rainfall. In Spain, chemical pollutant in the aquifers of rural and urban area was examined

(Fick et al. 2009) and 72 chemical compounds and 23 transformation products were present in the groundwater; all were from the WWTPs influent. Among pharmaceuticals, ciprofloxacin was detected with high relative concentration of 32.75 ng L^{-1}, probably from agricultural purposes or/and insufficient water treatment (Cabeza et al. 2008). On the other side in Catalonia, 18 types of sulfonamide were detected with concentration ranging from 0.01 to 3460.57 ng L^{-1}, mainly from livestock fertilizer (Garcia-Galan et al. 2010; Jurado et al. 2012).

Many studies have reported that groundwater contains not only antibiotics, but also pesticides, hormones, industrial compounds, and some personal care products, with much lower concentrations than found in WWTPs and rivers (Barnes et al. 2008; Cabeza et al. 2008; Garcia-Galan et al. 2010; Hu et al. 2010; Lapworth et al. 2012). The occurrence of pharmaceutical compounds in groundwater is reported by many researchers, and contamination is probably from landfill, insufficient treatment of wastewater, and to incorrect disposal of wastes from pharmaceutical factories and hospitals (Schwarzbauer et al. 2002; Asano and Cotruvo 2004). Some of the pharmaceutical compounds including antibiotics released into the water bodies and develop resistance power in the microorganism against other antimicrobial drugs (Voulvoulis et al. 2013; Sui et al. 2015). The physicochemical properties of some pharmaceutical compounds as well as the ecological properties of different water bodies controlled the persistence and occurrence of some pharmaceuticals (Gurr and Reinhard 2006; Liu et al. 2011). Some medicinal compounds need long time to be degraded by metabolic action within a patient, which suggests the potential accumulation of such compounds in animal tissue. Therefore, the bioaccumulation of pharmaceuticals and their potential is another factor for aquifers contamination (Lai et al. 2002).

Some of but not all pharmaceutical compounds discharged into the groundwater and get adsorbed into a solid phase like biota, sediments, or suspended solid particles, but the adsorption rate depends upon chemical properties of these compounds (Lai et al. 2000; Jones et al. 2002). A summary of some classes of antibiotics present in different environmental compartments is listed in Table 3.2.

3.4.5 Existence in Plant and Aquatic Animals

The pharmaceutical compounds from water compartment move to crops, vegetables, aquatic plants, and animals (Eggen et al. 2011; Dong et al. 2012; Na et al. 2013). In Napoli, ciprofloxacin was found to be accumulated in root of carrot (*Daucus* spp.) and barley (*Hordeum vulgare*), with bioaccumulation factor ⍰1 higher than that of leaf (Eggen et al. 2011). The uptake ability of plant compartment is different from part to part. In a study, Hu et al. (2010) reported the sequence of accumulation factor of antibiotics as root ⍰ stem⍰ leaf, and also attributed this variation to seasonal changes. In winter, bioaccumulation of antibiotics is more than summer. For instance, oxytetracycline, tetracycline, and chlortetracycline were bioaccumulated in coriander leaves at very high level in winter, with relative concentration ranges

Table 3.2 Distribution of different classes of antibiotics in the environmental compartments

Class and antibiotics	Concentration	Environmental compartments	Reference
Sulfonamides (Sulfamethazine)	0.005–50 μM	Marine sediment slurry	Hou et al. (2015)
Sulfonamides (Sulfamethazine)	20–100 mg L^{-1}	Soil treated with poultry manure	Awad et al. (2016)
Sulfonamides (Sulfadiazine)	0–100 mg kg^{-1} with liquid manure	Soil treated with or without pig liquid manure	Hammesfahr et al. (2011)
Quinolones and Fluoroquinolones (Ciprofloxacin)	5 and 50 mg kg^{-1}	Soil microcosms	Cui et al. (2014)
Tetracyclines (Tylosin)	N50 mg kg^{-1}	Soil	Demoling et al. (2009)
β-Lactams (Amoxicillin)	10–100 mg kg^{-1}	Soil treated with manure	Binh et al. (2007)
β-Lactams (PenicillinG) and Tetracyclines (Oxytetracycline)	80 μg mL^{-1}	River bacterial community under long term antibiotic stresses by treated wastewater from two antibiotic producing facilities	Li et al. (2011)
Glycopeptides (Vancomycin)	1–1000 mg L^{-1}	River sediments sampled 10 m upstream and 10 m downstream from a WWTP	Laverman et al. (2015)
Lincosamides (Lincomycin)	0.05–500 mg kg^{-1}	Two forest soils with different pH and clay content	Cermak et al. (2008)
Macrolides (Natamycin)	50–200 mg L^{-1}	Soil treated with manure	Hammesfahr et al. (2008)
Quinolones and Fluoroquinolones (Ciprofloxacin)	0–200 μg mL^{-1}	Salt marsh sediment	Cordova-Kreylos and Scow (2007)
Quinolones and Fluoroquinolones (Ciprofloxacin)	0.2–2 mg L^{-1}	Marine sediment with ciprofloxacin in the overlying water	Naslund et al. (2008)
Tetracyclines (Tylosin)	2000 mg kg^{-1}	Sandy soil	Muller et al. (2002)
Tetracyclines (Chlortetracycline)	0–100 mg kg^{-1}	Soil spiked with DOM extracted from pig manure	Liu et al. (2015)
Tetracyclines (Chlortetracycline)	1, 10, and 100 mg kg^{-1}	Soil microcosms	Yang et al. (2009)
Tetracyclines (Oxytetracycline)	10 mg kg^{-1}	Wheat rhizosphere soil	Liu et al. (2012)
Tetracyclines (Oxytetracycline)	5–200 mg kg^{-1}	Grass and agroforestry soils	Unger et al. (2013)
Tetracyclines (Oxytetracycline)	200 ppm	Agricultural soil spiked with swine manure	ECDC (2015)
Sulfonamides (Sulfadiazine)	10–100 mg kg^{-1}	Soil amended with glucose	Schauss et al. (2009)

(continued)

Table 3.2 (continued)

Class and antibiotics	Concentration	Environmental compartments	Reference
Sulfonamides (Sulfadiazine)	10 and 100 μg g^{-1}	Soil amended with pig manure	Schauss et al. (2009)
Sulfonamides (Sulfadiazine)	1–50 mg kg^{-1}	Soil amended with glucose	Underwood et al. (2011)
Sulfonamides (Sulfamethoxazole)	0.005–50 μM	Enrichment culture with groundwater from pristine zone of a sandy drinking-water aquifer (anaerobic and heterotrophic media)	Hund-Rinke et al. (2004)
Sulfonamides (Sulfamethoxazole)	240–520 μg L^{-1}	Microbial communities from an oligotrophic aquifer	Haack et al. (2012)
mixture of Sulfamethoxazole Erythromycin and Ciprofloxacin	0.33–3.33 μg L^{-1}	Pristine aquifer microbial community	Schauss et al. (2009)
Sulfonamides (Sulfadiazine)	10 and 100 mg g^{-1}	Bulk soil	Underwood et al. (2011)
A mixture of 16 antibiotics (Sulfonamides, Quinolones, Macrolides, Tetracycline)	0–1500 ng L^{-1}	Biofilms exposed to river waters from downstream sites	Proia et al. (2013)
Sulfonamides (Sulfamethoxazole)	20–500 mg kg^{-1}	Soil amended with manure from alfalfa or antibiotic-treated pig	Binh et al. (2007)
Sulfonamides (Sulfapyridine)	100–1000 mg kg^{-1}	Topsoil	Sengelov et al. (2003) and Schmitt et al. (2005)
Tetracyclines (Tetracycline)	5–500 mg kg^{-1}	Soil treated with pig manure	Brandt et al. (2009)
Tetracyclines (Oxytetracycline)	100–1000 mg kg^{-1}	Pristine topsoil amended with milled maize straw or glucose	Reddersen et al. (2002)
Macrolides (Natamycin)	50–200 mg L^{-1}	Bulk soil and rhizosphere soil suspensions on agar medium	Naslund et al. (2008)
Sulfonamides (Sulfamethoxazole)	20–500 mg kg^{-1}	Soil amended or non-amended with manure from alfalfa or antibiotic-treated pig	Binh et al. (2007)
Sulfonamides (Sulfachloropyridazine)	100 mg kg^{-1}	Soil amended with fresh pig slurry or with alfalfa manure	Demoling and Baath (2008)
Sulfonamides (Sulfadiazine)	1–100 mg kg^{-1}	Soil	Demoling and Baath (2008)
Tetracyclines (Tylosin)	50–1500 mg $kg-^{1}$	Soil with or without alfalfa	Demoling and Baath (2008)
Sulfonamides (Sulfadiazine)	10–100 mg kg^{-1}	Soil amended with Manure	Heuer et al. (2008)

(continued)

Table 3.2 (continued)

Class and antibiotics	Concentration	Environmental compartments	Reference
Quinolones and Fluoroquinolones (Ciprofloxacin)	Quinolones and Fluoroquinolones (Ciprofloxacin)	Interstitial water samples of wetlands	Heuer et al. (2008)
Aminoglycosides (Streptomycin)	400 mg L^{-1}	Activated sludge	Weber et al. (2014)
Sulfonamides (Sulfadiazine)	10–100 mg kg^{-1}	Soil amended with Manure	Tomlinson et al. (1966)
Sulfonamides (Sulfadimethoxine)	50–200 mg kg^{-1}	Soil	Kotzerke et al. (2008)
Tetracyclines (Oxytetracycline)	12.5–75 mg L^{-1}	Plus active cultures of the nitrifying bacteria Nitrosomonas and Nitrobacter	Toth et al. (2011)
Tetracyclines (Chlortetracycline)	50–200 μg kg^{-1} in poultry manure	Soil	Klaver and Matthews (1994)
Tetracyclines (Chlortetracycline)	1 mg L^{-1}	Groundwater	Ahmad et al. (2014)
Ciprofloxacin	0.16–21.74 μg kg^{-1}	River sediments	Hannaa et al. (2018)
Enrofloxacin	0.16–24.42 μg kg^{-}		
Levofloxacin	0.82–2.89 μg kg^{-1}		
Norfloxacin	0.14–2.20 μg kg^{-1}		
Chloramphenicol	0.98–1.53 μg kg^{-1}		
β-lactam (Amoxycillin)	0–16.7 ng L^{-1}	Rivers	Grenni et al. (2018)
Macrolide (Clarithromycin)	8.3–149.0 ng L^{-1}		
Sulfonamide (Sulfamrthoxazole)	950 ng L^{-1}	Groundwater	Adler et al. (2018)
Tetracyclines	100 mg kg^{-1}	Slurry and fermentation residues	Adler et al. (2018)

from 78–330, 1.9–5.6, and 92–481 μg kg^{-1}, respectively. The bioaccumulation range of sulfadoxine, sulfachloropyridazine, chloramphenicol, and sulfamethoxazole in radish leaves was found between 0.2–0.6, 0.1–0.5, 8–30, and 0.9–2.7 μg kg^{-1}, respectively. In addition, concentrations ranged from 1.7–3.6, 1.1, and 5–20 μg kg^{-1}in celery leaves for ofloxacin, pefloxacin, and lincomycin, respectively. Quinolones were also predominantly occurred in aquatic plants and aquatic animals with the corresponding concentrations ranged from 8.37–6532 μg kg^{-1} and 17.8–167 μg kg^{-1}, respectively. Similarly, macrolides were also detected at a non-detectable level of 182 μg kg^{-1}in China (Li et al. 2013). Contaminants from

sludge amended soils are transferred to the plant through uptake by root, retention by root, leaves uptake, translocation, and animal intake. Many scientists showed the adverse effect of antibiotics on plant activities like growth rate, nitrogen, fixation, heterocyst formation, and bioaccumulation (Forni et al. 2002; Migliore et al. 2003). From the plants they reach to animals and get accumulated into their body tissues and consequently lead to food chain contamination. Soil characteristics such as sorption kinetics, pH, and organic matter affect accumulation of pharmaceutical compound in the soil (Jjemba 2002). The leaching of pharmaceuticals to the groundwater can be prevented by sorption, and may reduce the effect of antibiotics on soil microbes (Kotzerke et al. 2008; Bialk-Bielinska et al. 2012), as well as, it plays an important role in drug plant uptake, which is determined upon the hydroponic conditions (Wu et al. 2015). Organic content of soli is another factor affection pharmaceuticals uptake by plants. The uptake ability is decreased when the organic component of soil is increased; however, under normal farming conditions, the potential uptake is low (Holling et al. 2012).

Chemical substances used in aquaculture may directly release to the surface water, whereas veterinary drugs in soil may have the potential to discharge to surface water or reach the aquifers (Tarazona and Vega 2002). The exposure route of pharmaceuticals and veterinary medicines in aquacultures are variable, one possible source is the food chain of animals that accumulate drugs in their tissues. Another possible route is the crop with significant levels of accumulated medicines from manure or soil; fish with accumulated drugs used for aquaculture treatment purposes and the main source is contaminated ground and wastewater. In a study, Guo et al. (2016) reported that tetracyclines are common veterinary medicines in livestocks in China, where it was found in animal manure with very high concentration of about 5775.6 mg kg^{-1}, indicating that the livestock is the main source of contamination of aquatic culture.

3.5 The Hazardous Impact of Antibiotics in the Environment

The presence and occurrence of pharmaceuticals in the environment have been shown through many studies. The rate of entrance in to the environment is more than the rate of removal and antibiotics are considered to be persistent of pseudo-persistent (Mackay et al. 2014). The toxic effect of antibiotics involves all biotic compartments including microbial community, animal, plant and humans (Madden et al. 2009). Generally, the sequences of the negative action of antibiotics occur at the molecular level firstly, then to the cellular levels, organism level, individual level, population level, community level, and finally at ecosystem level. The sensitivity at the DNA level is considered as the most accurate and effective bioindicator for toxicity of pollution in the soil. The genotoxicity and antibiotic stress could be determined by DNA damage and alteration in enzyme activities in various organisms (Halling-Sorensen 2000; Stuart et al. 2012).

In farmland soil, many metabolic processes in microorganisms could be affected or disrupted by antibiotics, such as nitrogen fixation and nutrients fluxes (Larsson 2014). The microorganisms in wastewater treatment plants showed a pattern of flexibility to overcome antibiotic hazards, mostly by horizontal gene transfer and their ability to accumulate mutations, because the concentration of residual drugs and their exposure route is quite different (Larsson 2014). The most important issue about the risk of antibiotics in the environment is the development of antibiotic resistance of pathogens in human and animals (Kümmerer 2004). Antibiotic resistance bacteria confer their resistance mainly from the environment rather than other sources, as the external environment creates a reservoir for resistance genes and provides a chance for genetic recombination other than in the pathogens inside the body. Misuse of antibiotics, wrong disposal through the aquatic environment, insufficient treatment of wastewater and sludge make the dissemination of resistance genes more frequent (Ashbolt et al. 2013). The occurrence of antibiotics in the environment is likely to have a toxic effect on the biota, for example, ciprofloxacin and ofloxacin were showed to be in fresh water with high concentration and can have most potent effect even at low concentration (Feitosa-Felizzola and Chiron 2009; Bengtsson-Palme and Larsson 2016).

Ciprofloxacin in fresh water with concentration of 5 μg L^{-1} inhibited the growth of cyanobacteria *Mycrocystisaeruginosa* (Robinson et al. 2005) and *Vibrio fischeri* and the algae *Psuedokirchunellasubcapitata*at concentrations of 0.9 μg L^{-1} and 4.47 μg L^{-1}, respectively. To some extent, some antibiotics target the intracellular components of plant cells, as tetracycline, fluoroquinolones, and macrolides target the chloroplast and inhibit the mitochondrial protein synthesis (Brain et al. 2008); fluoroquinolones, in particular, have a negative effect on the morphology and photosynthesis in plant (Aritstilde et al. 2010). Tetracycline is able to cause chromosomal aberration and has toxic effect on photosynthesis and growth (Xie et al. 2011). Similarly, β-lactam antibiotics have slightly affected on plants such as rice (*Oryzasativa* L), carrot (*Daucuscarota* L.), and Chainese cabbage (*B. chinesis* L.) (Grewal et al. 2006; Grzebelus and Skop 2014; Meng et al. 2014). Furthermore, the reductions in photosynthetic pigments, carotenoid, and chlorophylls were also caused by tetracyclin, erythromycin, and ciprofloxacin (Pomati et al. 2004; Quanten et al. 2007; Yaronskaya et al. 2007; Hillis et al. 2011). Ciprofloxacin strongly inhibits the net assimilation rate of foliage (*Triticum aestivum*) (Opris et al. 2013).

3.6 Conclusions

Pharmaceutical compounds are known to be important for human life and welfare and economy in modern communities, but contamination by antibiotics has been detected worldwide in natural environment (aquatic, soil, animal, and plant). However, antibiotics play a role in evolving the biological effects, which were estimated on all biota. Not all antibiotics are present in the environment at high levels to cause remarkable effects but low concentration is sometimes enough to cause effects on animals and plants. The biological and physicochemical properties and persistence decide the chronic and long-term effects of pharmaceuticals on the environment. Some of the pharmaceuticals are able to biodegrade easily, while others need more time to degrade and thus the probability of causing negative effects also varies. Likewise, the exposure route, dose of residues, time of exposure, and kinetic mechanisms are factors responsible for antibiotic-induced effects.

Antibiotics can reach the surface water, drinking water, groundwater, seawater, soil, plants, and aquatic organisms by variety of ways, but WWTPs are considered as the most significant source for dissemination of antibiotics. Pharmaceutical manufacturing plant, hospital effluent, and application of livestock manures are other means for antibiotic discharge. Landfill application of WWTPs sludge, sewage, and fertilizer are shown to be important sources for antibiotics entrance into the food chain through animals and agriculture practices. Several processes contribute to eliminate antibiotics in conventional and advanced WWTPs such as sorption, biodegradation, photodegradation, and oxidation; however, the efficiency of antibiotic removal in WWTPs is variable due to the ability of some pharmaceuticals for sorption onto biosolids or to be remaining at their aqueous phase. Many questions have risen regarding the identification of parent antibiotics or their transformation products with other toxic organic and inorganic substances, as the mixtures and their compositions. The presence of such mixtures with other toxic compounds in wastewater treatment plants has another effect on non-target organisms. The development of antibiotics and antibiotic resistance genes in the environment is gaining a great concern over the time, and their impact on ecosystem needs fast pace to understand the interaction of these compounds with the environmental constituents.

References

Adler N, Balzer F, Blondzik K et al (2018) Antibiotics and antibiotic resistances in the environment background, challenges and options for action. The Publisher German Environment Agency/UBA. ISSN 2363-829X. P:44.

Adzitey F (2015) Antibiotic classes and antibiotic susceptibility of bacterial isolates from selected poultry; a mini review. World Vet J 5(3):36–41

Agunbiade FO, Moodley B (2016) Occurrence and distribution pattern of acidic pharmaceuticals in surface water, wastewater, and sediment of the Msunduzi River, Kwazulu-Natal, South Africa. Environ Toxicol Chem 35:36–46

Ahmad M, Vithanage M, Kim K et al (2014) Inhibitory effect of veterinary antibiotics on denitrification in groundwater: a microcosm approach. Sci World J 879831

Al Aukidy M, Verlicchi P, Jelic A et al (2012) Monitoring release of pharmaceutical compounds: occurrence and environmental risk assessment of two WWTP effluents and their receiving bodies in the Po Valley, Italy. Sci Total Environ 438:15–25

Alidina M, Hoppe-Jones C, Yoon M et al (2014) The occurrence of emerging trace organic chemicals in wastewater effluents in Saudi Arabia. Sci Total Environ 478:152–162

Aminov RI (2010) A brief history of the antibiotic era: lessons learned and challenges for the future. Front Microbiol 1(134):1–7

Aritstilde L, Melis A, Sposito G (2010) Inhibition of photosynthesis by a fluoroquinolone antibiotic. Environ Sci Technol 44:1444–1450

Asano T, Cotruvo J (2004) Groundwater recharge with reclaimed municipal wastewater: health and regulatory considerations. Water Res 38:1941–1951

Ashbolt NJ, Amezquita A, Backhaus T et al (2013) Human health risk assessment (HHRA) for environmental development and transfer of antibiotic resistance. Environ Health Perspect 121:993–1001

aus der Beek T, Weber FA, Bergmann A et al (2016) Pharmaceuticals in the environment-global occurrences and perspectives. Environ Toxicol Chem 35:823–835

Awad YM, Ok YS, Igalavithana AD et al (2016) Sulphamethazine in poultry manure changes carbon and nitrogen mineralisation in soils. Chem Ecol 32:899–918

Baquero F, Martinez JL, Canton R (2008) Antibiotics and antibiotic resistance in water environments. Curr Opin Biotechnol 19:260–265

Barnes KB, Kolpin DW, Furlong ET et al (2008) A national reconnaissance of pharmaceuticals and other organic wastewater contaminants in the United States, I. Groundwater. Sci Total Environ 402:192–200

Bengtsson-Palme J, Larsson DGJ (2016) Concentrations of antibiotics predicted to select for resistant bacteria: proposed limits for environmental regulation. Environ Int:86. https://doi.org/10.1016/j.envint.2015.10.015

Benotti MJ, Trenholm RA, Vanderford BJ et al (2009) Pharmaceuticals and endocrine disrupting compounds in U.S. drinking water. Environ Sci Technol 43:597–603

Bester K (2003) Triclosan in a sewage treatment process – balances and monitoring data. Water Res 37:3891–3896

Bialk-Bielinska A, Maszkowska J, Mrozik W et al (2012) Sulfadimethoxine and sulfaguanidine: their sorption potential on natural soils. Chemosphere 86:1059–1060. https://doi.org/10.1016/j.chemosphere.20111111.11.058

Binh CTT, Heuer H, Gomes NCM et al (2007) Short-term effects of amoxicillin on bacterial communities in manured soil. FEMS Microbiol Ecol 62:290–302

Bird K, Boopathy R, Nathaniel R et al (2019) Water pollution and observation of acquired antibiotic resistance in Bayou Lafourche, a major drinking water source in Southeast Louisiana, USA. Environ Sci Pollut Res. https://doi.org/10.1007/s11356-018-4008-5

Blackwell PA, Boxall ABA, Kay P et al (2005) Evaluation of a lower tier exposure assessment model for veterinary medicines. J Agric Food Chem 53:2192–2201

Boxall ABA, Rudd MA, Brooks BW et al (2012) Pharmaceuticals and personal care products in the environment: what are the big questions? Environ Health Perspect 120:1221–1229

Brain RA, Hanson ML, Solomon KR et al (2008) Aquatic plants exposed to pharmaceuticals: effects and risks. Rev Environ Contam Toxicol 192:67–115

Brandt KK, Sjoholm OR, Krogh KA et al (2009) Increased pollution-induced bacterial community tolerance to sulfadiazine in soil hotspots amended with artificial root exudates. Environ Sci Technol 43:2963–2968

Brooks GF, Butel JS, Morse SA (2004) Jawetz, Melnick and Adelbergs medical microbiology, 23rd edn. McGraw Hill Companies, Singapore, p 877

Bürgmann H, Frigon D, Gaze WH et al (2018) Water and sanitation: an essential battlefront in the war on antimicrobial resistance. FEMS Microbiol Ecol 94:fiy101

Cabello FC (2006) Heavy use of prophylactic antibiotics in aquaculture: a growing problem for human and animal health and for the environment. Environ Microbiol 8:1137–1144. https://doi.org/10.1111/j.1462-2920.2006.01054.x

Cabeza Y, Candela L, Ronen D et al (2008) Monitoring the occurrence of emerging contaminant in the treated wastewater and groundwater between 2008 and 2010. The BaixLlobregat (Barcelona, Spain). J Hazard Mater 239–240:32–39

Calderon CB, Sabundayo BP (2007) Antimicrobial classifications: drugs for bugs. In: Schwalbe R, Steele-Moore L, Goodwin AC (eds) Antimicrobial susceptibility testing protocols. CRC Press/ Taylor and Frances Group. ISBN 978-0-8247-4100-6

Carballa M, Omil F, Lema JM et al (2004) Behavior of pharmaceuticals, cosmetics and hormones in a sewage treatment plant. Water Res 38:2918–2926

Carvalho IT, Santos L (2016) Antibiotics in the aquatic environments: a review of the European scenario. Environ Int 94:736–757

Cermak L, Kopecky J, Novotna J et al (2008) Bacterial communities of two contrasting soils reacted differently to lincomycin treatment. Appl Soil Ecol 40:348–358

Cha JM, Yang S, Carlson KH (2006) Trace determination of B-lactam antibiotics in surface water and urban wastewater using liquid chromatography combined with electrospray tandem mass spectrometry. J Chromatogr A 1115:46–57. https://doi.org/10.1016/j.chroma.2006.02.086

Chen H, Zhang M (2013) Occurrence and removal of antibiotic resistance genes in municipal wastewater and rural domestic sewage treatment systems in eastern China. Environ Int 55:9–14

Choquet-Kastylevsky G, Vial T, Descotes J (2002) Allergic adverse reactions to sulfonamides. Curr Allergy Asthma Rep 2:16–25

Christian T, Schneider RJ, Färber HA et al (2003) Determination of antibiotic residues in manure, soil, and surface waters. Acta Hydrochim Hydrobiol 31:36–44. https://doi.org/10.1002/aheh.200390014

Cordova-Kreylos AL, Scow KM (2007) Effects of ciprofloxacin on salt marsh sediment microbial communities. ISME J 1:585–595

Cui H, Wang SP, Fu J et al (2014) Influence of ciprofloxacin onmicrobial community structure and function in soils. Biol Fertil Soils 50:939–947

DeJongh CM, Kooij PJF, de Voogt P et al (2012) Screening and human health risk assessment of pharmaceuticals and their transformation products in Dutch surface waters and drinking water. Sci Total Environ 427:70–77

Demoling LA, Baath E (2008) No long-term persistence of bacterial pollution-induced community tolerance in tylosin-polluted soil. Environ Sci Technol 42:6917–6921

Demoling LA, Baath E, Greve G et al (2009) Effects of sulfamethoxazole on soil microbial communities after adding substrate. Soil Biol Biochem 41:840–848

Diaz-Cruz M, Lopez de Alda M, Barcelo D (2003) Environmental behaviour and analysis of veterinary and human drugs in soils, sedimnets and sludge. TrAC Trends Anal Chem 22:340

Díaz-Cruz MS, García-Galán MJ, Barceló D (2008) Highly sensitive simultaneous determination of sulfonamide antibiotics and one metabolite in environmental waters by liquid chromatography–quadrupole linear ion trap–mass spectrometry. J Chromatogr A 1193:50–59. https://doi.org/10.1016/j.chroma.2008.03.029

Diehl DL, Lapara TME (2010) Effect of temperature on the fate of genes encoding tetracycline resistance and the integrase of class 1 integrons within anaerobic and aerobic digesters treating municipal wastewater solids. Environ Sci Technol 44:9128–9133

Dolliver H, Gupta S (2008) Antibiotic losses in leaching and surface runoff from manure-amended agricultural land. J Environ Qual 37:1227–1237

Dong L, Gao J, Xie X et al (2012) DNA damage and biochemical toxicity of antibiotics in soil on the earthworm *Eiseniafetida*. Chemosphere 89:44–51

ECDC (2015) (European Centre for Disease Prevention and Control), EFSA (European Food Safety Authority) and EMA (European Medicines Agency). ECDC/EFSA/EMA first joint report on the integrated analysis of the consumption of antimicrobial agents and occurrence of antimicrobial resistance in bacteria from humans and food producing animals. EFSA J 13:4006. P114 http://www.efsa.europa.eu/sites/default/files/scientific_output/files/main_documents/4006.pdf. Accessed January 2017

Eggen T, Asp TN, Grave K et al (2011) Uptake and translocation of metformin, ciprofloxacin and narsin in forage and crop plants. Chemosphere 85:26–33

Eguchi K, Nagase H, Ozawa M et al (2004) Evaluation of antimicrobial agents for veterinary use in the ecotoxicity test using microalgae. Chemosphere 57:1733–1738

Etebul E, Arikekpar I (2016) Antibiotics: classification and mechanisms of action with emphasis on molecular perspectives. IJAMBR 4:90–101
Feitosa-Felizzola J, Chiron S (2009) Occurrence and distribution of selected antibiotics in a small Mediterranean stream (Arc River, Southern France). J Hydrol 364:50–57. https://doi.org/10.1016/j.jhydrol.2008.10.006
Ferber D (2003) Antibiotic resistance. WHO advises kicking the livestock antibiotic habit. Science 301:1027
Fick J, Söderström H, Lindberg RH et al (2009) Contamination of surface, ground, and drinking water from pharmaceutical production. Environ Toxicol Chem 28:2522–2527. https://doi.org/10.1897/09-073.1
Flaherty CM, Dodson SI (2005) Effects of pharmaceuticals on Daphnia survival, growth, and reproduction. Chemosphere 61:200–207. https://doi.org/10.1016/j.chemosphere.2005.02.016
Focazio MJ, Kolpin DW, Barnes KK et al (2008) A national reconnaissance for pharmaceuticals and other organic wastewater contaminants in the United States – (II) untreated drinking water sources. Sci Total Environ 402:201–216
Forni C, Cascone A, Fiori M et al (2002) Sulfadimethoxine and Azollafiliculoids Lam.: a model of drug remediation. Water Res 36:3398–3403
Frank U, Tacconelli E (2012) The Daschner guide to in-hopsital antibiotic therapy. European standards. Available online at: http://www.springer.com/978-3-642-18401-7. 300 p
Garcia-Galan MJ, Garrido T, Fraile J et al (2010) Simultaneous occurrence of nitrates and sulfonamide antibiotics in two ground water bodies of Catalonia (Spain). J Hydrol 383:93–101
Giger W, Alder AC, Golet EM et al (2003) Occurrence and fate of antibiotics as trace contaminants in waste waters, sewage sludges, and surface waters. Chimia Int J Chem 57:485–491
Ginebreda A, Muñoz I, de Alda ML et al (2010) Environmental risk assessment of pharmaceuticals in rivers: relationships between hazard indexes and aquatic macroinvertebrate diversity indexes in the Llobregat River (NE Spain). Environ Int 36:153–162. https://doi.org/10.1016/j.envint.2009.10.003
Göbel A, Thomsen A, McArdell CS et al (2005) Extraction and determination of sulfonamides, macrolides, and trimethoprim in sewage sludge. J Chromatogr A 1085:179–189. https://doi.org/10.1016/j.chroma.2005.05.051
Göbel A, Mcardell C, Joss A et al (2007) Fate of sulfonamides, macrolides, and trimethoprim in different wastewater treatment technologies. Sci Total Environ 372:361–371
Gothwal R, Shashidhar T (2014) Antibiotic pollution in the environment: a review. Clean Soil Air Water 42(9999):1–11
Grenni P, Ancona V, Caracciolo AB (2018) Ecological effects of antibiotics on natural ecosystems: a review. Microchem J 136:25–39
Grewal D, Gill R, Gosal SS (2006) Influence of antibiotic cefotaxime on somatic embryogenesis and plant regeneration in indica rice. Biotechnol J 1:1158–1162
Grzebelus E, Skop L (2014) Effect of β-lactam antibiotics on plant regeneration in carrot protoplast cultures. In Vitro Cell Dev Biol Plant 50:568–575
Guo XY, Hao LJ, Qiu PZ et al (2016) Pollution characteristics of 23 veterinary antibiotics in livestock manure and manure-amended soils in Jiangsu province, China. J Environ Sci Heal B 51(6):383–392
Gurr C, Reinhard M (2006) Harnessing natural attenuation of pharmaceuticals and hormones in rivers. Environ Sci Technol 40:2872–2876
Haack SK, Metge DW, Fogarty LR et al (2012) Effects on groundwater microbial communities of an engineered 30-day in situ exposure to the antibiotic sulfamethoxazole. Environ Sci Technol 46:7478–7486
Halling-Sorensen B (2000) Algal toxicity of antibacterial agent used in intensive farming. Chemosphere 40:731–739
Hammesfahr U, Heuer H, Manzke B et al (2008) Impact of the antibiotic sulfadiazine and pig manure on the microbial community structure in agricultural soils. Soil Biol Biochem 40:1583–1591

Hammesfahr U, Bierl R, Thiele-Bruhn S (2011) Combined effects of the antibiotic sulfadiazine and liquid manure on the soil microbial-community structure and functions. J Plant Nutr Soil Sci 174:614–623

Hannaa N, Sunb P, Sunc Q et al (2018) Presence of antibiotic residues in various environmental compartments of Shandong province in eastern China: its potential for resistance development and ecological and human risk. Environ Int 114:131–142

Heise J, Höltge S, Schrader S et al (2006) Chemical and biological characterization of non-extractable sulfonamide residues in soil. Chemosphere 65:2352–2357

Heuer H, Focks A, Lamshoft M et al (2008) Fate of sulfadiazine administered to pigs and its quantitative effect on the dynamics of bacterial resistance genes in manure and manured soil. Soil Biol Biochem 40:1892–1900

Higgins PG, Fluit AC, Schmitz FJ (2003) Fluoroquinolones: structure and target sites. Curr Drug Targets 4:181–190

Hillis DG, Fletcher J, Solomon KR et al (2011) Effects of ten antibiotics on seed germination and root elongation in three plant species. Arch Environ Contam Toxicol 60:220–232

Hocquet D, Muller A, Bertrand X (2016) What happens in hospitals does not stay in hospitals: antibiotic-resistant bacteria in hospital wastewater systems. J Hosp Infect 93:395–402

Holling CS, Bailey JL, Vanden Heuvel B, Kinney CA (2012) Uptake of human pharmaceuticals and personal care products by cabbage (*Brassica campestris*) from fortified and biosolids-amended soils. J Environ Monit 14:3029. https://doi.org/10.1039/c2em30456b

Hou L, Yin G, Liu M et al (2015) Effects of sulfamethazine on denitrification and the associated N_2O release in estuarine and coastal sediments. Environ Sci Technol 49:326–333

Hu J, Shi J, Chang H et al (2008) Phenotyping and genotyping of antibiotic-resistant *Escherichia Coli* isolated from a Natural River basin. Environ Sci Technol 42:3415–3420. https://doi.org/10.1021/es7026746

Hu X, Zhou Q, Lou Y (2010) Occurrence and source analysis of typical veterinary antibiotics in manure, soil, vegetables and groundwater from organic vegetable bases, Northern China. Environ Pollut 158:2992–2998

Huerta-Fontela M, Teresa Galceran M, Ventura F (2011) Occurrence and removal of pharmaceuticals and hormones through drinking water treatment. Water Res 45:1432–1442

Hund-Rinke K, Simon M, Lukow T (2004) Effects of tetracycline on the soil microflora: function, diversity, resistance. J Soils Sediments 4:11–16

Ji X, Shen Q, Liu F et al (2012) Antibiotic resistance gene abundances associated with antibiotics and heavy metals in animal manures and agricultural soils adjacent to feedlots in Shanghai; China. J Hazard Mater 235–236:178–185

Jia A, Wan Y, Xiao Y et al (2012) Occurrence and fate of quinolone and fluoroquinolone antibiotics in a municipal sewage treatment plant. Water Res 46:387–394

Jjemba P (2002) The potential impact of veterinary and human therapeutic agents in manure and biosolids on vegetation: a review. Agric Ecosyst Environ 93:267–278

Jjemba PK (2006) Excretion and ecotoxicity of pharmaceutical and personal care products in the environment. Ecotoxicol Environ Saf 63:113–130

Jones O, Voulvoulis N, Lester J (2002) Aquatic environmental assessment of the top 25 English prescription pharmaceuticals. Water Res 36:5013–5022

Ju F, Beck K, Yin X et al (2019) Wastewater treatment plant resistomes are shaped by bacterial composition, genetic exchange, and upregulated expression in the effuent microbiomes. ISME J 13:346–360

Jurado A, Vazquez-Sune E, Carrera J et al (2012) Emerging organic contaminants in groundwater in Spain: a review of sources, recent occurrence and fate in a European context. Sci Total Environ 440:82–94

Kanda R, Griffin P, James H et al (2003) Pharmaceutical and personal care products in Sewage treatment works. J Environ Monit 5:823–830

Kapoor G, Saigal S, Elongavan A (2017) Action and resistance mechanisms of antibiotics: A guide for clinicians. J Anaesthesiol Clin Pharmacol 33:300–305

Kemper N (2008) Veterinary antibiotics in the aquatic and terrestrial environment. Ecol Indic 8:1–13
Khan SJ, Ongerth JE (2002) Estimation of pharmaceutical residues in primary and secondary sewage sludge based on quantities of use and fugacity modelling. Water Sci Technol 46 (3):105–113
Klaver AL, Matthews RA (1994) Effects of oxytetracycline on nitrification in a model aquatic system. Aquaculture 123:237–247
Klein EY, Van Boeckel TP, Martinez EM et al (2018) Global increase and geographic convergence in antibiotic consumption between 2000 and 2015. Proc Natl Acad Sci 115(15):201717295. https://doi.org/10.1073/pnas.1717295115
Kotzerke A, Sharma S, Schauss K et al (2008) Alterations in soil microbial activity and N-transformation processes due to sulfadiazine loads in pig manure. Environ Pollut 153:315–322
Kumar K, Gupta SC, Baidoo SK et al (2005) Antibiotic uptake by plants from soil fertilized with animal manure. J Environ Qual 34:2082–2085
Kümmerer K (2004) Resistance in the environment. J Antimicrob Chemother 54:311–320. https://doi.org/10.1093/jac/dkh325
Kümmerer K (2009) Antibiotics in the aquatic environment–a review–part I. Chemosphere 75:417–434. https://doi.org/10.1016/j.chemosphere.2008.11.086
Lai K, Johnson K, Scrimshaw M et al (2000) Binding of waterborne steroid estrogens to solid phases in river and estuarine systems. Environ Sci Technol 34:3890–3894
Lai K, Scrimshaw M, Lester JN (2002) Prediction of bioaccumulation factors and body burden of natural and synthetic estrogens in aquatic organisms in the river systems. Sci Total Environ 289:159–168
Lapworth DJ, Baran N, Stuart ME et al (2012) Emerging organic contamination in groundwater: a review of sources, fate and occurrence. Environ Pollut 163:287–303
Larsson DGJ (2014) Antibiotics in the Environmnet. Upsala J Med Sci 119:108–112
Larsson DGJ, de Pedro C, Paxeus N (2007) Effluent from drug manufactures contains extremely high levels of pharmaceuticals. J Hazard Mater 148:751–755. https://doi.org/10.1016/j.jhazmat.2007.07.008
Laverman AM, Cazier T, Yan C et al (2015) Exposure to vancomycin causes a shift in the microbial community structure without affecting nitrate reduction rates in river sediments. Environ Sci Pollut Res Int 22:13702–13709
Li D, Yang M, Hu J (2008) Determination and fate of oxytetracycline and related compounds in oxytetrac. Environ Toxicol 27:80–86
Li D, Qi R, Yang M et al (2011) Bacterial community characteristics under long-term antibiotic selection pressures. Water Res 45:6063–6073
Li W, Shi Y, Gao L et al (2013) Occurrence and removal of antibiotics in a municipal wastewater reclamation plant in Beijing, China. Chemosphere 92:435–444
Li XW, Xie YF, Li CL et al (2014) Investigation of residual fluoroquinolones in a soil–vegetable system in an intensive vegetable cultivation area in Northern China. Sci Total Environ 468–469:258–264. https://doi.org/10.1016/j.scitotenv.2013.08.057
Lienert J, Bürki T, Escher BI (2007) Reducing micropollutants with source control: substance flow analysis of 212 pharmaceuticals in faeces and urine. Water Sci Technol J IntAssoc Water Pollut Res 56:87–96. https://doi.org/10.2166/wst.2007.560
Lindberg R, Olofsson U, Rendahl P et al (2006) Behavior of fluoroquinolones and trimethoprim during mechanical, chemical, and active sludge treatment of sewage water and digestion of sludge. Environ Sci Technol 40:1042
Lindberg RH, Fick J, Tysklind M (2010) Screening of antimycotics in Swedish sewage treatment plants–waters and sludge. Water Res 44:649–657. https://doi.org/10.1016/j.watres.2009.10.034
Lindsey ME, Meyer M, Thurman EM (2001) Analysis of trace levels of sulfonamide and tetracycline antimicrobials in groundwater and surface water using solid-phase extraction and liquid

chromatography/mass spectrometry. Anal Chem 73(19):4640–4646. https://doi.org/10.1021/ac010514w

Liu F, Wu J, Ying GG et al (2011) Changes in functional diversity of soil microbial community with addition of antibiotics sulfamethoxazole and chlortetracycline. Appl Microbiol Biotechnol 95:1615–1623

Liu W, Pan N, Chen W et al (2012) Effect of veterinary oxytetracycline on functional diversity of soil microbial community. Plant Soil Environ 58:295–301

Liu B, Li Y, Zhang X et al (2015) Effects of chlortetracycline on soil microbial communities: comparisons of enzyme activities to the functional diversity via Biolog Eco Plates™. Eur J Soil Biol 68:69–76

López-Serna R, Jurado A, Vázquez-Suñé E et al (2013) Occurrence of 95 pharmaceuticals and transformation products in urban groundwaters underlying the metropolis of Barcelona. Spain Environ Pollut 174:305–315. https://doi.org/10.1016/j.envpol.2012.11.022

Louvet JN, Giammarino C, Potier O et al (2010) Adverse effects of erythromycin on the structure and chemistry of activated sludge. Environ Pollut 158:688–693

Mackay D, Hughes DM, Luisa RM, Bonnell M (2014) The Role of Persistence in Chemical Evaluations. 10:588–594. https://doi.org/10.1002/ieam.1545

Madden JC, Enoch SJ, Hewitt M et al (2009) Pharmaceuticals in the environment: good practice in predicting acute ecotoxicological effects. Toxicol Lett 185:85–101

Madikizela LM, Tavengwa NT, Chimuka L (2017) Status of pharmaceuticals in African water bodies: occurrence, removal and analytical methods. J Environ Manag 193:211–220. https://doi.org/10.1016/j.jenvman.2017.02.022

Mahmood AR, Al-Haideri HH, Fikrat M, Hassan FM (2019) Detection of antibiotics in drinking water treatment plants in Baghdad City, Iraq. Hind Adv Pub Heal, Article ID 7851354, 10 pages. https://doi.org/10.1155/2019/7851354

Managaki S, Murata A, Takada H et al (2007) Distribution of macrolides, sulfonamides, and trimethoprim in tropical waters: ubiquitous occurrence of veterinary antibiotics in the Mekong Delta. Environ Sci Technol 41:8004–8010. https://doi.org/10.1021/es0709021

Martinez JL, Baquero F, Andersson DI (2007) Predicting antibiotic resistance. Nat Rev Microbiol 5:958–965

Matongo S, Birungi G, Moodley B et al (2015) Pharmaceutical residues in water and sediment of Msunduzi River, KwaZulu-Natal, South Africa. Chemosphere 134:133–140

McManus PS, Stockwell VO, Sundin GW et al (2002) Antibiotic use in plant agriculture. Annu Rev Phytopathol 40:443–465

Meakins N, Bubb J, Lester JN (1994) The fate and behaviour of organic micropollutants during wastewater treatment processes: a review. Int J Environ Pollut 4:27

Meng Q, Liu Z, Zhang Y et al (2014) Effects of antibiotics on in vitro-cultured cotyledons. Vitr Cell Dev Biol Plant 50:436–441

Michael I, Rizzo L, McArdell CS et al (2013) Urban wastewater treatment plants as hotspots for the release of antibiotics in the environment: a review. Water Res 47:957–995

Migliore L, Cozzolino S, Fiori M (2003) Phytotoxicity to and uptake of enerofloxacin in crop plants. Chemosphere 52:1233–1244

Mompelat S, Le Bot B, Thomas O (2009) Occurrence and fate of pharmaceutical products and by-products, from resource to drinking water. Environ Int 35:803–814. https://doi.org/10.1016/j.envint.2008.10.008

Mompelat S, Thomas O, Le Bot B (2011) Contamination levels of human pharmaceutical compounds in French surface and drinking water. J Environ Monit 13:2929–2939

Moore D (2015). Antibiotic classification and mechanism. http://www.orthobullets.com/basic-science/9059/antibiotic-classification-and-mechanism

Muller AK, Westergaard K, Christensen S et al (2002) The diversity and function of soil microbial communities exposed to different disturbances. Microb Ecol 44:49–58

Munir M, Wong K, Xagoraraki I (2011) Release of antibiotic resistant bacteria and genes in the effuent and biosolids of five wastewater utilities in Michigan. Water Res 45:681–693

Na G, Fang X, Cai Y et al (2013) Occurrence, distribution, and bioaccumulation of antibiotics in coastal environment of Dalian, China. Mar Pollut Bull 69:233–237
Naslund J, Hedman JE, Agestrand C (2008) Effects of the antibiotic ciprofloxacin on the bacterial community structure and degradation of pyrene in marine sediment. Aquat Toxicol 90:223–227
Opris O, Soran ML, Coman V (2013) Determination of some frequently used antibiotics in waste water using solid phase extraction followed by high performance liquid chromatography with diode array and mass spectrometry detection. Cent Eur J Chem 11:1343–1351
Orya J, Bricheux G, Togola A et al (2016) Ciprofloxacin residue and antibiotic-resistant biofilm bacteria in hospital effluent. Environ Pollut 214:635–645
Östman M, Lindberg RH, Fick J et al (2017) Screening of biocides, metals and antibiotics in Swedish sewage sludge and wastewater. Water Res 115:318–328
Park J, Kim MH, Choi K et al (2007) Environmental risk assessment of pharmaceuticals: model application for estimating pharmaceutical exposures in the Han River Basin, Re-06. Korea Environment Institute, p 201
Pauwels B, Verstraete W (2006) The treatment of hospital wastewater: an appraisal. Water Health 4:405–416
Pei R, Kim SC, Carlson KH et al (2006) Effect of river landscape on the sediment concentrations of antibiotics and corresponding antibiotic resistance genes (ARG). Water Res 40:2427–2435. https://doi.org/10.1016/j.watres.2006.04.017
Peng X, Yu Y, Tang C et al (2008) Occurrence of steroid estrogens, endocrine-disrupting phenols, and acid pharmaceutical residues in urban riverine water of the Pearl River Delta. South China Sci Total Environ 397:158–166
Peng X, Tang C, Yu Y et al (2009) Concentrations, transport, fate, and releases of polybrominated diphenyl ethers in sewage treatment plants in the Pearl River Delta, South China. Environ Int 35:303–309
Peng X, Zhang K, Tang C et al (2011) Distribution pattern, behavior and fate of antibacterials in urban aquatic environments in South China. J Environ Monit 13:446–454
Peterson LR (2008) Currently available antimicrobial agents and their potential for use as monotherapy. Clin Microbial Infect 14:30–45
Poirel L, Brinas L, Verlinde A et al (2005) BEL-1, a novel clavulanic acid-inhibited extended-spectrum beta-lactamase, and the class 1 integron In120 in *Pseudomonas aeruginosa*. Antimicrob Agents Chemother 49:3743–3748
Polk RE, Johnson CK, McClish D et al (2004) Predicting hospital rates of fluoroquinolone-resistant *Pseudomonas Aeruginosa* from fluoroquinolone use in US hospitals and their surrounding communities. Clin Infect Dis 39:497–503. https://doi.org/10.1086/422647
Pomati F, Netting AG, Calamari D et al (2004) Effects of erythromycin, tetracycline and ibuprofen on the growth of *Synechocysris*sp. and *Lemma minor*. Aquat Toxicol 67:387–396
Proia L, Lupini G, Osorio V et al (2013) Response of biofilm bacterial communities to antibiotic pollutants in a Mediterranean river. Chemosphere 92:1126–1135
Pruden A, Larsson DGJ, Amezquita A et al (2013) Management options for reducing the release of antibiotics and antibiotic resistance genes to the environment. Environ Health Perspect 121:878–885
Quanten L, Chaerle L, Noben JP et al (2007) Effects of tetracycline on wild-type and inducible P35So IPT-5/TETR transgenic tobacco plants. Physiol Plant 130:290–300
Reddersen K, Heberer T, Dünnbier U (2002) Identification and significance of phenazone drugs and their metabolites in ground and drinking water. Chemosphere 49:539–544
Reinthaler FF, Posch J, Feierl G et al (2003) Antibiotic resistance of *E. Coli* in sewage and sludge. Water Res 37:1685–1690. https://doi.org/10.1016/S0043-1354(02)00569-9
Renew JE, Huang CH (2004) Simultaneous determination of fluoroquinolone, sulfonamide, and trimethoprim antibiotics in wastewater using tandem solid phase extraction and liquid chromatography–electrospray mass spectrometry. J Chromatogr A 1042:113–121. https://doi.org/10.1016/j.chroma.2004.05.056
Rizzo L, Manaia C, Merlin C et al (2013) Urban wastewater treatment plants as hotspots for antibiotic resistant bacteria and genes spread into the environment: a review. Sci Total Environ 447:345–360

Roberts PH, Thomas KV (2006) The occurrence of selected pharmaceuticals in wastewater effluent and surface waters of the lower Tyne catchment. Sci Total Environ 356:143–153. https://doi.org/10.1016/j.scitotenv.2005.04.031

Robinson AA, Belden JB, Lydy MJ (2005) Toxicity of fluoroquinolone antibiotics to aquatic organisms. Environ Toxicol Chem 24:423–430. https://doi.org/10.1897/04-210R.1

Rodriguez-Mozaz S, Chamorro S, Marti E et al (2015) Occurrence of antibiotics and antibiotic resistance genes in hospital and urban wastewaters and their impact on the receiving river. Water Res 69:234–242

Rodriguez-Rojas A, Rodriguez-Beltran J, Couce A et al (2013) Antibiotics and antibiotic resistance: a better fight against evolution. Int J Med Microbiol 303:293–297

Russell AD (2004) Types of antibiotics and synthetic antimicrobial agents. In: Denyer SP (ed) Hugo and Russells pharmaceutical microbiology, 7th edn. Blackwell Science, London, pp 152–186

Sarmah AK, Meyer MT, Boxall ABA (2006) A global perspective on the use, sales, exposure pathways, occurrence, fate and effects of veterinary antibiotics (VAs) in the environment. Chemosphere 65:725–759

Schauss K, Focks A, Leininger S et al (2009) Dynamics and functional relevance of ammonia-oxidizing archaea in two agricultural soils. Environ Microbiol 11:446–456

Schlegel HG (2003) General microbiology, 7th edn. Cambridge University Press, Cambridge, p 676

Schmitt H, Haapakangas H, Van Beelen P (2005) Effects of antibiotics on soil microorganisms: time and nutrients influence pollution-induced community tolerance. Soil Biol Biochem 37:1882–1892

Schwarzbauer J, Heim S, Brinker S et al (2002) Occurrence and alteration of organic contaminants in seepage and leakage water from a waste deposit landfill. Water Res 36:2275–2287

Segura PA, Takada H, Correa JA et al (2015) Global occurrence of anti-infectives in contaminated surface waters: impact of income inequality between countries. Environ Int 80:1386–1398. https://doi.org/10.1016/j.envint.2015.04.001

Sengelov G, Agerso Y, Halling-Sorensen B et al (2003) Bacterial antibiotic resistance levels in Danish farmland as a result of treatment with pig manure slurry. Environ Int 28:587–595

Singer H, Muller S, Tixier C et al (2002) Triclosan: occurrence and fate of a widely used biocide in the aquatic environment: field measurements in wastewater treatment plants, surface waters, and lake sediments. Environ Sci Technol 36:4998

Singer RS, Finch R, Wegener HC et al (2003) Antibiotic resistance – the interplay between antibiotic use in animals and human beings. Lancet Infect Dis 3:47–51

Singer AC, Howard BM, Johnson AC et al (2008) Meeting report: risk assessment of tamiflu use under pandemic conditions. Environ Health Perspect 116:1563–1567

Singer AC, Shaw H, Rhodes V et al (2016) Review of antimicrobial resistance in the environment and its relevance to environmental regulators. Front Microbiol 7:1–22. https://doi.org/10.3389/fmicb.2016.01728

Slatore CG, Tilles SA (2004) Sulfonamide hypersensitivity. Immunol Allergy Clin North Am 24:477–490

Smith DL, Harris AD, Johnson JA et al (2002) Animal antibiotic use has an early but important impact on the emergence of antibiotic resistance in human commensal bacteria. Proc Natl Acad Sci USA 99:6434–6439

Stackelberg PE, Gibs J, Furlong ET et al (2007) Efficiency of conventional drinking-water-treatment processes in removal of pharmaceuticals and other organic compounds. Sci Total Environ 377:255–272

Standley LJ, Rudel RA, Swartz CH et al (2008) Wastewater-contaminated groundwater as a source of endogenous hormones and pharmaceuticals to surface water ecosystems. Environ Toxicol Chem 27:2457–2468. https://doi.org/10.1897/07-604.1

Stuart M, Lapworth D, Crane E et al (2012) Review of risk from potential emerging contaminants in UK groundwater. Sci Total Environ 416:1–21

Suarez S, Carballa M, Omil F et al (2008) Chemlnform Abstract: how are pharmaceutical and personal care products (PPCPs) removed from Urban wastewaters? Rev Environ Sci Bio/Technol 7:125–138

Sui Q, Cao X, Lu S et al (2015) Occurrence, sources and fate of pharmaceuticals and personal care products in the groundwater: a review. Emerg Con 1:14–24
Szczepanowski R, Linke B, Krahn I et al (2009) Detection of 140 clinically relevant antibiotic-resistance genes in the plasmid metagenome of wastewater treatment plant bacteria showing reduced susceptibility to selected antibiotics. Microbiology 155:2306–2319
Tarazona J, Vega M (2002) hazard and risk assessment of chemicals for terres. Toxicology 181–182:187–191
Taylor NM, Davies RH, Ridley A et al (2008) A survey of fluoroquinolone resistance in *Escherichia Coli* and thermophilic campylobacter spp. on poultry and pig farms in great Britain. J Appl Microbiol 105:1421–1431. https://doi.org/10.1111/j.1365-2672.2008.03877.x
Teijon G, Candela L, Tamoh K et al (2010) Occurrence of emerging contaminants, priority substances (2008/105/CE) and heavy metals in treated wastewater and groundwater at Depurbaix facility (Barcelona, Spain). Sci Total Environ 408:3584–3595. https://doi.org/10.1016/j.scitotenv.2010.04.041
Ternes T, Joss A (2006) Human pharmaceuticals, hormones and fragrances: the challenge of micropollutants in urban water management. IWA Publishing, p 406
Thiele-Bruhn S (2003) Pharmaceutical antibiotic compounds in soils – a review. J Plant Nutr Soil Sci 166:145–167. https://doi.org/10.1002/jpln.200390023
Tomlinson TG, Boon AG, Trotman CNA (1966) Inhibition of nitrification in activated sludge process of sewage disposal. J Appl Bacteriol 29:266–291
Toth JD, Feng Y, Dou Z (2011) Veterinary antibiotics at environmentally relevant concentrations inhibits soil iron reduction and nitrification. Soil Biol Biochem 43:2470–2472
Underwood JC, Harvey RW, Metge DW et al (2011) Effects of the antimicrobial sulfamethoxazole on groundwater bacterial enrichment. Environ Sci Technol 45:3096–3101
Unger IM, Goyne KW, Kennedy AC et al (2013) Antibiotic effects on microbial community characteristics in soils under conservation management practices. Soil Sci Soc Am J 77:100–112
USP (2003) Veterinary pharmaceutical information monographs – antibiotics. J Veter Pharmacol Therap 26(s2):1–271
Valcarcel Y, Alonso SG, Rodriguez-Gil JL et al (2011) Detection of pharmaceutically active compounds in the rivers and tap water of the Madrid Region (Spain) and potential ecotoxicological risk. Chemosphere 84:1336–1348
van Hoek AHAM, Mevius D, Guerra B et al (2011) Acquired antibiotic resistance genes: an overview. Front Microbiol 2:203. https://doi.org/10.3389/fmicb.2011.00203
Verlicchi P, Al Aukidy M, Zambello E (2012) Occurrence of pharmaceutical compounds in urban wastewater: removal, mass load and environmental risk after a secondary treatment – a review. Sci Total Environ 429:123
Verlicchi P, Zambello E, Al Aukidy M (2013) Analysis, removal, effects and risk of pharmaceuticals in the water cycle: occurrence and transformation in the environment. In: Petrovic M, Perez S, Barcelo D (eds) Comprehensive analytical chemistry, vol 62. Elsevier, p 231
Verlicchi P, Al Aukidy M, Zambello E (2015) What have we learned from worldwide experiences on the management and treatment of hospital effluent? – an overview and a discussion on perspectives. Sci Total Environ 514:467–491
Vila J, Martí S, Sánchez-Céspedes J (2007) Porins, efflux pumps and multidrug resistance in Acinetobacterbaumannii. J Antimicrob Chemother 59(6):1210–1215
Voulvoulis N, Barcelo D, Verlicchi P (2013) Pharmaceutical Residues in Sewage Treatment Works and their Fate in the Receiving Environment. 2015. Issues in Environmental Science and Technology No. 41 Print ISBN: 978-1-78262-189-8 PDF eISBN: 978-1-78262-234-5 ISSN 1350-7583A catalogue record for this book is available from the British Libraryr The Royal Society of Chemistry p 312
Voulvoulis N, Barcelo D, Verlicchi P (2016) Pharmaceutical residues in sewage treatment works and their fate in the receiving environment. 2015. Issues in environmental science and technology no. 41 Print ISBN: 978-1-78262-189-8 PDF eISBN: 978-1-78262-234-5 ISSN 1350-7583A catalogue record for this book is available from the British Library, The Royal Society of Chemistry, p 312

Waggot A (1981) Trace organic substances in the river Lee (Great Britain). In: Cooper WJ (ed) Chemistry in water reuse, 1st edn. Ann Arbour Science, Ann Arbor, pp 55–99
Walsh C (2003) Antibiotics: actions, origins, resistance, 1st edn. ASM Press, Washington, DC, p 345
Watkinson A, Murby E, Costanzo S (2007) Removal of antibiotics in conventional and advanced wastewater treatment: implications for environmental discharge and wastewater recycling. Water Res 41:4164–4176
Weber KP, Petersen EJ, Bissegger S et al (2014) Effect of gold nanoparticles and ciprofloxacin on microbial catabolism: a community-based approach. Environ Toxicol Chem 33:44–51
White D, Cox E (2013) Fighting the impact of antibiotic-resistance. FDA Consumer Health Information. www.fda.gov/downloads/. . ./UCM350090.pdf. Accessed 4 Sept 2016
WHO (2019) Essential medicine and Health products. https://www.who.int/medicines/news/2019/WHO_releases2019AWaRe_classification_antibiotics/en/
Wu G, Liu Z, Chen T et al (2015) Elevation-dependent variations of tree growth and intrinsic water-use efficiency in Schrenk spruce (*Picea schrenkiana*) in the western Tianshan Mountains. China Front Plant Sci 6:309. https://doi.org/10.3389/fpls.2015.00309
Xie X, Zhou Q, Bao Q et al (2011) Genotoxicity of tetracycline as an emerging pollutant on root meristem cells of wheat (*Triticum Aestivum* L.). Environ Toxicol 26:417–423. https://doi.org/10.1002/tox.20567
Xu WH, Zhang G, Zou SC et al (2007) Determination of selected antibiotics in the Victoria Harbour and the Pearl River, South China using high performance liquid chromatography–electrospray ionization tandem mass spectrometry. Environ Pollut 145:672–679
Xu J, Xu Y, Wang H et al (2015) Occurrence of antibiotics and antibiotic resistance genes in a sewage treatment plant and its effluent-receiving river. Chemosphere 119:1379–1385
Yang QX, Zhang J, Zhu KF et al (2009) Influence of oxytetracycline on the structure and activity of microbial community in wheat rhizosphere soil. J Environ Sci 21:954–959
Yang Y, Li B, Zou S et al (2014) Fate of antibiotic resistance genes in sewage treatment plant revealed by metagenomic approach. Water Res 62:97–106
Yaronskaya EB, Gritskevich ER, Trukhanovets NL et al (2007) Effects of kinetin on early stages of chlorophyll biosynthesis in streptomycin-treated barley seedling. Russ J Plant Physiol 54:388–395
Yasojima M, Nakada N, Komori K et al (2006) Occurrence of levofloxacin, clarithromycin and azithromycin in wastewater treatment plant in Japan. Water Sci Technol 53:227
Yiruhan Y, Wang Q, Mo C et al (2010) Determination of four fluoroquinolone antibiotics in tap water in Guangzhou and Macao. Environ Pollut 158:2350–2358
Yoneyama H, Katsumata R (2006) Antibiotic resistance in bacteria and its future for novel antibiotic development. Biosci Biotechnol Biochem 70:1060–1075
Zhang D, Lin L, Luo Z et al (2011) Occurrence of selected antibiotics in Jiulongjiang River in various seasons, South China. J Environ Monit 13:1953–1960. https://doi.org/10.1039/c0em00765j
Zhang CH, Tang JW, Wang LL et al (2015) Occurrence of antibiotics in water and sediment from Zizhuyuan Lake. Pol J Environ Stud 24(4):1831–1836
Zhengqi Y, Weinberg HS, Meyer MT (2004) Occurrence of antibiotics in drinking water. The University of North Carolina at Chapel Hill U.S. Geological Survey Kansas p 138–142
Zhou L, Ying G, Liu S et al (2013) Occurrence and fate of eleven classes of antibiotics in two typical wastewater treatment plants in South China. Sci Total Environ 452–453:365–376
Zuccato E, Castiglioni S, Bagnati R et al (2010) Source, occurrence and fate of antibiotics in the Italian aquatic environment. J Hazard Mater 179:1042–1048
Zwiener C, Frimmel F (2003) Short-term testes with a pilot sewage plant and biofilm reactors for the biological degradation of the pharmaceutical compounds clofibric acid, ibuprofen, and diclofenac. Sci Total Environ 309:201–211

Chapter 4
Application of Artificial Intelligence in Predicting Groundwater Contaminants

Sushant K. Singh, Ataollah Shirzadi, and Binh Thai Pham

4.1 Introduction

Groundwater constitutes 95% of all global freshwater, and more than 7 billion individuals depend on it for their domestic needs (NGWA 2020). Groundwater is naturally characterized by certain physical, chemical, and biological properties. Its major chemical constituents, whose concentrations may range between 1 mg/L and 1000 mg/L, are sodium, calcium, magnesium, bicarbonate, sulfate, chloride, and silica (Chapman 1996; Şen 2014). Secondary and elemental constituents of groundwater include iron, aluminum, potassium, carbonate, nitrate, fluoride, boron, and selenium, which may be found in concentrations between 0.01 mg/L and 10 mg/L (Chapman 1996). Several other elements occur in slight levels (0.0001 mg/L to 0.1 mg/L) and form the minor elemental constituents of groundwater; these include arsenic, barium, bromide, cadmium, chromium, cobalt, copper, iodide, lead, lithium, manganese, nickel, phosphate, strontium, uranium, and zinc (Chapman 1996; Schwarzenbach et al. 2010; Şen 2014). These natural groundwater constituents may be altered by processes that are physical (dispersion and filtration), geochemical (complexation, ionic strength, acid-base, oxidation–reduction, precipitation solution, and adsorption–desorption), or biochemical (decay, respiration, and cell synthesis) (Chapman 1996; Schwarzenbach et al. 2010). These processes are influenced by a variety of anthropogenic activities that lead to groundwater pollution, including

S. K. Singh (✉)
Artificial Intelligence & Analytics | Health Care and Life Sciences, Virtusa Corporation, New York, NY, USA

A. Shirzadi
Department of Rangeland and Watershed Management, Faculty of Natural Resources, University of Kurdistan, Sanandaj, Iran

B. T. Pham
Institute of Research and Development, Duy Tan University, Da Nang, Vietnam

A. Singh et al. (eds.), *Water Pollution and Management Practices*,
https://doi.org/10.1007/978-981-15-8358-2_4

industrial, agricultural, and mining development, as well as urbanization and groundwater-resource management (Chapman 1996; Schwarzenbach et al. 2010). Among these activities, unsewered sanitation, wellhead contamination, accidental spillage, agrochemicals, irrigation, and saline intrusion have serious impacts on the quality of groundwater used for drinking purposes (Chapman 1996, Schwarzenbach et al. 2010). Other activities responsible for groundwater contamination are land and stream discharge of sewage and effluent, sewage oxidation of lagoons, sewer leakage, landfilling, solid-waste disposal, highway drainage soak-away, process water/effluent lakes, tank and pipeline leakage, well-disposal effluent, aerial fallout, sludge and slurry disposal, wastewater irrigation, mine-drainage discharge, process water/sludge lagoons, solid mine tailings, oilfield brine disposal, hydraulic disturbance, and recovering water levels (Chapman 1996; Schwarzenbach et al. 2010).

These activities can be triggered by a number of social, economic, environmental, and political actions, as well as by variations in climate and socioeconomic level (Singh 2017; Vörösmarty et al. 2000; Alcamo et al. 2007). It has been consistently documented that anthropogenic activities are the major contributors to groundwater pollution; for example, human-created compounds (such as Aldrin, Chlordane, DDT, Dieldrin, Dioxins, Endrin, Furans, Heptachlor, Hexachlorobenzene, Mirex, PCBs, and Toxaphene, popularly known as the "Dirty Dozen") were particularly targeted by the 2004 Stockholm Convention on Persistent Organic Pollutants (Schweitzer and Noblet 2018). According to the World Health Organization (WHO), nearly 11% of the global population still lacks a primary drinking-water service, and approximately 29% of the worldwide population relies on unsafe drinking water, which led to the loss of 1.2 million lives (WHO 2017; Ritchie and Roser 2019). According to the International Agency for Research on Cancer (IARC), both anthropogenically and naturally occurring chemicals in groundwater are categorized according to their possible carcinogenic effects on humans into either Group 1 (carcinogenic), Group 2A (probably carcinogenic), Group 2B (possibly carcinogenic), Group 3 (not carcinogenic), or Group 4 (probably not carcinogenic) (WHO 2017). Arsenic and fluoride are the two most common naturally occurring chemicals. Nitrate (in both its NO_3^- and NO_2^-forms) is one of the agriculturally produced compounds posing a significant threat to human health (WHO 2017). Groundwater contamination is a global public health challenge, and the prediction of groundwater contaminant presence would help in creating proactive mitigation policies (Burri et al. 2019; Singh 2018). In the United States (U.S.) alone, more than 100,000 lifetime cancer cases may be linked to carcinogenic chemicals in drinking water (Evans et al. 2019).

The use of various machine learning (ML), deep learning (DL), and artificial intelligence (AI) techniques in developing prediction models based on environmental data is highly trusted among scientists (Singh et al. 2018). However, the application of these techniques to the development of groundwater contaminant-prediction models is still in its rudimentary stages. Considering the great expense of environmental data collection and laboratory analysis, ML, DL, and AI techniques could be extremely useful for developing systems supportive of decision-making in environmental-contamination monitoring. For example, it would cost nearly

$40,000 to collect 100 groundwater samples, and the mean cost per sample to analyze metals would be around $260 (USEPA 1997). A detailed description of possible challenges in the collection and analysis of water-quality data can be found here (Crocker and Bartram 2014). Additionally, the Environmental Protection Agency provides comprehensive guidance on the possible costs associated with collection, monitoring, and testing of water samples. Costs may vary according to the instruments and laboratory facilities used, but even so, they may continue to be an obstacle to conducting such research in low-income regions where groundwater contamination is a severe problem.

This chapter focuses only on arsenic, fluoride, and nitrate because of their serious human health impacts and global presence in groundwater. Considering the obstacles mentioned in the previous section, the chapter is structured in three sections, the first of which highlights the prevalence of these groundwater contaminants and their health impacts. The second section covers the application of various AI techniques to the prediction of arsenic, fluoride, and nitrate contamination of groundwater. The final section entails a discussion of the advantages of certain techniques and concludes by recommending future directions of exploration and practical application.

4.2 Groundwater Contamination

4.2.1 Arsenic

Arsenic, a metalloid and the world's most 53rd-most abundant element, occupies just 0.0001% of the earth's crust and occurs in unpolluted groundwater in concentrations below 10 μg/L (Braman 1975). Out of arsenic's five valence states, its inorganic form, arsenite, is the most toxic; in groundwater, however, the organic form of arsenic, arsenate, predominates (Braman 1975; Oremland and Stolz 2003). Because of its adverse health effects on human beings, the Environmental Protection Agency has categorized arsenic as a "Group A" element. Currently, more than 100 countries are experiencing geogenic arsenic contamination of groundwater, and more than 296 million individuals' lives are at risk (Fig. 4.1) (Chakraborti et al. 2017).

Approximately 66% of European, 59% of Asian, 37% of American, 24% of African, and 1% of Oceanian countries are impacted by arsenic contamination of groundwater (Singh 2017; Murcott 2012). It has been more than two centuries since arsenic contamination was first testified to in Germany in 1885 and more than one since arsenic-induced health effects were discovered in 1917 in the province of Cordoba, Argentina (Chakraborti et al. 2017; Murcott 2012; Nordstrom 2002). Over time, advancements in arsenic research confirmed that prolonged exposure to arsenic in water, food, soil, or air could lead to a variety of health problems, including both carcinogenic and non-carcinogenic diseases (Chakraborti et al. 2017; Mazumder 2008; Chakraborti et al. 2003; Bhattacharya et al. 2010). These health issues are summarized in Table 4.1.

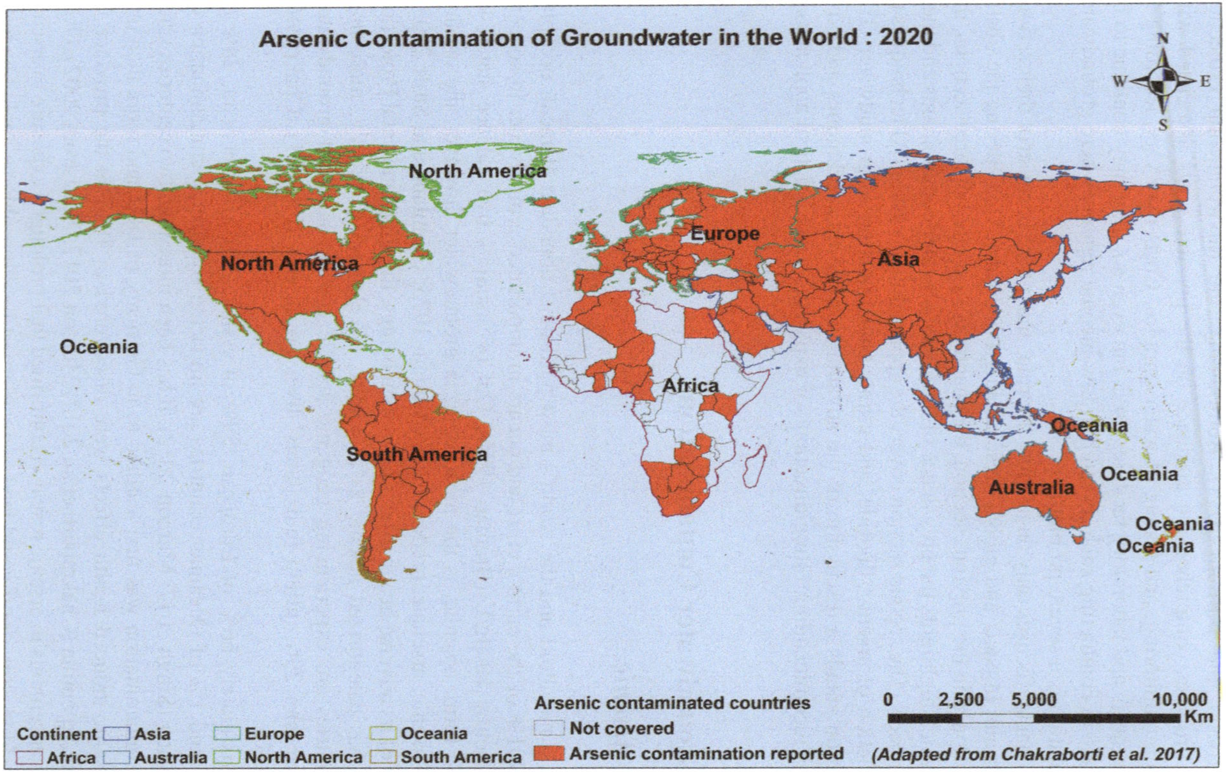

Fig. 4.1 Arsenic contamination of groundwater in the world. (Adapted from Chakraborti et al. 2017)*
*This is the most updated map as of today based on the available data through published sources

Table 4.1 A summary of the signs and symptoms of arsenic poisoning

Effects	Symptoms	Reference
Dermal	**Acute signs and symptoms**	Chakraborti et al. (2017), Tchounwou et al. (2019), Mitra et al. (2020), Chakraborti et al. (2018), Sinha and Prasad (2020)
	Delayed appearance of Mee's lines in nail beds	
	Dermatitis	
	Melanosis	
	Vesiculation	
	Chronic signs and symptoms	
	Hyperpigmentation	
	Pigment changes on the face, neck, and back—"Raindrop" appearance	
	Skin lesions	
	Hyperkeratosis	
	Desquamation	
Gastrointestinal	**Acute signs and symptoms**	Chakraborti et al. (2017), Tchounwou et al. (2019), Mitra et al. (2020), Chakraborti et al. (2018) and Sinha and Prasad (2020)
	Garlic odor on breath	
	Severe abdominal pain	
	Nausea and vomiting	
	Thirst	
	Dehydration	
	Anorexia	
	Heartburn	
	Bloody or rice-water diarrhea	
	Dysphagia	
	Chronic signs and symptoms	
	Esophagitis	
	Gastritis	
	Colitis	
	Abdominal discomfort	
	Anorexia	
	Malabsorption	
	Weight loss	
Cardiovascular	**Acute signs and symptoms**	Chakraborti et al. (2017), Tchounwou et al. (2019), Mitra et al. (2020), Chakraborti et al. (2018) and Sinha and Prasad (2020)
	Hypotension	
	Shock	
	Ventricular arrhythmia	
	Congestive heart failure	
	Irregular pulse	
	T-wave inversion	
	Persistent prolongation of QT interval	
	Chronic signs and symptoms	
	Arrhythmia	
	Pericarditis	

(continued)

Table 4.1 (continued)

Effects	Symptoms	Reference
	Blackfoot disease (gangrene with spontaneous amputation)	
	Raynaud's syndrome	
	Acrocyanosis (intermittent)	
	Ischemic heart disease	
	Cerebral infarction	
	Carotid atherosclerosis	
	Hypertension	
	Microcirculation abnormalities	
Respiratory	**Acute signs and symptoms** Irritation of nasal mucosa Pharynx, larynx, and bronchi Pulmonary edema Tracheobronchitis Bronchial pneumonia Nasal septum perforation **Chronic signs and symptoms** Rhino-pharyngo-laryngitis Tracheobronchitis Pulmonary insufficiency (emphysematous lesions) Chronic restrictive/obstructive diseases	Chakraborti et al. (2017), Tchounwou et al. (2019), Mitra et al. (2020), Chakraborti et al. (2018) and Sinha and Prasad (2020)
Neurological	**Acute signs and symptoms**	Chakraborti et al. (2017), Tchounwou et al. (2019), Mitra et al. (2020), Chakraborti et al. (2018) and Sinha and Prasad (2020)
	Sensorimotor peripheral axonal neuropathy (paresthesia, hyperesthesia, neuralgia)	
	Neuritis	
	Autonomic neuropathy with unstable blood pressure, anhidrosis, sweating, and flushing	
	Leg/muscular cramps	
	Lightheadedness	
	Headache	
	Weakness	
	Lethargy	
	Delirium	
	Encephalopathy	
	Hyperpyrexia	
	Tremors	
	Disorientation	
	Seizure	
	Stupor	
	Paralysis	
	Coma	

(continued)

Table 4.1 (continued)

Effects	Symptoms	Reference
	Chronic signs and symptoms	
	Neuropathy	
	Polyneuritis and motor paralysis	
	Hearing loss	
	Mental retardation	
	Encephalopathy, symmetrical peripheral polyneuropathy (sensori-motor, resembling Landry–Guillain–Barre syndrome)	
	Electromyography abnormalities	
	Peripheral neuropathy of sensory and motor neurons	
Hepatic	**Acute signs and symptoms**	Chakraborti et al. (2017), Tchounwou et al. (2019), Mitra et al. (2020), Chakraborti et al. (2018) and Sinha and Prasad (2020)
	Elevated liver enzyme levels	
	Fatty infiltration	
	Congestion	
	Central necrosis	
	Cholangitis	
	Cholecystitis	
	Chronic signs and symptoms	
	Enlarged, tender liver	
	Elevated hepatic enzyme levels	
	Cirrhosis	
	Portal hypertension without cirrhosis	
	Fatty degeneration	
Renal	**Acute signs and symptoms**	Chakraborti et al. (2017), Tchounwou et al. (2019), Mitra et al. (2020), Chakraborti et al. (2018) and Sinha and Prasad (2020)
	Hematuria	
	Oliguria	
	Proteinuria	
	Leukocyturia	
	Glycosuria	
	Uremia	
	Acute tubular necrosis	
	Renal cortical necrosis	
Hematological	**Acute signs and symptoms**	Chakraborti et al. (2017), Tchounwou et al. (2019), Mitra et al. (2020), Chakraborti et al. (2018) and Sinha and Prasad (2020)
	Anemia	
	Leukopenia	
	Thrombocytopenia	
	Bone marrow suppression	
	Disseminated intravascular coagulation	
	Chronic signs and symptoms	

(continued)

Table 4.1 (continued)

Effects	Symptoms	Reference
	Bone marrow hypoplasia	
	Anemia	
	Aplastic anemia	
	Leukopenia	
	Thrombocytopenia	
	Impaired folate metabolism	
	Karyorrhexis	
Endocrinological	**Chronic signs and symptoms**	Chakraborti et al. (2017), Mitra et al. (2020), Chakraborti et al. (2018) and Sinha and Prasad (2020)
	Diabetes mellitus	
Other	**Acute sign and symptoms**	Chakraborti et al. (2017), Chakraborti et al. (2018) and Sinha and Prasad (2020)
	Rhabdomyolysis	
	Conjunctivitis	
	Chronic signs and symptoms	
	Lens opacity	
Carcinogenic	Bladder cancer	Chakraborti et al. (2017), Tchounwou et al. (2019), Mitra et al. (2020), Chakraborti et al. (2018) and Sinha and Prasad (2020)
	Lung cancer	
	Kidney cancer	
	Liver cancer	
	Skin cancer	
Reproductive and developmental	Fetal growth	Chakraborti et al. (2017), Mitra et al. (2020), Chakraborti et al. (2018) and Sinha and Prasad (2020)
	Fetal death	
	Pregnancy complications	
	Premature delivery	
	Gonadal dysfunction	
	Reduction of testosterone and gonadotrophins	
	Apoptosis and necrosis	

4.2.2 *Fluoride*

Fluoride, the world's 13th most abundant element, occupies 0.06–0.09% of the earth's crust. Being highly reactive, it changes its elemental form rapidly and is converted into both organic and inorganic compounds. In groundwater, it is present in concentrations lower than 1 mg/L. At the 1.5 mg/L level, fluoride improves bone growth and helps protect against dental caries (cavities). Around the world, over 200 million individuals across 84 countries are at risk due to fluoride-contaminated drinking water (Fig. 4.2) (Jha et al. 2013; Ali et al. 2016).

Ironically, human beings require a low concentration of fluoride, for example, between 0.5 and 1.5 mg/L, to strengthen bones and teeth. However, higher doses of fluoride, imbibed through water, air, food, or any other source, can lead to a variety

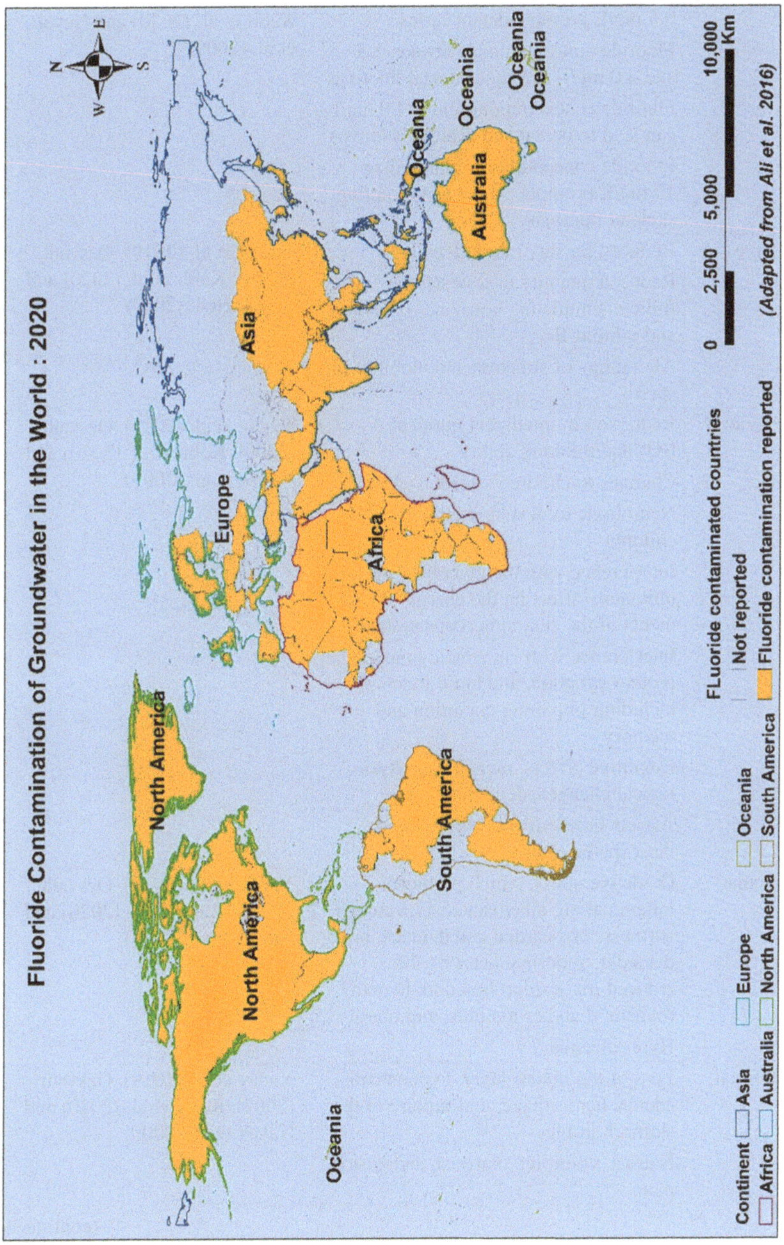

Fig. 4.2 Fluoride contamination of groundwater in the world. (Adapted from Ali et al. 2016, Yadav et al. 2019)*
*This is the most updated map as of today based on the available data through published sources

Table 4.2 A summary of the sign and symptoms of fluoride poisoning

Effects	Symptoms	Reference
Fluorosis	Fluoride concentrations lower than 0.5 mg/L promote dental caries	Ali et al. (2016), Ozsvath (2009), Kabir et al. (2020) and Nayak et al. (2009)
	Fluoride concentrations between 1.5 and 4.0 mg/L can cause dental fluorosis	
	Fluoride concentrations above 4.0 mg/L can lead to dental and skeletal fluorosis	
	Fluoride concentrations higher than 10 mg/L promote cancer and crippling skeletal fluorosis	
Reproductive system	Reduced fertility rate in females	Yadav et al. (2019), Ozsvath (2009), Kabir et al. (2020) and Nayak et al. (2009)
	Reduced quantity of testosterone, follicle-stimulating hormones (FSH), and inhibin-B	
	Alterations of structure and mobility of sperm	
Neurobehavioral effects	Reduction in intelligent quotient (IQ) and thinking ability	Yadav et al. (2019), Ozsvath (2009), Kabir et al. (2020) and Nayak et al. (2009)
	Changes to children's mental capacity	
	Neurotoxic to developing brains of children	
	Interference with the glycolysis cycle, ultimately affecting the energy requirements of the central nervous system	
	Interference with enzymatic function, protein structure, and brain function, including impairing cognition and memory	
	Cognitive effects, mental and physiological changes, dementia	
	Affects biochemical activities and visuospatial abilities	
Cardiovascular system	Oxidative stress, which promotes inflammation, atherosclerosis, vascular stiffness, myocardial cell damage, bradycardia, abnormal heart rhythms, reduced myocardial function, hypothyroidism, diabetes mellitus, and obesity	Yadav et al. (2019), Ozsvath (2009), Kabir et al. (2020) and Nayak et al. (2009)
	Hypocalcemia	
Gastrointestinal effects	Loss of the mucus layer, hyperanemia, edema, hemorrhage, and rupture of the stomach lining	Yadav et al. (2019), Ozsvath (2009), Kabir et al. (2020) and Nayak et al. (2009)
	Nausea, vomiting, diarrhea, and gastric pain	

(continued)

Table 4.2 (continued)

Effects	Symptoms	Reference
Endocrine system	Structural changes and dysfunctions in the thyroid gland	Yadav et al. (2019), Ozsvath (2009), Kabir et al. (2020) and Nayak et al. (2009)
	Increased parathyroid and calcitonin activity, secondary hyperparathyroidism, and impaired glucose tolerance	
Urinary/renal system	Increased risk of kidney stones and non-carcinogenic effects on kidney	Yadav et al. (2019), Ozsvath (2009), Kabir et al. (2020) and Nayak et al. (2009)
	Metabolic, histopathological, and pathological changes in the glomeruli	
Other	Damage of human soft tissues, muscles, erythrocytes, gastrointestinal mucosa, ligaments, spermatozoa, and thyroid glands	Kabir et al. (2020)
	Brittle bones	
	Stunted growth	
	Alzheimer's disease	
	Encumbered movement due to bony and dental lesions	
	Decreased secretion of the pituitary lactotrophic hormone	
	Alterations in levels of volatile fatty acids	
	Loss of appetite and body weight	
	Delays in postpartum estrus	
	Muscle weakness and associated muscle loss	
	Increased respiration and heart rates in animals	
	Decreased milk and wool production in livestock	

of health issues, as summarized in Table 4.2 (Chakraborti et al. 2016; Grandjean 2019; Kabir et al. 2020).

4.2.3 *Nitrate*

Nitrate—used here to refer to the complete class of compounds known as nitrates—is a naturally occurring salt. Nitrate is highly soluble, part of the nitrogen cycle, and an essential component of fertilizers and explosives (Gamble 2019). Fertilizers are the primary source of nitrate contamination, introducing it to groundwater via agricultural runoff (Zhou 2015; Keeney 1989). The WHO sets the maximum level of nitrate contamination of groundwater at 50 mg/L for NO_3^- and 11.3 mg/L for NO_3^-N (Ward et al. 2018; Ward et al. 2005). High groundwater concentrations of nitrate can adversely impact both human health and environmental health (Zhou

2015; Sousa et al. 2014). Nitrate contamination has been detected in nearly 60 countries (Fig. 4.3).

Newly born babies (those of less than 6 months of age) are the primary victims of nitrate pollution. They may suffer from methemoglobinemia, popularly known as "blue baby syndrome," if they consume water with a nitrate concentration greater than 10 mg/L. The adverse human health impacts of nitrate first surfaced in 1945 (Zhou 2015) and are summarized in Table 4.3.

4.3 Application of Artificial Intelligence in Groundwater Contaminant Prediction

Artificial intelligence (AI), a subset of machine learning (ML), has been used to identify and resolve an array of challenges across various domains, including information technology, business, banking, retail, pharmacy, healthcare, insurance, and life sciences, among others. However, the application of AI to the identification, prediction, and monitoring of environmental problems is a recent phenomenon. Environmental research is challenged by a lack of ground data, i.e., collected from fields through surveys and laboratory analysis; therefore, it depends heavily on remote sensing images. Geographical information system (GIS) tools are used to extract information from these remote sensing images, simulate the imaged areas, and develop prediction models. Machine learning, deep learning, and reinforcement learning (RL) help simulate human knowledge of a specific domain and assist in developing intelligent systems that can learn from historical data and predict patterns in unseen data (Chau 2006). Various basic and advanced AI techniques have been used to develop prediction models of landslide susceptibility and groundwater potentiality (Pham et al. 2019; Pham et al. 2018; Phong et al. 2019; Rizeei et al. 2019; Chen et al. 2019; Sameen et al. 2019, 2020), for example. The application of AI techniques in predicting groundwater contaminants, in contrast, is still at a primitive level. The handful of studies that are available on the application of these techniques in predicting groundwater contamination by arsenic, fluoride, and nitrate are summarized below in Tables 4.4, 4.5, and 4.6, respectively.

4.4 Application of AI in Predicting Arsenic Contamination of Groundwater

Data-analysis techniques for predicting groundwater contamination of arsenic were first reported in 2006, when researchers developed a probabilistic statistical model to predict the probability that arsenic concentrations would exceed the level of 5 μg/L in the drinking-water wells of the New England of the USA (Ayotte et al. 2006). A logistic regression model of geologic and anthropogenic sources of arsenic,

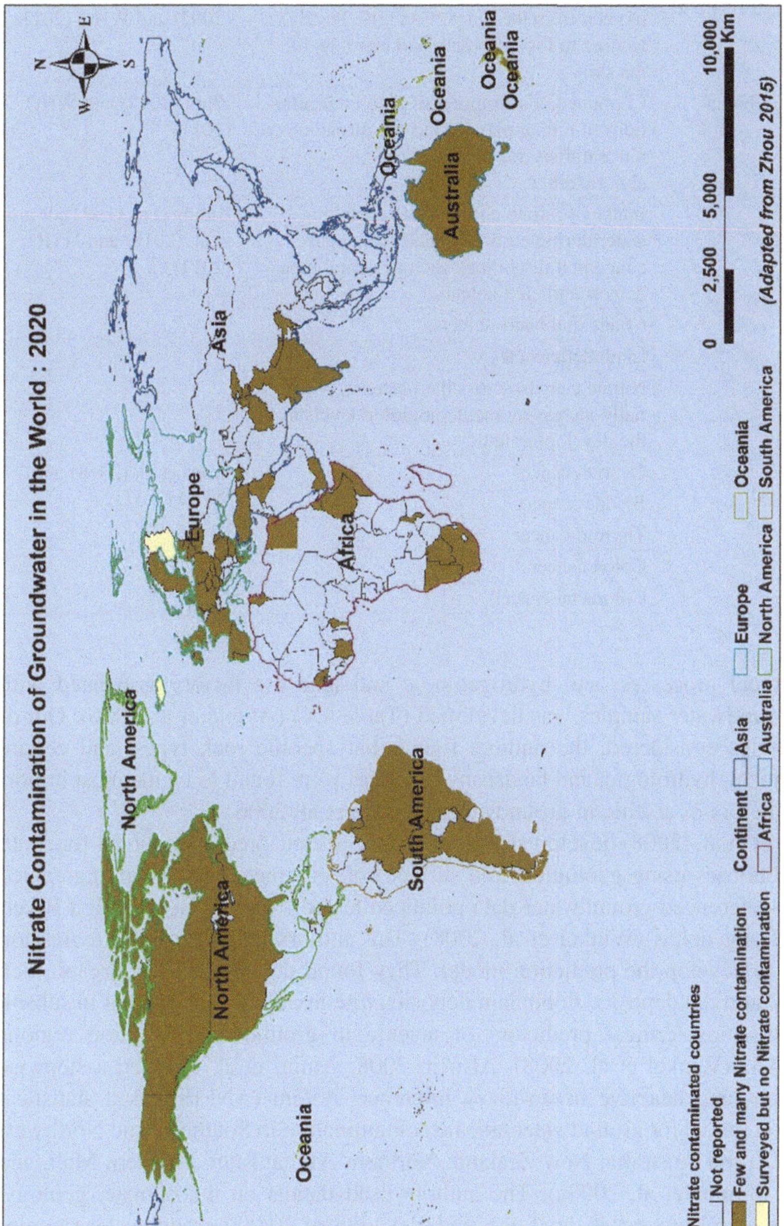

Fig. 4.3 Nitrate contamination of groundwater in the world. (Adapted from Zhou 2015)*
*This is the most updated map as of today based on the available data through published sources

Table 4.3 A summary of the sign and symptoms of nitrate poisoning

Effects	Symptoms	Reference
Methemoglobinemia	Red blood cells become unable to carry oxygen from the lungs. Can arise hastily, leading to loss of breath and blueness of the skin.	Zhou (2015), Ward et al. (2005) and WHO (2011)
Gastric problems	Continued consumption of high concentrations of nitrate may lead to the formations of nitrosamines that cause gastric abnormalities.	Zhou (2015) and WHO (2011)
Pregnancy	Intake of nitrate-contaminated drinking water during early pregnancy can cause congenital disabilities such as neural tube defects and cleft palates.	Bundy et al. (1994), Ward et al. (2018) and WHO (2011)
	Congenital heart defects.	
	Limb deficiencies.	
	Nitrate can also cross the placenta, potentially increasing methemoglobin levels in the developing fetus.	
Cancer	Ovarian cancer	Ward et al. (2018) and WHO (2011)
	Bladder cancer	
	Thyroid cancer	
	Colon cancer	
	Colorectal cancer	

geochemical processes, and hydrogeologic and land-use factors associated with 2470 groundwater samples was developed (Table 4.4) (Ayotte et al. 2006). Out of 50 variables considered, the authors found that specific rock types, and certain geochemical, hydrologic, and landscape variables were found to be the most important predictors of arsenic in groundwater (Ayotte et al. 2006).

Winkel et al. (2008) developed a groundwater-arsenic prediction model for south Asian countries using geological and surface soil parameters for 1756 aggregated and geo-referenced groundwater data points collected from the Bengal, Red River, and Mekong deltas (Winkel et al. 2008). The authors used a logistic regression method to develop the prediction model. They found deltaic deposits, organic-rich deposits, alluvial deposits, floodplain deposits, fine-textured soils, and silt in subsoil to be the most critical predictors of arsenic in groundwater in these regions (Table 4.4) (Winkel et al. 2008). Also in 2008, Amini et al. reported a stepwise regression- and adaptive neuro-fuzzy inference system (ANFIS)-based statistical prediction model for groundwater arsenic contamination in Southeast and Northwest China, Central Australia, New Zealand, Northern Afghanistan, Northern Mali, and Zambia (Amini et al. 2008a). The authors used details on the climate, geology, hydrology, soil properties, land use, and elevation of 20,000 groundwater samples across the studied regions and found that in those regions, evapotranspiration, temperature, precipitation, distance from volcanoes, drainage code, subsoil silt content, topsoil silt content, distance from volcanic rocks, subsoil carbon-to-nitrogen

Table 4.4 Artificial intelligence techniques used to predict arsenic contamination of groundwater

Case study	Predictors	Method	Reference
1	Avalon belt	Logistic regression	Ayotte et al. (2006)
	Bronson Hill belt		
	Eugeosyncline sequence		
	Mesozoic basin		
	Waits River basin		
	Pelitic rocks (Bronson Hill)		
	Peraluminous granite (New Hampshire Maine sequence)		
	Mafic rocks (Narragansett Basin)		
	Concord granite (Dc1m, granite)		
	Madrid Fm. (DSm, metamorphic)		
	Rindgemere Fm., lower member (DSrb, metamorphic)		
	Berwick Fm., calcareous member (SObc, Calcpelite)		
	Eliot Fm., Calef member (SOec, metamorphic)		
	Kittery Fm. (SOk, metamorphic)		
	Perry Mountain Fm. (Sp, metamorphic)		
	Rangeley Fm., lower part, (Srl, metamorphic)		
	Sangerville Fm. (Sspm, metamorphic)		
	Waterville Fm. (Sw, metamorphic)		
	Kittery Fm. (SZk, metamorphic)		
	Massabesic gneiss complex (Zmz, granite)		
	Stream sediment arsenic, (ln) mg kg^{-1}		
	Pleistocene marine inundation intrusive granitic pluton category (within 3 km of pluton)		
	Developed land flag (cut point of 33%) elevation (1:24,000 scale DEM, m)		
	Population density (person km^{-2})		
	Precipitation, mm $year^{-1}$		
	Water bodies (% area in 1000 m radius buffer)		
2	Easting	Regression-kriging	Lado et al. (2008)
	Northing		
	Depth		
	Arsenic		
	NH_4^+ (ammonium cation)		
	Fe (Iron)		
	Mn (manganese)		
	Pb (Lead)		

(continued)

Table 4.4 (continued)

Case study	Predictors	Method	Reference
	Alkalinity		
	Digital elevation model		
	Slope		
	Topographic wetness index		
	Hydrologic flux index		
	Topographic convergence index		
	Geological formation		
3	Elevation	Logistic regression	Winkel et al. (2008)
	Slope		
	Distance to rivers		
	Evapotranspiration		
	Precipitation		
	Runoff		
	Temperature		
	Irrigated areas		
	Topsoil clay content		
	Topsoil silt content		
	Topsoil sand content		
	Subsoil clay content		
	Subsoil silt content		
	Subsoil sand content		
	Subsoil cation exchange capacity		
	Soil drainage code		
	Coarse textured soils		
	Medium textured soils		
	Fine-textured soils		
	Subsoil nitrogen content		
	Subsoil organic carbon content		
	Subsoil pH		
	Pre-Holocene deposits		
	Organic-rich deposits		
	Other Holocene deposits		
	Tidal deposits		
	Deltaic deposits		
	Flood plain deposits		
	Alluvial deposits		
4	Elevation	Regression analysis	Amini et al. (2008a)
	Slope	Adaptive neuro-fuzzy inferencing	
	Geology age		
	Evapotranspiration		
	Precipitation		

(continued)

Table 4.4 (continued)

Case study	Predictors	Method	Reference
	Runoff		
	Temperature		
	Irrigated areas		
	Topsoil C/N ratio		
	To soil clay content		
	Topsoil silt content		
	Topsoil sand content		
	Subsoil C/N ratio		
	Subsoil clay content		
	Subsoil silt content		
	Subsoil sand content		
	Subsoil cation exchange capacity		
	Soil drainage code		
	Subsoil nitrogen content		
	Subsoil organic carbon content		
	Subsoil pH		
	Distance from volcanoes		
	Distance from volcanic rocks		
	Distance to rivers		
5	pH	Multiple linear regression	Cho et al. (2011)
	Electrical conductivity	Principal component analysis	
	Total dissolved solids	Artificial neural network	
	Temperature	Principal component analysis and artificial network analysis	
	Redox potential		
6	Pre-Holocene deposits	Logistic regression	Yang et al. (2014)
	Deltaic deposits		
	Organic-rich deposits		
	Alluvial deposits		
	Tidal deposits		
	Coarse-textured soil		
	Medium-textured soil		
	Fine-textured soil		
	Sand content		
	Silt content		
	Clay content		
	Slope		
	Topographic wetness index		
	Density of rivers		
	Distance to rivers		

(continued)

Table 4.4 (continued)

Case study	Predictors	Method	Reference
7	pH	Back-propagation artificial neural network	Park et al. (2016)
	Electrical conductivity	Support vector machine	
	Total dissolved solids		
	Temperature		
	Redox potential		
	Well depths		
8	Distance to river ganga	Regression	Singh et al. (2016)
	Distance to river Sone		
	Depth of hand pumps		
	Elevation of hand pumps		
	Surface elevation		
	Slope		
	Flow direction		
	Flow accumulation		
	Distance to a drainage point		
	Latitude		
	Longitude		
9	Volcano-sedimentary schist (Birimian)	Logistic regression	Bretzler et al. (2017)
	Volcanite: Basalt, andesite, rhyolite (Birimian)		
	Orthogneiss (Birimian)		
	Granite		
	Tonalite		
	Distance to faults		
	Distance to mineral deposits (metal ores)		
	Distance to granitoid rocks		
	Drainage direction		
	Flow accumulation		
10	pH	Logistic regression	Podgorski et al. (2017)
	Depth		
	Aridity		
	Fluvisols		
	Irrigated area		
	Slope		
	Holocene fluvial sediments		
	Soil organic carbon		
	Soil pH		
	Slope		

(continued)

Table 4.4 (continued)

Case study	Predictors	Method	Reference
11	Casing diameter	Boosted regression tree	Erickson et al. (2018)
	Type of aquiferin which well is located		
	Stratigraphic unit at the top of the open interval		
	Stratigraphic unit at the bottom of the open interval		
	Percent of excessively well-drained soil		
	Percent of moderately well-drained soil		
	Percent of poorly drained soil		
	Percent of somewhat excessively drained soil		
	Percent of somewhat poorly drained soil		
	Percent of very poorly drained soil within		
	Percent of well-drained soil within		
	Percent of soil classified as hydrologic group AD		
	Percent of soil classified as hydrologic group B		
	Percent of soil classified as hydrologic group BD		
	Percent of soil classified as hydrologic group C		
	Percent of soil classified as hydrologic group CD		
	Percent of soil classified as hydrologic group D		
	Nearest major river		
	Estimated mean annual recharge		
	Depth to the water table, meters below land surface		
	Aluminum content in soil C horizon, weight percent		
	Barium content in soil C horizon, mg/kg		
	Beryllium content in soil C horizon, mg/kg		
	Strontium content in soil C horizon, mg/kg		
	Thorium content in soil C horizon, mg/kg		
	Thallium content in soil C horizon, mg/kg		
	Uranium content in soil C horizon, mg/kg		
	Lanthanum content in soil C horizon, mg/kg		
	Cadmium content in soil C horizon, mg/kg		
	Cobalt content in soil C horizon, mg/kg		
	Tungsten content in soil C horizon, mg/kg		
	Zinc content in soil C horizon, mg/kg		
	Sodium content in soil C horizon, weight percent		

(continued)

Table 4.4 (continued)

Case study	Predictors	Method	Reference
	Phosphorus content in soil C horizon, mg/kg		
	Rubidium content in soil C horizon, mg/kg		
	Total clay content in soil C horizon, weight percent		
	Lead content in soil C horizon, mg/kg		
	Nickel content in soil C horizon, mg/kg		
	Top bedrock unit		
	Stream order number of the closest stream to well		
	Distance to nearest stream indicated by a stream order number, meters		
	Calculated distance from screened interval to clay layer, meters		
	Land surface elevation, meters (point data used for initial model development)		
	Land surface elevation, meters (point data used for predictive map development)		
	Completed well depth, meters		
	Casing depth, meters		
	Length of the screened interval, meters		
	Eastward-measured distance		
	Northward-measured distance		
	Calculated elevation at the bottom well, meters		
	Area- and depth-weighted average percent of sand		
	Area- and depth-weighted average percent of silt		
	Area- and depth-weighted average percent of clay		
	Percent of soil classified as hydrologic group A		
	Land use code		
	Distance to the nearest major river, meters		
	Bismuth content in soil C horizon, mg/kg		
	Calcium content in soil C horizon, weight percent		
	Iron content in soil C horizon, weight percent		
	Potassium content in soil C horizon, weight percent		
	Lithium content in soil C horizon, mg/kg		

(continued)

Table 4.4 (continued)

Case study	Predictors	Method	Reference
	Magnesium content in soil C horizon, weight percent		
	Dolomite content in soil C horizon, weight percent		
	Copper content in soil C horizon, mg/kg		
	Manganese content in soil C horizon, mg/kg		
	Molybdenum content in soil C horizon, mg/kg		
	Total thickness of glacial deposits, meters		
	Thickness of coarse-grained sediment within the glacial deposits, meters		
	Calculated thickness of fine-grained sediment within the glacial deposits, meters		
	Texture-based estimated equivalent vertical hydraulic conductivity of the glacial deposits, meters per day		
	Texture-based estimated equivalent transmissivity of the glacial deposits, square meters per day		
	Texture-based estimated equivalent horizontal hydraulic conductivity of the glacial deposits, meters per day		
	Specific-capacity-based transmissivity of coarse-grained sediment within the glacial deposits, square meters per day		
	Specific-capacity-based horizontal hydraulic conductivity of coarse-grained sediment within the glacial deposits, meters per day		
	Predicted nitrate concentration, mg/L		

ratio, subsoil pH, topsoil clay content, and subsoil clay content were the strongest predictors of arsenic in groundwater (Amini et al. 2008a). Cho et al. (2011) have developed multiple linear regression, principal component regression, and artificial neural network models, as well as models based on a combination of principal components and artificial neural networks, using just 141 groundwater samples collected from Cambodia, Laos, and Thailand. The significance of the geochemical parameters used to develop these ML models varied between each model; however, all variables were found to predict arsenic in the studied groundwater (Cho et al. 2011). In a relatively recent study, the authors utilized the conductivity, temperature, redox, pH, well depth, and TDS of 350 groundwater samples collected from Cambodia, Laos, and Thailand and to develop artificial neural network (ANN) and support vector machine (SVM) ML models (Park et al. 2016). In another study, the authors developed logistic regression and boosted regression tree models using

Table 4.5 Artificial intelligence techniques used to predict fluoride contamination of groundwater

Case Study	Predictors	Method	Reference
1	Actual evapotranspiration	Logistic regression	Podgorski et al. (2018)
	Calcisols		
	Cropland		
	Felsic igneous rocks		
	Mafic igneous rocks		
	Noncarb, sedimentary rocks		
	Potential evapotranspiration		
	Precipitation		
	Sand fraction		
	Silt fraction		
	Slope		
2	Elevation	Multiple linear regression	Amini et al. (2008b)
	Slope	Adaptive neuro-fuzzy inference system	
	Evapotranspiration		
	Precipitation		
	Temperature		
	Runoff		
	Irrigation		
	Topsoil clay content		
	Topsoil silt content		
	Topsoil sand content		
	Topsoil C/N ratio		
	Subsoil clay content		
	Subsoil silt content		
	Subsoil sand content		
	Subsoil C/N ratio		
	Subsoil cation exchange capacity		
	Soil drainage code		
	Subsoil nitrogen content		
	Subsoil organic carbon content		
	Subsoil pH		
	Distance from felsic volcanic rocks		
	Distance from mafic volcanic rocks		
	Distance from other volcanic rocks		
	Distance from intrusive felsic rocks		

(continued)

Table 4.5 (continued)

Case Study	Predictors	Method	Reference
	Distance from intrusive mafic rocks		
	Distance from metamorphic rocks		
	Sedimentary rocks		
	Distance from normal faults		
	Distance from rivers		
3	Electrical conductivity	Multilayer perceptron	Barzegar et al. (2017)
	pH	Support vector machine	
	Ca^{2+}	Extreme learning machine	
	Mg^{2+}		
	Na^{+}		
	K^{+}		
	HCO_3^{-}		
	CO_3^{2-}		
	Cl^{-}		
	$SO^{2-}{}_4$		
	F^{-}		
4	Electrical conductivity	Fuzzy logic	Nadiri et al. (2013)
	pH	Artificial neural network	
	Ca^{2+}	Neuro-fuzzy	
	Mg^{2+}	Committee machine with artificial intelligence	
	Na^{+}		
	K^{+}		
	HCO_3^{-}		
	CO_3^{2-}		
	Cl^{-}		
	SO_4^{2-}		
	F^{-}		
	SiO_2		
5	Ca^{2+}	Instance-based *k*-nearest neighbors	Khosravi et al. (2019)
	Mg^{2+}	Locally weighted learning	
	Na^{+}	K-star algorithm	
	K^{+}	Regression by discretization	
	HCO_3^{-}		
	CO_3^{2-}		
	Cl^{-}		
	SO_4^{2-}		
	F^{-}		

Table 4.6 Artificial intelligence techniques used to predict nitrate contamination of groundwater

Case Study	Predictors	Method	Reference
1	Urban land	Multiple linear regression	Knoll et al. (2019)
	Arable land	Classification and regression trees	
	Grassland	Random forest	
	Forest	Boosted regression trees	
	Special crops		
	N-surplus on agricultural land		
	Hydrogeological units		
	Seepage water rate		
	Groundwater recharge rate		
	Nitrate concentration in seepage water		
	Nitrate concentration in groundwater recharge		
	Soil groups		
	Field capacity		
	Humus content in topsoil		
2	Surface water flow direction	Classification and regression trees	Rodriguez-Galiano et al. (2018)
	Drop raster	Random forest	
	Groundwater table depth	Support vector machines	
	Vadose zone thickness		
	Transmissivity		
	Module of hydraulic gradient		
	Maximum level of photosynthetic activity in canopy		
	Time of maximum photosynthesis in canopy		
	Cumulated NDVI for the post-maximum month		
	Overall population (based on census)		
	Distance from population centroids		
	Land cover reclassified according to its potential impact on nitrate pollution		
	Distance to irrigation canals with water-quality problems		
	Distance to cemeteries		
	Density of industries and facilities, extended to a radius of 1 km		
	Density of industries and facilities, extended to a radius of 3 km		
	Density of industries and facilities, extended to a radius of 5 km		
	Livestock density within 1 km radius of the livestock farms		

(continued)

Table 4.6 (continued)

Case Study	Predictors	Method	Reference
	Livestock density within 3 km radius of the livestock farms		
	Livestock density within 5 km radius of the livestock farms		
3	Depth of groundwater	Boosted regression trees	Sajedi-Hosseini et al. (2018)
	Net recharge	Multivariate discriminant analysis	
	Aquifer media	Support vector machine	
	Soil media		
	Topography		
	Impact of vadose zone		
	Hydraulic conductivity		
	Topographic wetness index		
	Distance to river		
	Slope percent		
	Drainage density		
	Soil type		
	Elevation		
	Land use		
	Lithology		
4	Calendar year of sample	Censored ML regression	Messier et al. (2019)
	Calendar month of sample	Random forest	
	Longitude value of well location		
	Latitude value of well location		
	Log of modeled mean well depth		
	Standard deviation of modeled well depth		
	Geometric mean of modeled well depth		
	Back log-transformed mean of modeled well depth		
	Back log-transformed variance of modeled well depth		
	Proportion of 1992 agricultural lands within the well buffer		
	Annual county-level nitrogen fertilizer apportioned in 1992 to agrarian lands within the well buffer		
	Physiographic region of well		
	pH		
	Hydrologic soil group A		
	Hydrologic soil group B		

(continued)

Table 4.6 (continued)

Case Study	Predictors	Method	Reference
	Hydrologic soil group C		
	Hydrologic soil group D		
	Available water capacity		
	Histosol soil type		
	Topographic wetness index		
	Slope		
	Nitrogen load		
	Septic system density		
	Decaying contribution of nitrate from wastewater treatment plants		
	Decaying contribution of nitrate from cattle		
	Decaying contribution of nitrate from poultry		
	Decaying contribution of nitrate from lagoons		
	Decaying contribution of nitrate from wastewater treatment residual		
	National land cover		
	STATSGO-based mean water table depth		
5	Land cover/land use	Multiple linear regression	Ouedraogo et al. (2019)
	Population density	Random forest regression	
	Nitrogen application		
	Climate class		
	Type of region		
	Rainfall class		
	Depth to groundwater		
	Aquifer type		
	Soil type		
	Unsaturated zone		
	Topography/slope		
	Recharge		
	Hydraulic conductivity		

domestic and public supply data from 5219 water sources and incorporated 76 variables, including hydrologic, geologic, meteorological, geochemical, and land-use factors (Ayotte et al. 2016). In a regional study in rural India, the authors developed a simple regression model from 88 groundwater samples using hydrogeological and topographic features (Singh et al. 2016). They found that the depth of tube wells, flow direction of surface water, distance to rivers, and distance to drainage point were the most important predictors of arsenic in groundwater (Singh et al. 2016). The authors of another study developed a logistic regression model using ten geohydrological and topographic features of 1498 water-sample collection points in rural Burkina Faso (Bretzler et al. 2017). They found that volcanic-sedimentary schist (Birimian), volcanite (basalt, andesite, rhyolite (Birimian)), granite, distance to mineral deposits, and distance to granitoid rocks were the most important arsenic-contamination predictors in the studied region (Bretzler et al. 2017). Podgorski and his colleagues have developed a logistic regression model using 12 geochemical predictors and 1200 groundwater samples in Pakistan (Podgorski et al. 2017). They found that in there, potential evapotranspiration, precipitation, aridity, irrigated area, slope, soil organic carbon, soil pH, and Holocene fluvial sediments were the most important predictors of arsenic presence (Podgorski et al. 2017). In a recent study, the authors developed a boosted regression tree machine learning model using 74 features of physical or geochemical triggers of arsenic contamination from 3283 water samples in the north-central USA (Erickson et al. 2018). They found that 33 of the 74 variables served as essential arsenic predictors in the model (Erickson et al. 2018).

It is apparent from this collection of studies and models that hydrogeological and topographic factors play significant roles in predicting groundwater arsenic contamination and that logistic regression has been a preferred method of developing arsenic-prediction models.

4.5 Application of AI in Predicting Fluoride Contamination of Groundwater

Amini et al. (2008a, b) developed the first global model of fluoride contamination of groundwater. The model is based on 60,000 groundwater samples collected from 25 countries around the world. Thirty-one variables were used to develop probabilistic models, applying linear regression and adaptive neuro-fuzzy inference system (ANFIS) techniques (Table 4.5) (Amini et al. 2008b).

Nadiri et al. (2013), using a minimal dataset of the hydro-chemical properties of 132 groundwater samples collected over 4 years (2004–2008), developed ML models: a Sugenofuzzylogic, a Mamdanifuzzylogic, an ANN, a neuro-fuzzy, and a committee machine with artificial intelligence (CMAI) model (Nadiri et al. 2013). In 2017, Barzegar et al. (2017) incorporated additional samples into the same dataset to develop extreme learning machine (ELM), multilayer perceptron (MLP), and SVM models. Recently, Podgorski et al. (2018) have developed multinomial logistic regression and random forest models based on geological, climatic, and soil features of 12,600 groundwater samples collected from all over India. They found 15 of 25 variables to be the most crucial fluoride predictors in the studied area (Podgorski et al. 2018). In another recent study, the authors developed instance-based k-nearest neighbors, locally weighted learning, M5P, and regression-by-discretization models on a small dataset of 143 groundwater samples using hydro-chemical properties of the water samples (Khosravi et al. 2019).

It is readily apparent that all of these models except for one are based on a minimal dataset (Barzegar et al. 2017; Khosravi et al. 2019; Nadiri et al. 2013) and on the hydro-chemical properties alone of collected water samples. Considering the numerical predictors, it is also evident that various linear regression machine learning techniques were applied. However, when it comes to a multidimensional case (Podgorski et al. 2018), the preferred method is logistic regression, a binary classification technique.

4.6 Application of AI in Predicting Nitrate Contamination of Groundwater

Only a few studies on prediction of nitrate contamination of groundwater use AI techniques (Table 4.6). In one such study, Sajedi-Hosseini et al. (2018) developed boosted regression tree, multivariate discriminant analysis, and support vector machine models based on the hydrologic and topographic parameters of 102 water samples collected from Iran.

In a recent study, the authors developed random forest, classification and regression trees, and support vector machine models using hydrogeological, hydrological, census, and Normalized Difference Vegetation Index (NDVI) data extracted from remote sensing images for 110 water samples collected in Spain (Rodriguez-Galiano et al. 2018). In another, the authors developed a random forest regression model examining the land use, soil type, hydrogeology, topography, climatology, type of region, and nitrogen fertilizer application factors of 250 groundwater samples collected in Africa (Ouedraogo et al. 2019). Comparatively, more comprehensive fluoride prediction models have been developed by Messier et al. (2019); these apply random forest, gradient boosted machine, support vector machine, neural network, and kriging techniques to a larger dataset of 22,000 samples collected from private wells in the U.S. state of North Carolina, and they consider nearly 120 variables (Table 4.5). Additionally, Knoll et al. (2019) have developed multiple linear regression, classification and regression trees, random forest, and boosted regression tree models on 1890 groundwater samples collected from the German state of Hesse. The authors employed the hydrogeological, hydrologic, land use, and soil properties of the sampled sites (Knoll et al. 2019).

It is evident that nitrate prediction models have been developed using various hydrogeologic, land use, and topographic properties, and that the random forest method has been the most preferable option.

4.7 Discussion and Conclusions

According to the best of the knowledge gained from the literature review, certain artificial intelligence algorithms have been suggested and developed for classification tasks related to the mapping of environmental problems. According to their distribution-probability functions, the developed algorithms can be classified into three categories: (i) Bayes-based algorithms (e.g., naïve Bayes and Bayes net), (ii) functional algorithms (e.g., logistic regression, artificial neural network, support vector machine), and (iii) decision-tree algorithms (e.g., random forest). The diversity of these algorithms is referred to as their applicability in classification problems. Spatial prediction of environmental issues is related to two issues, which are (i) model selection and (ii) factor/predictor selection and conditioning. In model selection, users peruse existing models in search of one that is fitted to a training dataset. There are neither standards or guidelines to clarify the optimal model for a given task (Svetnik et al. 2003). This means that users may need to find the applicable model through a trial-and-error approach, tuning parameters and running algorithms on the training dataset and then comparing results to select the best fit. What variables are appropriate to consider as predictors may vary according to data availability, and differences in selected variables may create uncertainties during the modeling process (Shirzadi et al. 2019). For example, a given LR model may return different results in one case study than it would in another. This difference depends on the fact that the variables defined in each case study are different, causing the results to differ as well (Bui et al. 2018; Tien Bui et al. 2019).

Moreover, for a given case study, two models may return unlike results because of differences in their probability-distribution functions. An LR model, for example, might be fitted to the training dataset while an RF model might not. In other words, the variables in a single studied area will be constant for all applied models, but in this situation, the models would return different results. For this reason, a model's capacity to accurately predict is simply referred to as its ability to explain causal processes, and there is enough room within the discipline for both standards of evaluation (Shmueli 2010).

As per the above discussion and literature review, although some algorithms have been developed for the classification task, few have been proposed or used for groundwater contamination, including that of arsenic, fluoride, and nitrate. As we have mentioned, the LR model has been more heavily relied upon for mapping arsenic and fluoride contamination of groundwater. Although LR was employed as a statistical machine learning technique in the 1990s to assess groundwater nitrate vulnerability (Eckhardt and Stackelberg 1995; Tesoriero and Voss 1997), it has been more recently used to map arsenic (Twarakavi and Kaluarachchi 2005; Venkataraman 2010; Winkel et al. 2008) and fluoride-contamination resources (Barzegar et al. 2017; Podgorski et al. 2018; Singh et al. 2013). The applicability and usability of the LR model that makes it a promising alternative solution to the mapping of groundwater contamination is derived from several advantages, including the ability to mathematically compute empirical weights for each variable, to statistically remove the variables that lack predictive ability (via multi collinearity tests), and to consider which variables most significantly affecting the results (at a 95% confidence interval) (Focazio 2002). Additionally, the obtained weights, which are computed based on observed data, lead to the production of a reasonable, reliable vulnerability map that can be used as an appropriate tool by water-resource managers (Mair and El-Kadi 2013). The LR model has been considered a benchmark soft computing model (Chapi et al. 2017; Miraki et al. 2019; Tien Bui et al. 2019) whose performance has been confirmed in studies of groundwater-potential mapping (Elío et al. 2017; Mair and El-Kadi 2013; Ozdemir 2011; Rizeei et al. 2018).

The current study also concludes that for mapping nitrate contamination of groundwater, among the applied machine learning algorithms of recent years, the random forest approach has predominated. The RF is a tree-based algorithm that is extremely flexible (Muchlinski et al. 2016). It exchanges a high degree of variance between each tree for a low bias in predicting the outcome variable. If the assumptions of the modeling process, including linearity of variables, collinearity, and homoscedasticity, are not remarkable, other methods that are not regression-based may provide better estimations. However, if the training dataset of an event is rare and crucial, the flexibility of algorithms such as RF can outperform LR models (Muchlinski et al. 2016). A study carried out by Rodriguez-Galiano et al. (2014) indicated that the RF model had superior goodness-of-fit and prediction accuracy, when compared with the LR model, for nitrate vulnerability mapping of the Vega de Granada aquifer in the southern part of Iberia (southeast of Granada City). The authors stated that the application of RF models is simple and that their results are interpretable. Eventually, they summarized the advantages of the RF model for nitrate vulnerability mapping and groundwater studies as follows:

- It can learn the complicated, nonlinear relationship between the independent and dependent variables.
- The different types of variables can be incorporated for analysis because the RF model does not need any assumption, whereas the LR model does.
- A large number of training datasets can be efficiently handled by the RF model without variable deletion.
- It can estimate the predictive power of variables.
- It can produce an internal unbiased estimate of the prediction (out-of-bag) error.
- It is partly robust and sturdily resistant to outliers and spurious datasets.
- The computation time of RF model is lower than that of other machine learning methods, including ANN or SVM. In other words, it is a high-predictive-accuracy algorithm.
- Concerning noise, it is most robust than other machine learning models.

According to the above, it can be claimed that AI techniques for classification tasks are diverse, and that therefore, they should be tested on a given groundwater-contamination dataset and their results compared. The application of various AI techniques in predicting the presence of groundwater contaminants is a promising approach but a recent phenomenon. Groundwater contamination can be triggered by a myriad of environmental as well as anthropogenic processes. Geographic, hydrogeologic, geological, hydrochemical, land use, and socioeconomic factors at play in areas with contaminated water have vital roles in inducing the release of specific contaminants into the groundwater. The above review also shows that contamination is a regional phenomenon, so global prediction models may not be appropriate for most groundwater contaminants. An array of ML techniques, including but not limited to regression and classification algorithms has been used to develop such prediction models. However, there still exists a gap in the selection of appropriate algorithms. It is also apparent that many such prediction models were developed from miniscule sample sizes. Larger sample sizes will be required in the development of robust prediction models. Studies also suggest that since environmental data are highly nonlinearly correlated and multidimensional, it would be advantageous to explore nonlinear classifiers and hybrid machine learning models. Accurate and robust prediction models would help in predicting potentially contaminated areas, obviating costly expenditures on laboratory analytical water testing. Furthermore, environmental managers, policymakers, and local administrations could use these models to develop proactive mitigation strategies. These models will empower us to save millions of human lives by preventing their exposure to these contaminants, especially arsenic and nitrate, in the first place.

References

Alcamo J, Martina F, Michael M (2007) Future long-term changes in global water resources driven by socio-economic and climatic changes. Hydrolog Sci J 52(2):247–275

Ali S, Thakur SK, Sarkar A, Shekhar S (2016) Worldwide contamination of water by fluoride. Environ Chem Letters 14(3):291–315

Amini M, Abbaspour KC, Berg M, Winkel L, Hug SJ, Hoehn E, Yang H, Johnson CA (2008a) Statistical modeling of global geogenic arsenic contamination in groundwater. Environ Sci Technol 42(10):3669–3675

Amini M, Mueller K, Abbaspour KC, Rosenberg T, Afyuni M, Møller KN, Sarr M, Johnson CA (2008b) Statistical modeling of global geogenic fluoride contamination in groundwaters. Environ Sci Technol 42(10):3662–3668

Ayotte JD, Nolan BT, Nuckols JR, Cantor KP, Robinson GR, Baris D, Hayes L, Karagas M, Bress W, Silverman DT (2006) Modeling the probability of arsenic in groundwater in New England as a tool for exposure assessment. Environ Sci Technol 40(11):3578–3585

Ayotte JD, Nolan BD, Gronberg JO (2016) Predicting arsenic in drinking water wells of the Central Valley, California. Environ Sci Technol 50(14):7555–7563

Barzegar R, Moghaddam AA, Adamowski J, Fijani E (2017) Comparison of machine learning models for predicting fluoride contamination in groundwater. Stochastic Environ Res Risk Assess 31(10):2705–2718

Bhattacharya P, Samal AC, Majumdar J, Santra SC (2010) Arsenic contamination in rice, wheat, pulses, and vegetables: a study in an arsenic affected area of West Bengal, India. Water Air Soil Pollut 213(1–4):3–13

Braman RS (1975) Arsenic in the environment. ACS Publications

Bretzler A, Lalanne F, Nikiema J, Podgorski J, Pfenninger N, Berg M, Schirmer M (2017) Groundwater arsenic contamination in Burkina Faso, West Africa: predicting and verifying regions at risk. Sci Total Environ 584:958–970

Bui DT, Panahi M, Shahabi H, Singh VP, Shirzadi A, Chapi K, Khosravi K, Chen W, Panahi S, Li S (2018) Novel hybrid evolutionary algorithms for spatial prediction of floods. Sci Rep 8(1):1–14

Bundy LG, Knobeloch L, Webendorfer B, Jackson GW, Shaw BH (1994) Nitrate in Wisconsin groundwater: sources and concerns. University of Wisconsin–Extension

Burri NM, Weatherl R, Moeck C, Schirmer (2019) A review of threats to groundwater quality in the anthropocene. Sci Total Environ 684:136–154

Chakraborti D, Mukherjee SC, Pati S, Sengupta MK, Rahman MM, Chowdhury UK, Lodh D, Chanda CR, Chakraborti AK, Basu GK (2003) Arsenic groundwater contamination in Middle Ganga Plain, Bihar, India: a future danger? Environ Health Perspect 111(9):1194–1198

Chakraborti D, Rahman MM, Chatterjee A, Das D, Das B, Nayak B, Pal A, Chowdhury UK, Ahmed S, Biswas BK (2016) Fate of over 480 million inhabitants living in arsenic and fluoride endemic Indian districts: magnitude, health, socio-economic effects and mitigation approaches. J Trace Ele Med Biol 38:33–45

Chakraborti D, Singh SK, Rashid HM, Rahman MM (2017) Arsenic: occurrence in groundwater. Encyclopedia of environmental health, vol 2. Elsevier, Burlington, pp 1–17. https://doi.org/10.1016/B978-0-12-409548-9.10634-7

Chakraborti D, Singh SK, Rahman MM, Dutta RN, Mukherjee SC, Pati S, Kar PB (2018) Groundwater arsenic contamination in the Ganga River basin: a future health danger. Int J Environ Res Pub Health 15(2):180. https://doi.org/10.3390/ijerph15020180

Chapi K, Singh VP, Shirzadi A, Shahabi H, Bui DT, Pham BT, Khosravi K (2017) A novel hybrid artificial intelligence approach for flood susceptibility assessment. Environ Model Softw 95:229–245

Chapman DV (1996) Water quality assessments: a guide to the use of biota, sediments, and water in environmental monitoring

Chau K (2006) A review on integration of artificial intelligence into water quality modelling. Marine Pollut Bullet 52(7):726–733

Chen W, Pradhan B, Li S, Shahabi H, Rizeei HM, Hou E, Wang S (2019) Novel hybrid integration approach of bagging-based fisher's linear discriminant function for groundwater potential analysis. Nat Resour Res 28(4):1239–1258

Cho KH, Sthiannopkao S, Pachepsky YA, Kim KW, Kim JH (2011) Prediction of contamination potential of groundwater arsenic in Cambodia, Laos, and Thailand using artificial neural network. Water Res 45(17):5535–5544

Crocker J, Bartram J (2014) Comparison and cost analysis of drinking water quality monitoring requirements versus practice in seven developing countries. Int J Environ Res Pub Health 11 (7):7333–7346

Eckhardt DAV, Stackelberg PE (1995) Relation of ground-water quality to land use on Long Island, New York. Groundwater 33(6):1019–1033

Elío J, Crowley Q, Scanlon R, Hodgson J, Long S (2017) Logistic regression model for detecting radon prone areas in Ireland. Sci Total Environ 599:1317–1329

Erickson ML, Elliott SM, Christenson CA, Krall AL (2018) Predicting geogenic arsenic in drinking water wells in glacial aquifers, North-Central USA: accounting for depth-dependent features. Water Resour Res 54(12):172–187

Evans S, Campbell C, Naidenko OV (2019) Cumulative risk analysis of carcinogenic contaminants in United States drinking water. Heliyon 5(9):e02314

Focazio MJ (2002) Assessing ground-water vulnerability to contamination: providing scientifically defensible information for decision makers. Vol. 1224: US Department of the Interior, US Geological Survey

Gamble A (2019) Ullmann's encyclopedia of industrial chemistry. Charleston Advisor 20(4):46–50

Grandjean P (2019) Developmental fluoride neurotoxicity: an updated review. Environ Health 18 (1):1–17

Jha SK, Singh RK, Damodaran T, Mishra VK, Sharma DK, Rai D (2013) Fluoride in groundwater: toxicological exposure and remedies. J Toxicol Environ Health B 16(1):52–66

Kabir H, Gupta AK, Tripathy S (2020) Fluoride and human health: systematic appraisal of sources, exposures, metabolism, and toxicity. Crit Rev Environ Sci Technol 50(11):1116–1193

Keeney DR (1989) Sources of nitrate to ground water. In: Developments in agricultural and managed forest ecology. Elsevier, pp 23–34

Khosravi K, Barzegar R, Miraki S, Adamowski J, Daggupati P, Alizadeh MR, Pham BT, Alami MT (2019) Stochastic modeling of groundwater fluoride contamination: introducing lazy learners. Groundwater. https://doi.org/10.1111/gwat.12963

Knoll L, Breuer L, Martin B (2019) Large scale prediction of groundwater nitrate concentrations from spatial data using machine learning. Sci Total Environ 668:1317–1327
Lado LR, Polya D, Winkel L, Berg M, Hegan A (2008) Modelling arsenic hazard in Cambodia: a geostatistical approach using ancillary data. Appl Geochem 23(11):3010–3018
Mair A, El-Kadi AI (2013) Logistic regression modeling to assess groundwater vulnerability to contamination in Hawaii, USA. J Conta Hydrol 153:1–23
Mazumder GDN (2008) Chronic arsenic toxicity & human health. Ind J Med Res 128(4):436–447
Messier KP, Wheeler DC, Flory AR, Jones RR, Patel D, Nolan BT, Ward MH (2019) Modeling groundwater nitrate exposure in private wells of North Carolina for the agricultural health study. Sci Total Environ 655:512–519
Miraki S, Zanganeh SH, Chapi K, Singh VP, Shirzadi A, Shahabi H, Pham BT (2019) Mapping groundwater potential using a novel hybrid intelligence approach. Water Resour Manag 33 (1):281–302
Mitra A, Chatterjee S, Gupta DK (2020) Environmental arsenic exposure and human health risk. In: Fares A, Singh SK (eds) Arsenic water resources contamination: challenges and solutions. Springer, Cham, pp 103–129
Muchlinski D, David S, Jingrui H, Matthew K (2016) Comparing random forest with logistic regression for predicting class-imbalanced civil war onset data. Polit Anal 24(1):87–103
Murcott S (2012) Arsenic contamination in the world. An international sourcebook
Nadiri AA, Fijani E, Tsai FTC, Moghaddam AA (2013) Supervised committee machine with artificial intelligence for prediction of fluoride concentration. J Hydroinf 15(4):1474–1490
Nayak B, Roy MM, Das B, Pal A, Sengupta MK, De SP, Chakraborti D (2009) Health effects of groundwater fluoride contamination. Clinical Toxicol 47(4):292–295
NGWA (2020) Groundwater facts. https://www.ngwa.org/what-is-groundwater/About-groundwater/groundwater-facts. Accessed 4 Feb 2020
Nordstrom DK (2002) Worldwide occurrences of arsenic in ground water. Ame Assoc Advance Sci 296:2143
Oremland RS, Stolz JF (2003) The ecology of arsenic. Science 300(5621):939–944
Ouedraogo I, Defourny P, Vanclooster M (2019) Application of random forest regression and comparison of its performance to multiple linear regression in modeling groundwater nitrate concentration at the African continent scale. Hydrogeol J 27(3):1081–1098
Ozdemir A (2011) Using a binary logistic regression method and GIS for evaluating and mapping the groundwater spring potential in the Sultan Mountains (Aksehir, Turkey). J Hydrol 405 (1–2):123–136
Ozsvath DL (2009) Fluoride and environmental health: a review. Rev Environ Sci Biotechnol 8 (1):59–79
Park Y, Ligaray M, Kim YM, Kim JH, Cho KH, Sthiannopkao S (2016) Development of enhanced groundwater arsenic prediction model using machine learning approaches in Southeast Asian countries. Desalinat Water Treat 57(26):12227–12236
Pham BT, Prakash I, Dou J, Singh SK, Trinh PT, Tran HT, Le TM, Tran VP, Khoi DK, Shirzadi A (2018) A novel hybrid approach of landslide susceptibility modeling using rotation forest ensemble and different base classifiers. Geocarto Int:1–38
Pham BT, Abolfazl J, Prakash I, Singh SK, Quoc NK, Bui DT (2019) Hybrid computational intelligence models for groundwater potential mapping. Catena 182:101–104
Phong TV, Phan TT, Prakash I, Singh SK, Shirzadi A, Chapi K, Ly HB, Ho LS, Quoc NK, Pham BT (2019) Landslide susceptibility modeling using different artificial intelligence methods: a case study at Muong Lay district, Vietnam. Geocarto Int:1–24
Podgorski JE, Eqani SAMAS, Khanam T, Ullah R, Shen H, Berg M (2017) Extensive arsenic contamination in high-pH unconfined aquifers in the Indus Valley. Sci Adv 3(8):e1700935
Podgorski JE, Labhasetwar P, Saha D, Berg M (2018) Prediction modeling and mapping of groundwater fluoride contamination throughout India. Environ Sci Technol 52(17):9889–9898
Ritchie H, Roser M (2019) Clean water. Our world in data

Rizeei HM, Azeez OS, Pradhan B, Khamees HH (2018) Assessment of groundwater nitrate contamination hazard in a semi-arid region by using integrated parametric IPNOA and data-driven logistic regression models. Environ Monit Assess 190(11):633

Rizeei HM, Pradhan B, Saharkhiz MA, Lee S (2019) Groundwater aquifer potential modeling using an ensemble multi-adoptive boosting logistic regression technique. J Hydrol 579:124–172

Rodriguez-Galiano V, Mendes MP, Garcia-Soldado MJ, Chica-Olmo M, Ribeiro L (2014) Predictive modeling of groundwater nitrate pollution using random Forest and multisource variables related to intrinsic and specific vulnerability: a case study in an agricultural setting (Southern Spain). Sci Total Environ 476:189–206

Rodriguez-Galiano VF, Luque-Espinar JA, Chica-Olmo M, Mendes MP (2018) Feature selection approaches for predictive modelling of groundwater nitrate pollution: an evaluation of filters, embedded and wrapper methods. Sci Total Environ 624:661–672

Sajedi-Hosseini F, Malekian A, Choubin B, Rahmati O, Cipullo S, Coulon F, Pradhan B (2018) A novel machine learning-based approach for the risk assessment of nitrate groundwater contamination. Sci Total Environ 644:954–962

Sameen MH, Pradhan B, Lee S (2019) Self-learning random forests model for mapping groundwater yield in data-scarce areas. Natu Resour Res 28(3):757–775

Sameen MI, Pradhan B, Lee S (2020) Application of convolutional neural networks featuring Bayesian optimization for landslide susceptibility assessment. Catena 186:104249

Schwarzenbach RP, Thomas E, Thomas BH, Gunten UV, Wehrli B (2010) Global water pollution and human health. Ann Rev Environ Resour 35:109–136

Schweitzer L, Noblet J (2018) Water contamination and pollution. In: Green chemistry. Elsevier, pp 261–290

Şen Z (2014) Practical and applied hydrogeology. Elsevier

Shirzadi A, Solaimani K, Roshan MH, Kavian A, Chapi K, Shahabi H, Keesstra S, Ahmad BB, Bui DT (2019) Uncertainties of prediction accuracy in shallow landslide modeling: sample size and raster resolution. Catena 178:172–188

Shmueli G (2010) To explain or to predict? Stat Sci 25(3):289–310

Singh SK (2017) Conceptual framework of a cloud-based decision support system for arsenic health risk assessment. Environ Sys Decis 37(4):435–450. https://doi.org/10.1007/s10669-017-9641-x

Singh, SK (2018) Application of artificial intelligence in environmental modelling and sustainability. In Euro Scicon Conference on Applied Science, Biofuels & Petroleum Engineering, Athens, Greece, November 12–13, 2018

Singh SK, Srivastava PK, Pandey AC (2013) Fluoride contamination mapping of groundwater in Northern India integrated with geochemical indicators and GIS. Water Sci Technol Water Supp 13(6):1513–1523

Singh SK, Brachfeld SA, Taylor RW (2016) Evaluating hydrogeological and topographic controls on groundwater arsenic contamination in the Middle-Ganga plain in India: towards developing sustainable arsenic mitigation models. In: Emerging issues in groundwater resources. Springer, pp 263–287

Singh SK, Taylor RW, Rahman MM, Pradhan B (2018) Developing robust arsenic awareness prediction models using machine learning algorithms. J Environ Manag 211:125–137. https://doi.org/10.1016/j.jenvman.2018.01.044

Sinha D, Prasad P (2020) Health effects inflicted by chronic low-level arsenic contamination in groundwater: a global public health challenge. J Appl Toxicol 40(1):87–131

Sousa MR, Rudolph DL, Frind E (2014) Threats to groundwater resources in urbanizing watersheds: the Waterloo Moraine and beyond. Can Water Resour J 39(2):193–208

Svetnik V, Liaw A, Tong C, Culberson JC, Sheridan RP, Feuston BP (2003) Random forest: a classification and regression tool for compound classification and QSAR modeling. J Informat Comp Sci 43(6):1947–1958

Tchounwou PB, Yedjou CG, Udensi UK, Pacurari M, Stevens JJ, Patlolla AK, Noubissi F, Kumar S (2019) State of the science review of the health effects of inorganic arsenic: perspectives for future research. Environ Toxicol 34(2):188–202

Tesoriero AJ, Voss FD (1997) Predicting the probability of elevated nitrate concentrations in the Puget Sound Basin: implications for aquifer susceptibility and vulnerability. Groundwater 35 (6):1029–1039

Tien Bui D, Shahabi H, Omidvar E, Shirzadi A, Geertsema M, Clague JJ, Khosravi K, Pradhan B, Pham BT, Chapi K (2019) Shallow landslide prediction using a novel hybrid functional machine learning algorithm. Remote Sen 11(8):931

Twarakavi NKC, Kaluarachchi JJ (2005) Aquifer vulnerability assessment to heavy metals using ordinal logistic regression. Groundwater 43(2):200–214

USEPA (1997) Ambient ground water quality monitoring cost analysis. United States Environmental Protection Agency (USEPA)

Venkataraman K (2010) Occurrence, distribution, and speciation of arsenic in the Southern High Plains Aquifer System. Texas Tech University, Lubbock, TX

Vörösmarty CJ, Green P, Salisbury J, Lammers RB (2000) Global water resources: vulnerability from climate change and population growth. Science 289(5477):284–288

Ward MH, DeKok TH, Levallois P, Brender J, Gulis G, Nolan BT, Van Derslice J (2005) Workgroup report: drinking-water nitrate and health – recent findings and research needs. Environ Health Perspec 113(11):1607–1614

Ward M, Jones R, Brender J, de Kok T, Weyer P, Nolan B, Villanueva C, van Breda S (2018) Drinking water nitrate and human health: an updated review. Int J Environ Res Pub Health 15 (7):1557

WHO (2011) Nitrate and nitrite in drinking-water: background document for development of WHO guidelines for drinking-water quality. World Health Organization (WHO)

WHO (2017) Progress on drinking water, sanitation and hygiene: 2017 update and SDG baselines. World Health Organization (WHO) and the United Nations Children's Fund (UNICEF)

Winkel L, Berg M, Amini M, Hug SJ, Johnson CA (2008) Predicting groundwater arsenic contamination in Southeast Asia from surface parameters. Nat Geosci 1(8):536

Yadav KK, Kumar S, Pham QB, Gupta N, Rezania S, Kamyab H, Yadav S, Vymazal S, Kumar V, Tri DQ (2019) Fluoride contamination, health problems and remediation methods in Asian groundwater: a comprehensive review. Ecotoxicol Environ Saf 182:109362

Yang N, Winkel LHE, Johannesson KH (2014) Predicting geogenic arsenic contamination in shallow groundwater of South Louisiana, United States. Environ Sci Technology 48 (10):5660–5666

Zhou Z (2015) A global assessment of nitrate contamination in groundwater. International Groundwater Resources Assessment Center. Internship report

Chapter 5
Arsenic Contamination of Groundwater and Its Mitigation Strategies

Surabhi Awasthi, Reshu Chauhan, Sanjay Dwivedi, Sudhakar Srivastava, and Rudra Deo Tripathi

5.1 Introduction

Contamination of arsenic (As) in groundwater and their resources globally affects human health. The problem of As contamination is becoming more worsen due to day by day increasing concentration of As and discovery of new As-contaminated areas. In recent years, many As-contaminated regions have been identified, but still systematic evaluation and monitoring of some areas that are at high risk remains to be carried out. Contamination of groundwater through geogenic and anthropogenic activities is the main concern for environment and human health (Fig. 5.1). Across the world in several countries including India, millions of people are using As-contaminated water beyond the permissible limit (10 μg/L). Many countries such as Bangladesh, Hungary, Taiwan, Argentina, China, Chile, Mexico, USA, Nepal, and India have crossed the safe limit of As in drinking water guided by WHO (IWA 2016; Singh 2017). Various standards of As have been accepted in several countries ranging from 5 mg/L in the United States to 50 mg/L in most developing countries (Ahmed 2003; Singh and Stern 2017). Ravenscroft et al. (2009) reported that millions of inhabitants are consuming contaminated drinking water having As more than the permissible limit, i.e., 10 μg/L.

Arsenic is a naturally occurring toxic metalloid with atomic number 33, atomic mass 75, and four oxidation states (−3, 0, +3, and + 5) (Awasthi et al. 2017). Arsenic

Authors Surabhi Awasthi and Reshu Chauhan have equally contributed to this chapter.

S. Awasthi · R. Chauhan · S. Dwivedi · R. D. Tripathi (✉)
CSIR – National Botanical Research Institute, Council of Scientific and Industrial Research, Lucknow, Uttar Pradesh, India

S. Srivastava
Institute of Environment and Sustainable Development, Banaras Hindu University, Varanasi, India

A. Singh et al. (eds.), *Water Pollution and Management Practices*,
https://doi.org/10.1007/978-981-15-8358-2_5

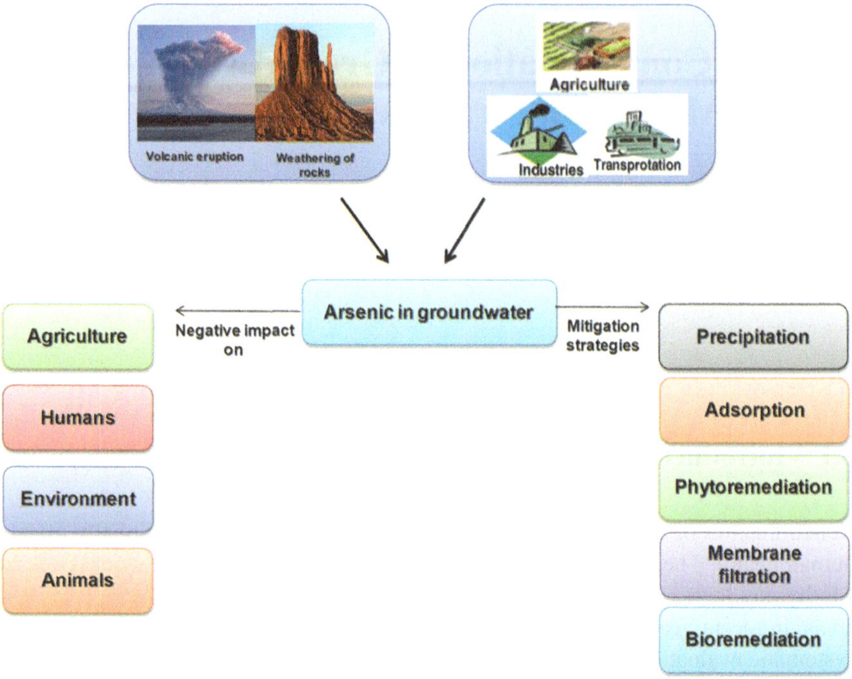

Fig. 5.1 Geogenic and anthropogenic contamination of arsenic in groundwater and its mitigation strategies

exists in >200 mineral forms, including arsenites, arsenates, arsenides, sulfides, sulfosalts, metal alloys, and native elements. Among these, arsenate minerals are scorodite, beudantite, yukonite, and sulfide viz., arsenopyrite, pyrite, loellingite, and realgar (Srivastava et al. 2015). Arsenite (AsIII) and arsenate (AsV) are the most prevalent and more toxic inorganic forms, while monomethylarsine (MMA), dimethylarsine (DMA), and trimethylarsine (TMA) are less toxic organic forms of As (Srivastava et al. 2015; Awasthi et al. 2017). Other organic forms of As include arsenosugars, arsenobetaine, and arsenocholine, which are less toxic or non-toxic and also found in seafood (Srivastava et al. 2015). Drinking of As-polluted water and its utilization for cooking and irrigation of crops poses severe threat to public health. Long-term human exposure of As from food and drinking water resulted in several skin diseases, diabetes, increased blood pressure, and cancer of lung, skin, bladder, and kidney (WHO 2011; USEPA 2013; Santra et al. 2013). To avoid the human exposure of As contamination, the hand pumps polluted with >50 mg/L of As were colored red in India and Bangladesh and the hand pumps with less contamination >50 μg/L were colored blue in India and with green color in Bangladesh (Nickson et al. 2007; Milton et al. 2007). Besides this precaution, several people still have to depend on As-contaminated drinking water (red-painted hand pumps) because of the water scarcity (Singh and Vedwan 2015; Singh and Stern 2017). The concentration of As is also increasing in groundwater due to the geochemical and physical conditions of aquifers and the water–rock interactions for the mobilization and accumulation of As in water. The order of As minerals dissolution in groundwater observed in the series of arsenics >arsenolite>orpiment>realgar>arsenopyrite>tennantite (Islam et al. 2013). The present chapter summarizes the possible sources of As contamination in groundwater and presents an overview of strategies for mitigation of As toxicity and to reduce the As level in drinking water and groundwater.

5.2 Global Arsenic Contamination of Groundwater

Contamination of As in groundwater and drinking water is a public health issue and adversely affects millions of people globally. This leads to a marked increase in cancer risk (Chakraborti et al. 2017). The International Agency for Research on Cancer (IARC) and the US Environmental Protection Agency (EPA) declared As and its compounds as class 1 human carcinogen (WHO 2004). The Agency of Toxic Substances and Disease Registry (ATSDR) marked As at number 1 position among 20 top hazardous substances. The Joint FAO/WHO Expert Committee on Food Additives (JECFA) analyzed the impact of As on human health and reported exceeded concentration of As (50–100 μg/L) in drinking water of many regions of the world (WHO 2019). Across the world, highly As-contaminated areas have been reported mainly in large deltas, e.g., Bengal delta (Chakraborti et al. 2010) and along river basins (Table 5.1) such as Duero Cenozoic Basin in Spain, Danube river basin in Hungry, Zenne river basin in Belgium Hetao river basin in Mongolia, Tulare Lake in the USA, and Paraiba do Sul delta in Brazil (Gómez et al. 2006; Nriagu et al. 2007; Khan and Ho 2011; Cutler et al. 2013; Mirlean et al. 2014). For technical and financial support in many states, a large number of National Rural Drinking Water Programmes (NRDWP) have been sponsored by the government (Ministry of Drinking Water and Sanitation) for safe drinking water. Up to 67% of fund was provided under NRDWP with priority to As- and fluoride-contaminated areas to tackle water quality problems. Advances have been made to the reduction of As exposure, e.g., by removal of As from drinking water and by providing residents with other resource of drinking water.

5.3 Sources of As and Its Impact on Human Health

Several geogenic and anthropogenic activities are being reported for increased As pollution in groundwater. Himalayan mountains and Shillong plateau are considered as main sources of As contamination in Gangetic river basin and delta sediments. Additionally, the Gondwana coal region, Bihar mica belt, the pyrite-bearing region in Vindhyan range, Sone river valley gold belt, and sulfide regions of eastern

Table 5.1 Arsenic contamination of groundwater in different countries

Country	As concentration (μg/L)	References
Bangladesh	≥1-5500	Ahmad et al. (2018)
Vietnam	1-3050	Rahman et al. (2009)
Thailand	1-5000	Jones et al. (2009)
Chile	1-900	Ferreccio and Sancha (2006)
Afghanistan	10-500	Nriagu et al. (2007)
West Bengal, India	10-3200	Chakraborti et al. (2010)
West Bengal, India	> 3600	Mishra et al. (2016)
Chhattisgarh, India	> 4500	
Mekong Delta, Cambodia	1-900	Sthiannopkao et al. (2008)
Brazil	0.4-350	Khan and Ho, (2011)
China	52-4440	Rahman et al. (2009)
Pakistan	≥906	Mukherjee et al. (2006), Khan and Ho (2011)
Taiwan	10-1820	Nriagu et al. (2007)
USA	10-2600	Cutler et al.(2013)
Argentina	4-5300	Smedley et al. (2005)
Nepal	10-2620	Shrestha (2012)

Himalayas are other geological sources for groundwater As contamination (Acharya et al. 1999; Bhattacharya et al. 2013). Mining, manufacturing, and processing of metallic ores using As-containing sulfides are main anthropogenic activities that contribute to groundwater As contamination. Anthropogenic activities that lead to As contamination affect the quality of surface water and discharge and runoff of groundwater (Khatri and Tyagi 2015). The excess As in groundwater is due to predominating sulfidic minerals of As such as pyrite and arsenopyrite and their association with other ore deposits (Borba et al. 2003). As-containing mineral, arsenopyrite (FeAsS), abundantly exists in anaerobic environments and in the other rock-forming minerals such as carbonate, phosphate, sulfide, silicate, and oxide (Smedley and Kinniburgh 2002). Arsenic in groundwater is attributed to many geochemical processes, including desorption of As from oxide and hydroxides, oxidation of As-bearing sulfides, dissolution of As (reductive)-bearing oxides and hydroxides, release of As from geothermal water as well as leaching of As. According to McArthur et al. (2001), the reductive dissolution of As-containing iron minerals in aquifer sediments chemically or with the help of microbes is the major cause of As release.

Drinking water contaminated with As severely affects human health and is a major environmental cause of cancer (Tripathi et al. 2007). In 1980, the International Agency for Research on Cancer (IARC) has listed As as a human carcinogen. Chronic exposure to As causes severe harm to internal organs such as digestive,

Table 5.2 Impact of arsenic on human health

	Symptoms	References
Respiratory	Laryngitis, tracheal bronchitis, rhinitis, pharyngitis, shortness of breath, perforation of nasal septum	Chakraborti et al. (2017)
Gastrointestinal	Heartburn, nausea, abdominal pain	Jain et al. (2016)
Dermal	Hyperpigmentation, abnormal skin thickening, narrowing of small arteries leading to numbness (Raynaud's disease), squamous and basal-cell cancer	Banerjee et al. (2011); Jain et al. (2016)
Cardiovascular	Heart attack, cardiac arrhythmias, thickening of blood vessels, loss of circulation leading to gangrene of extremities, hypertension	Wade et al. (2015)
Hematological	Anemia, low white-blood-cell count (leucopenia)	Correia et al. (2009)
Renal	Hematuria, proteinuria, shock, dehydration, cortical necrosis, cancer of kidneys and bladder	Zheng et al. (2014)
Reproductive	Spontaneous abortions, still-births, congenital malformations of fetus, low birth weight	Jain et al. (2016)

respiratory, neural, circulatory, and renal systems (ATSDR 2000; WHO/IPCS 2001) (Table 5.2). Ferreccio et al. (2000) reported the positive correlation between ingestion of inorganic As and lung cancer in humans in Chile. Besides the exposure of As-contaminated water, the dietary consumption of As-contaminated crops, vegetables, and spices are another major source of As exposure (Upadhyay et al. 2018). Among As-contaminated cereal crops, rice is the widely consumed and staple food for the large population of the world. Rice accumulates relatively high amount of As than other cereals (Mitra et al. 2017). People regularly exposed to As for more than 5 years may suffer with cancers of the hepatic, pulmonary, hematological, cardiovascular, renal, immunological, and neurological systems (Mazumder 2000; Chakraborti et al. 2017). Exposure of As may also cause spontaneous abortion in pregnant women (Chakraborti et al. 2016, 2017) and infants as well as children are more sensitive for the adverse effects of As (Das et al. 2009).

5.4 Mitigation Strategies of Arsenic

The mitigation measures for As removal are ranging from decreasing the level of As within the aquifer and dilution of the As contaminants by artificial recharge, blending with As free water, etc. Installation of As treatment unit and other resources of As free water are the two major means for As mitigation in hotspots of As contamination (Bundschuh et al. 2010). The common strategies adopted for As removal are based on the principles of co-precipitation, adsorption, oxidation, coagulation, flocculation, and filtration (Bundschuh et al. 2010). Sorghum biomass, sedges, cellulose, milled bones, keratin-rich biomass lettuce biomass, and cysteine-rich biomass are also used for As removal. Pond sand filters and sono filters are cost-effective household technologies that have been developed for As removal.

Advanced technologies such as phytoremediation, bioremediation, and artificially constructed wetlands are also effective strategies for As remediation (Bundschuh et al. 2010). Apart from this, deep tube wells, artificial groundwater recharge, surface water sources, rainwater harvesting systems, and digging of wells gained a marked degree of success in As amelioration technologies (Kabir and Howard 2007; Shibasaki et al. 2007; Shafiquzzaman et al. 2009; Bundschuh et al. 2010; Mosler et al. 2010). In southern parts of Bangladesh and many states of India, the rainwater harvesting is still a common practice. In Mizoram, almost 90% households use rainwater for drinking and cooking as a potential strategy to minimize arsenic toxicity. Government agencies such as Central Groundwater Board-Mid Eastern Region (CGWB-MER), Public Health Engineering Department (PHED) as well as UNICEF have started As mitigation programs (CGWB and NIH 2010).

Government of India (GoI) initiated As mitigation technologies include arsenic removal plants (ARP), new hand pumps (NHP), arsenic treatment units (ATU), and new tube wells with stand post (NTWSP) (CGWB and NIH 2010). Phytoremediation is also an effective strategy for As and fluoride removal. Plants have evolved an extraordinary potential to remediate As through strategies including uptake, repression, sequestration into vacuoles or extrusion. Arsenic removal plants will not be effective until they are not managed and maintained properly and there is an urgent need of political and people's participation. These proposed arsenic mitigation interventions will benefit millions of people, whether directly or indirectly (CGWB and NIH 2010).

5.4.1 Precipitation Processes

For As removal, precipitation is an effective method. Precipitation involves coagulation, coagulation-assisted microfiltration, filtration, and enhanced lime softening. Coagulation of As with salts of iron and aluminum and softening with lime is the most effective treatment. Adsorption co-precipitation with hydrolyzing metals such as iron and aluminum is the effective technique for groundwater As removal. Sedimentation is followed by rapid sand filtration or microfiltration to remove the precipitates (Mishra et al. 2016). In this method of As removal, oxidation of AsIII to AsV is necessary to improve the efficiency of this method. Hypochlorite and permanganate are commonly used for the oxidation of As. The examples of major techniques used for As removal based on precipitation process are fill and draw treatment unit, bucket treatment unit, iron As treatment unit, and tube well-attached As treatment.

5.4.1.1 Fill and Draw Units

This treatment unit is based on the precipitation method. Fill and draw treatment unit has good capacity to store water with slightly tapered bottom. The water contaminated with As is filled in the container along with addition of oxidant and coagulant. The water is mixed with the help of a manually operable mixer and then the unit is left overnight for sedimentation of precipitated As to occur. Then, next day settled water is taken out with the help of a pipe near bottom and is passed through sand bed before using this as drinking water (Ahmed 2001).

5.4.1.2 Bucket Treatment Unit

The bucket treatment unit operates on the principle of coagulation, co-precipitation, and adsorption. In this technique, one bucket serves the purpose of mixing As-contaminated water with chemicals like potassium permanganate and aluminum sulfate, and coagulation is promoted to enhance the sedimentation rate. The second bucket collects settled water. The water is finally filtered with a cloth and passed through another bucket containing sand filters and water is then used for drinking (Tahura et al. 2001).

5.4.1.3 Iron As Treatment Unit

This process involves the oxidation of soluble As forms into the insoluble forms followed by removal through filtration. Arsenic, which is usually present in reduced arsenite (As^{+3}) form, is oxidized to arsenate (As^{+5}) along with the oxidation of ferrous ions to ferric ions. The As^{+5} is adsorbed onto iron hydroxide and the precipitated As is then removed from the water through filteration (Akhter et al. 2015).

5.4.1.4 Tube Well-Attached As Treatment Unit

In this method, the arsenic removal plant is attached directly to the tube well. Arsenic removal in this technique utilizes principles of coagulation, sedimentation, and filtration. For the coagulation and sedimentation, sodium hypochlorite and aluminum alum are commonly used. This method has been found to remove As up to about 90% in the villages of West Bengal, India.

5.4.2 *Adsorptive Processes*

Adsorption involves the removal of As by surface chemical reaction that includes passage of water through a contact bed. In India and Bangladesh where the problem of groundwater As contamination is severe, sorptive media based on activated alumina are being extensively used for adsorption process for water treatment of As. In adsorption treatment no chemicals are used and the process based solely adsorption on the active surface of the media. The removal of As from natural water by adsorption method, and the use of granular ferric hydroxide as an adsorbent are highly effective (Mohan and Pittman 2007). Sono 3-Kolshi filter containing sand, brick chips, zero valent iron fillings, and wood coke are also good example of adsorbent used for As removal.

5.4.3 Membrane Processes

Membrane processes for As removal include nano-filtration, ultra-filtration, electro-dialysis, and reverse osmosis which use synthetic membranes for removal of many contaminants including As. The dramatic improvement in membrane technologies for water purification and treatment is due to its low energy cost, ease of scaling up, and high efficiency and stability over the past two decades. Membranes remove As through electric repulsion, filtration, and adsorption of arsenic-bearing compounds. Several cost-effective As removal filters have been developed by different national research organizations of India. The Indian Institute of Technology, Bombay (IITB) has developed a cost-effective, robust, iron-based As removal filter. Defence Research and Development Organization (DRDO) developed an As removal filter based on co-precipitation and adsorption. The Indian Institute of Technology, Kharagpur has developed a laterite-based As filter which is eco-friendly and ultra-low cost-effective. Agharkar Research Institute (ARI), Pune has developed a plant for As treatment. These As removal tools are efficient for As removal in lab conditions as well as in As-contaminated fields (Mishra et al. 2016). The use of membranes for the removal of contaminant like As has attracted attention as this possesses potential to be easily applicable even at personal home level. The membranes can also utilize biological functional components like specific transporter proteins to enhance the rate and efficiency of filtration (Werber et al. 2016; Ling et al. 2017).

5.4.4 Phytoremediation

The biological methods that include phytoremediation and bioremediation are ecofriendly and cost-effective for protecting human health and environment from toxic metal contamination. Phytoremediation involves the use of green plants for removal of contaminants. In phytoremediation, plant removes heavy metals by using one of these mechanisms, such as phytodegradation, phytoextraction, rhizofiltration, phytostabilization, and phytovolatilization (Kumar et al. 2020). There is an immense natural diversity in the As response among different plant species. Few plant species have a great potential of phytoremediation strategies as they are enriched with mechanisms for As detoxification and hyperaccumulation. A plant species is recognized as As hyperaccumulator if it accumulates more than 1000 μg/g As. Several aquatic macrophytes and wetland plants grown in As-contaminated areas are reported to hyperaccumulate As (Robinson et al. 2006). *Pteris vittata* is an excellent As hyperaccumulator and reported to accumulate As up to 22,630 μg g^{-1}in 6 weeks (Ye et al. 2011). Aldrich et al. (2007) reported that mesquite plant is a potential candidate for the phytoremediation of As-contaminated regions.

5.4.5 *Bioremediation*

Potential of living organisms to mitigate As contamination is known as bioremediation. Microorganisms have ability to grow and survive in the heavy metal (As)-contaminated areas. Some autotrophs and heterotrophs are reported to use As as their source of energy (Oremland 2009). Microbes methylate or biotransform the toxic form of As into less toxic form of As and thus can be used to amelioration of As from the contaminated environments. That microorganisms play an important role in the biogeochemical cycling of metal(oid) in the aquatic environment and have potential applications in bioremediation. It is also hypothesized that microorganisms are able to produce As-containing minerals or arsenosugars and thus can transform As into its less toxic form. Another important strategy of As removal includes plant-microbe interaction (Awasthi et al. 2018). The rhizospheric As-resistant microbes have been reported to play a vital role in plant growth promotion and phytoextraction of As from contaminated sites. Additionally, the ecological and socio-behavioral factors of As-affected areas, awareness about As-induced toxicity, and health risks should be of prime concern before design and implementation of any As mitigation proposals/policies. We must understand that so far there is no available treatment for As toxicity. The only solution to this ailment is non-As-contaminated drinking water, food, and essential vitamins and minerals. People living in the villages should be encouraged to include fresh fruits and vegetables in their diets due to their high nutritive value. They should be aware about the right cooking methods as over cooking can demolish essential nutrients in fruits and vegetables. In this reference, government should recruit food technologists, nutritionist, or medical personnel to aware villagers.

5.5 Conclusions and Future Prospects

Day by day, the addition of new As-affected areas due to the geogenic and anthropogenic activities has changed the present scenario of groundwater As contamination in India. Thus, it is of prime concern to explore the real picture of As contamination and its mitigation strategies to overcome the problem. In recent years, several As removal devices have been developed by different organizations that are proved to be efficient tools for As removal from groundwater. Deeper aquifers with no future risk of As contamination are helpful to supply As-free groundwater, thus exploration of deeper aquifers can provide a sound solution. Training and awareness program for As hazards and use of As removal tools to the user community would be helpful to minimize the exposure of As to human health. Considering the severity of the problem on a global scale, the awareness of population and implementation of facilities by setting proper guidelines is important for maintenance and mitigation of As problem. The government should monitor industrial and agriculture activities which contribute majorly in As pollution.

Acknowledgement SA is thankful to the Council of Scientific & Industrial Research for the award of Senior Research Fellowship [No. 31/08(332)/2017-EMR1]. RC is thankful to the Department of Science and Technology (DST) New Delhi, India for the award of DST-INSPIRE Junior/Senior Research Fellowship (No. DST/INSPIRE Fellowship/2013/321). RDT is thankful to CSIR, New Delhi for the Emeritus Scientist Scheme [No. 21(0978)/13/EMR-II].

References

Acharya SK, Chakraborty P, Lahiri S, Raymahashay BC, Guba S, Bhowmik A (1999) Arsenic poisoning in the Ganges delta. Nature 401:545–546

Agency for Toxic Substances and Disease Registry – ATSDR (2000) Toxicological profile for arsenic. U.S. Department of Health and Human Services, Public Health Service, Atlanta

Ahmad SA, Khan MH, Haque M (2018) Arsenic contamination in groundwater in Bangladesh: implications and challenges for healthcare policy. Risk Manag Healthc Policy 11:251

Ahmed MF (2001, May) An overview of arsenic removal technologies in Bangladesh and India. In: Proceedings of BUET-UNU international workshop on technologies for arsenic removal from drinking water, Dhaka, pp 5–7

Ahmed MF (2003) Arsenic contamination: Bangladesh perspective. ITN-Bangladesh, Centre for Water Supply and Waste Management, Dhaka

Akhter T, Abedin MZ, Basak JK, Akhter F (2015) Design and development of arsenic and iron removal unit for drinking water: a sustainable approach in environment. Asia Pac J Energy Environ 2:67–74

Aldrich MV, Peralta-Videa JR, Parsons JG, Gardea-Torresdey JL (2007) Examination of arsenic (III) and (V) uptake by the desert plant species mesquite (*Prosopis* spp.) using X-ray absorption spectroscopy. Sci Total Environ 379:249–255

Awasthi S, Chauhan R, Srivastava S, Tripathi RD (2017) The journey of arsenic from soil to grain in rice. Front Plant Sci 8:1007

Awasthi S, Chauhan R, Dwivedi S, Srivastava S, Srivastava S, Tripathi RD (2018) A consortium of alga (*Chlorella vulgaris*) and bacterium (*Pseudomonas putida*) for amelioration of arsenic toxicity in rice: a promising and feasible approach. Environ Exper Bot 150:115–126

Banerjee N, Nandy S, Kearns JK, Bandyopadhyay AK, Das JK, Majumder P, Basu S, Banerjee S, Sau TJ, States JC, Giri AK (2011) Polymorphisms in the TNF-α and IL10 gene promoters and risk of arsenic-induced skin lesions and other nondermatological health effects. Toxicol Sci 121:132–139

Bhattacharya P, Mukherjee A, Mukherjee AB (2013) Groundwater arsenic in India: source, distribution, effects and alternate safe drinking water sources. In: Reference module in earth systems and environmental sciences. Elsevier, Amsterdam, p 09342

Borba RP, Figueiredo BR, Rawlins B, Matschullat J (2003) Geochemical distribution of arsenic in waters, sediments and weathered gold mineralized rocks from Iron Quadrangle, Brazil. Environ Geol 44:39–52

Bundschuh J, Litter M, Ciminelli VS, Morgada ME, Cornejo L, Hoyos SG, Hoinkis J, Alarcon-Herrera MT, Armienta MA, Bhattacharya P (2010) Emerging mitigation needs and sustainable options for solving the arsenic problems of rural and isolated urban areas in Latin America–A critical analysis. Water Res 44:5828–5845

CGWB and NIH (2010) Mitigation and remedy of groundwater arsenic menace in India: a vision document. India, National Institute of Hydrology (NIH), Roorky and Central Ground Water Board (CGWB), Government of India

Chakraborti D, Rahman MM, Das B, Murrill M, Dey S, Mukherjee SC, Dhar RK, Biswas BK, Chowdhury UK, Roy S, Sorif S (2010) Status of groundwater arsenic contamination in Bangladesh: a 14-year study report. Water Res 44:5789–5802

Chakraborti D, Rahman MM, Ahamed S, Dutta RN, Pati S, Mukherjee SC (2016) Arsenic groundwater contamination and its health effects in Patna district (capital of Bihar) in the middle Ganga plain, India. Chemosphere 152:520–529

Chakraborti D, Rahman MM, Das B, Chatterjee A, Das D, Nayak B, Pal A, Chowdhury UK, Ahmed S, Biswas BK, Sengupta MK (2017) Groundwater arsenic contamination and its health effects in India. Hydrogeol J 25:1165–1181

Correia N, Carvalho C, Friões F, Araújo JP, Almeida J, Azevedo A (2009) Haemolytic anaemia secondary to arsenic poisoning: a case report. Cases J 2:7768

Cutler WG, Brewer RC, El-Kadi A, Hue NV, Niemeyer PG, Peard J, Ray C (2013) Bioaccessible arsenic in soils of former sugar cane plantations, Island of Hawaii. Sci Total Environ 442:177–188

Das B, Rahman MM, Nayak B, Pal A, Chowdhury UK, Mukherjee SC, Saha KC, Pati S, Quamruzzaman Q, Chakraborti D (2009) Groundwater arsenic contamination, its health effects and approach for mitigation in West Bengal, India and Bangladesh. Water Qual Expo Health 1:5–21

Ferreccio C, Sancha AM (2006) Arsenic exposure and its impact on health in Chile. J Health Popul Nutr 24(2):164–175

Ferreccio C, González C, Milosavjlevic V, Marshall G, Sancha AM, Smith AH (2000) Lung cancer and arsenic concentrations in drinking water in Chile. Epidemiology 11:673–679

Gómez JJ, Lillo J, Sahún B (2006) Naturally occurring arsenic in groundwater and identification of the geochemical sources in the Duero Cenozoic Basin, Spain. Environ Geol 50:1151–1170

Islam ABMR, Maity JP, Bundschuh J, Chen CY, Bhowmik BK, Tazaki K (2013) Arsenic mineral dissolution and possible mobilization in mineral–microbe–groundwater environment. J Hazard Mater 262:989–996

IWA (2016) Executive summary of arsenic contamination in the world. http://www.iwawaterwiki.org/xwiki/bin/view/Articles/ExecutiveSummaryofArsenicContaminationintheWorld

Jain J, Bajpai S, Gauba P (2016) Adverse Health effects of Arsenic toxicity. J Civil Eng Environ Technol 3:679

Jones H, Visoottiviseth P, Bux MK, Földényi R, Kováts N, Borbély G, Galbács Z (2009) Arsenic pollution in Thailand, Bangladesh, and Hungary. In: Reviews of environmental contamination and toxicology. Arsenic pollution and remediation: an international perspective, vol 197. Springer, New York, pp 163–187

Kabir A, Howard G (2007) Sustainability of arsenic mitigation in Bangladesh: results of a functionality survey. Int J Environ Health Res 17:207–218

Khan MA, Ho YS (2011) Arsenic in drinking water: a review on toxicological effects, mechanism of accumulation and remediation. Asian J Chem 23:1889

Khatri N, Tyagi S (2015) Influences of natural and anthropogenic factors on surface and groundwater quality in rural and urban areas. Front Life Sci 8:23–39

Kumar M, Yadav A, Ramanathan AL (2020) Arsenic contamination in environment, ecotoxicological and health effects, and bioremediation strategies for its detoxification. In: Bioremediation of industrial waste for environmental safety. Springer, Singapore, pp 245–264

Ling S, Qin Z, Huang W, Cao S, Kaplan DL, Buehler MJ (2017) Design and function of biomimetic multilayer water purification membranes. Sci Adv 3(4):e1601939

Mazumder DG (2000) Diagnosis and treatment of chronic arsenic poisoning. United Nations synthesis report on arsenic in drinking water. http://www.who.int/water_sanitation_health/dqw/arsenicum4.pdf

McArthur JM, Ravenscroft P, Safiulla S, Thirlwall MF (2001) Arsenic in groundwater: testing pollution mechanisms for sedimentary aquifers in Bangladesh. Water Resour Res 37:109–117

Milton AH, Smith W, Dear K, Ng J, Sim M, Ranmuthugala G, Lokuge K, Caldwell B, Rahman A, Rahman H, Shraim A (2007) A randomised intervention trial to assess two arsenic mitigation options in Bangladesh. J Environ Sci Health Part A 42:1897–1908

Mirlean N, Baisch P, Diniz D (2014) Arsenic in groundwater of the Paraiba do Sul delta, Brazil: an atmospheric source? Sci Total Environ 482:148–156

Mishra S, Dwivedi S, Kumar A, Chauhan R, Awasthi S, Tripathi RD, Mattusch J (2016) Current status of ground water arsenic contamination in India and recent advancements in removal techniques from drinking water. Int J Plant Environ 2:1–2

Mitra A, Chatterjee S, Moogouei R, Gupta DK (2017) Arsenic accumulation in rice and probable mitigation approaches: a review. Agronomy 7:67

Mohan D, Pittman CU Jr (2007) Arsenic removal from water/wastewater using adsorbents-a critical review. J Hazard Mater 142:1–53

Mosler HJ, Blöchliger OR, Inauen J (2010) Personal, social, and situational factors influencing the consumption of drinking water from arsenic-safe deep tubewells in Bangladesh. J Environ Manag 91:1316–1323

Nickson R, Sengupta C, Mitra P, Dave SN, Banerjee AK, Bhattacharya A, Basu S, Kakoti N, Moorthy NS, Wasuja M, Kumar M (2007) Current knowledge on the distribution of arsenic in groundwater in five states of India. J Environ Sci Health Part A 42:1707–1718

Nriagu JO, Bhattacharya P, Mukherjee AB, Bundschuh J, Zevenhoven R, Loeppert RH (2007) Arsenic in soil and groundwater: an overview. Trace Metals Contaminant Environ 9:3–60

Oremland RS (2009) December. Arsenic as an energy source for microbial growth. In AGU Fall Meeting Abstracts

Rahman MM, Naidu R, Bhattacharya P (2009) Arsenic contamination in groundwater in the Southeast Asia region. Environ Geochem Health 31:9–21

Ravenscroft P, Brammer H, Richards KS (2009) Arsenic pollution: a global synthesis. Wiley Blackwell, Chichester

Robinson B, Kim N, Marchetti M, Moni C, Schroeter L, van den Dijssel C, Milne G, Clothier B (2006) Arsenic hyperaccumulation by aquatic macrophytes in the Taupo Volcanic Zone, New Zealand. Environ Exp Bot 58:206–215

Santra SC, Samal AC, Bhattacharya P, Banerjee S, Biswas A, Majumdar J (2013) Arsenic in food chain and community health risk: a study in Gangetic West Bengal. Procedia Environ Sci 18:2–13

Shafiquzzaman M, Azam MS, Mishima I, Nakajima J (2009) Technical and social evaluation of arsenic mitigation in rural Bangladesh. J Health Popul Nutr 27:674

Shibasaki N, Lei P, Kamata A (2007) Evaluation of deep groundwater development for arsenic mitigation in western Bangladesh. J Environ Sci Health Part A 42:1919–1932

Shrestha R (2012) Arsenic contamination of groundwater in Nepal: good public health intention gone bad. Inquiries J 4(09):1–3

Singh SK (2017) Conceptual framework of a cloud-based decision support system for arsenic health risk assessment. Environ Syst Decis 37:435–450

Singh SK, Stern EA (2017) Global arsenic contamination: living with the poison nectar. Environ Sci Policy Sustain Dev 59:24–28

Singh SK, Vedwan N (2015) Mapping composite vulnerability to groundwater arsenic contamination: an analytical framework and a case study in India. Nat Hazards 75:1883–1908

Smedley PL, Kinniburgh DG (2002) A review of the source, behaviour and distribution of arsenic in natural waters. Appl Geochem 17:517–568

Smedley PL, Kinniburgh DG, Macdonald DMJ, Nicolli HB, Barros AJ, Tullio JO, Pearce JM, Alonso MS (2005) Arsenic associations in sediments from the loess aquifer of La Pampa, Argentina. Appl Geochem 20:989–1016

Srivastava S, Srivastava AK, Sablok G, Deshpande TU, Suprasanna P (2015) Transcriptomics profiling of Indian mustard (*Brassica juncea*) under arsenate stress identifies key candidate genes and regulatory pathways. Front Plant Sci 19(6):646. https://doi.org/10.3389/fpls.2015.00646

Sthiannopkao S, Kim KW, Sotham S, Choup S (2008) Arsenic and manganese in tube well waters of Prey Veng and Kandal Provinces, Cambodia. Appl Geochem 23:1086–1093

Tahura S, Shahidullah SM, Milton TRAH, Bhuiyan RH (2001) Evaluation of an arsenic removal household device: bucket treatment unit (BTU). Technologies for Arsenic Removal from Drinking Water:158–170

Tripathi RD, Srivastava S, Mishra S, Singh N, Tuli R, Gupta DK, Maathuis FJM (2007) Arsenic hazards: strategies for tolerance and remediation by plants. Trends Biotechnol 25(4):158–165

Upadhyay MK, Yadav P, Shukla A, Srivastava S (2018) Utilizing the potential of microorganisms for managing arsenic contamination: a feasible and sustainable approach. Front Environ Sci 6:24

USEPA (2013) Arsenic in drinking water. http://water.epa.gov/lawsregs/rulesregs/sdwa/arsenic/index.cfm

Wade TJ, Xia Y, Mumford J, Wu K, Le XC, Sams E, Sanders WE (2015) Cardiovascular disease and arsenic exposure in Inner Mongolia, China: a case control study. Environ Health 14:35

Werber JR, Osuji CO, Elimelech M (2016) Materials for next-generation desalination and water purification membranes. Natur Rev Mat 1:1–15

WHO (2001) Environmental health criteria, 224, arsenic and arsenic compounds, 2nd edn. World Health Organization, Geneva, pp 385–392

WHO (2011) Guidelines for drinking-water quality, vol 4. World Health Organization, Geneva

World Health Organization (2019) Safer water, better health. WHO, Geneva

Ye WL, Khan MA, McGrath SP Zhao FJ (2011) Phytoremediation of arsenic contaminated paddy soils with *Pteris vittata* markedly reduces arsenic uptake by rice. Environ Pollut 159:3739–3743

Zheng L, Kuo CC, Fadrowski J, Agnew J, Weaver VM, Navas-Acien A (2014) Arsenic and chronic kidney disease: a systematic review. Curr Environ Health Rep 1(3):192–207

Chapter 6
Wastewater Reuse in Peri-Urban Agriculture Ecosystem: Current Scenario, Consequences, and Control Measures

Prince Kumar Singh and Rajesh Kumar Sharma

6.1 Introduction

India is now the second most populated country in the world that sustains more than 16% of the world's population with only around 4% of the world's freshwater (The World Bank Group 2016). In India, the agriculture sector has been the largest consumer of water, although in next two decades share of water allocated to irrigation will be lowered to 10–15% (CWC 2000). Water is a vital natural resource that is an essential requirement for sustaining the life. It is investigated that 2.4 billion people in the world are unable to assess clean water, while 946 million of people are compelled to drink contaminated water and have unsafe sanitary practices (WHO 2014). The first evidence of wastewater reuse was found among the ancient Greeks, where flushed wastewater from public toilets was stored in several storage chambers through a sewerage line system (Jaramillo and Restrepo 2017).

Industrial or municipal wastewater reuse in agriculture is a common practice in suburban areas of developing countries (Urie 1986; Jeong et al. 2016; Jaramillo and Restrepo 2017), including India (Singh et al. 2004; Sharma et al. 2007; Kumar and Tortajada 2020). In this changing scenario, reutilization of domestic and industrial wastewater in agriculture for irrigating the plants such as crops, vegetables, etc., appears to be a valuable option. Besides being the source of irrigation, wastewater contains appreciable amounts of many plant nutrients such as macronutrients, micronutrients, organic matters, etc., but along with this it also contains harmful contents such as heavy metals and different emerging contaminants such as pharmaceuticals and personal care products, etc., which pose health risk to the living organisms (Kibuye et al. 2019). Heavy metals are more toxic due to its prevalence, non-biodegradable, and persistent nature.

P. K. Singh · R. K. Sharma (✉)
Centre of Advanced Study in Botany, Institute of Science, Banaras Hindu University, Varanasi, Uttar Pradesh, India

A. Singh et al. (eds.), *Water Pollution and Management Practices*,
https://doi.org/10.1007/978-981-15-8358-2_6

Overpopulation, industrialization, rapid urbanization, unplanned land use pattern, overexploitation of groundwater, chemical spills, storage tank leakage, unmanaged transportation, overuse, and surface runoff of agriculture fertilizers, etc., are the main cause of wastewater generation (Xia et al. 2017). Only one-tenth part of total generated sewage is treated and only one-third part of total urban household are connected to closed sewerage system (Sugam and Ghosh 2013). Most of the industries are not capable to treat wastewater due to their higher cost of operation and conventional chemicals. According to CPCB (2013), the total amount of sewage generation from 35 metropolitan cities (population one million and above) is 15,644 MLD and the treatment capacity is for 8040 MLD, i.e., only 51%.

Reuse of wastewater increased in the developed countries than developing countries due to the availability of more resources or facilities, e.g., in Europe and United States wastewater reuse increased 10–29% per year, and in Australia it increased up to 41% per year (Aziz and Farissi 2014).

For treatment of wastewater, types of sustainable infrastructure continuously increase by using several approaches such as physico-chemical approaches (sedimentation, chemical precipitation, adsorption, ion exchange, coagulation, catalytic removal, and nanotechnology) and biological or green approaches (different types of bioreactors, trickling filters and rotating biological contactor). By emphasizing these several techniques, we are expected to understand that how the wastewater is easily collected through drainage system, well treated, discharged, and reutilized. But unfortunately, in present time progress toward these several sustainable approaches is not well evenly distributed among all the nations (Rarasati et al. 2017).

There are several factors which affect the wastewater irrigation in agriculture field, e.g., chances of availability of freshwater or groundwater through tube well or canals for irrigation at affordable rates, consistency and reliability of wastewater generation through drainage system, level of nutrients in wastewater, acidity, alkalinity or salinity level of wastewater, contamination level of industrial effluents in wastewater, etc. Instead of freshwater irrigation, agriculture field irrigated by wastewater due to freshwater scarcity lead to food chain contamination, i.e., heavy metals transfer from soil to food plants (Fig. 6.1). Mobility of heavy metal depends upon its bioavailability in soil (present in different chemical forms) as well as its translocation and distribution varies by the species and population of plants (Liu et al. 2007; Sharma et al. 2020a).

Knowledge of the long-term impact of wastewater irrigation on metal or metalloid dynamics in soil-plant system should be improved for maximizing the benefits of wastewater irrigation as a viable source. Muchuweti et al. (2006) suggested drip irrigation method as a suitable and eco-friendly approach for mitigating the negative effect of wastewater irrigation on soil properties whereas flood irrigation in agriculture field badly affected the soil qualities. In present time, it becomes necessary to think about the existing urban wastewater disposal infrastructure, wastewater agriculture practices, quality of water consumed and its health implications, and the level of institutional awareness on wastewater-related issues (Rutkowski et al. 2007). By re-engineering the whole treatment plant structure, energy can be saved in wastewater treatment plants for increasing its reliability and efficiency. Combined heat

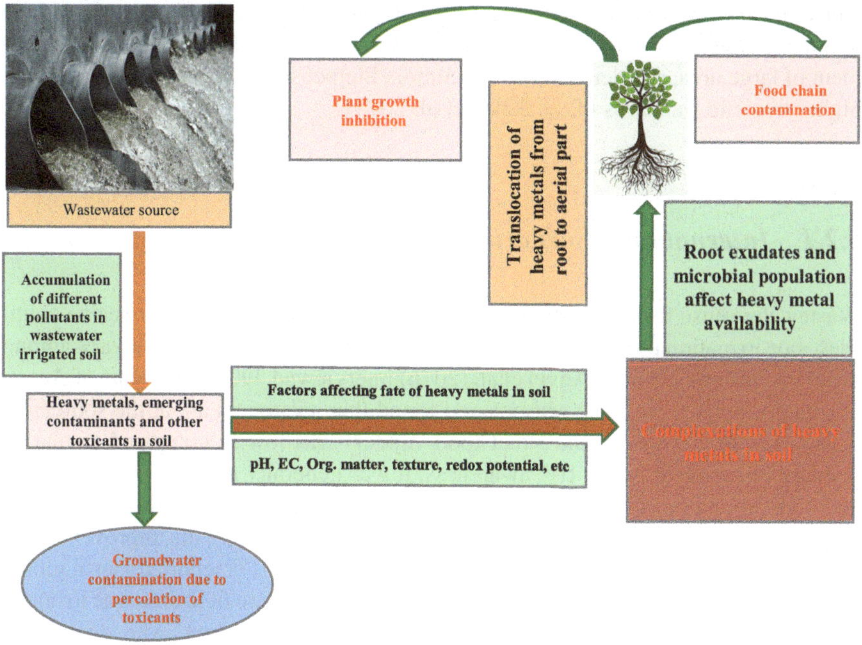

Fig. 6.1 Threats to food chain due to wastewater irrigation in agriculture field. (Modified from Khalid et al. 2018)

and power techniques (CHP) play an important role as a source of energy, which use its own energy in the form of heat and electricity. Waste materials released from the treatment plant can be used for generation of biogas and energy using CHP techniques.

With the above context, the present chapter is concerned with the different aspects of wastewater reuse both at national and international level. The chapter further gives comprehensive accounts on the impact of wastewater irrigation on soil and groundwater quality which adversely affect food chain. Moreover, this chapter also proposes several control measures for wastewater treatment and management strategies to minimize the human health risks associated with the reuse of wastewater.

6.2 Current Scenario of Wastewater Reuse

Treatment of wastewater and its reuse is not a new concept. Several decades ago, untreated domestic wastewater was applied in agriculture but as time passes different technologies evolved and treatment of wastewater becomes possible. Present knowledge of wastewater treatment continued up to twentieth century in the USA, Australia, Europe, and other parts of the world without taking serious public health concerns and negative environmental impacts. However, in the present time these treatment systems are not commonly accepted due to their drawbacks like requirement of large area, problems in field operation, high cost, and difficult to achieve the higher hygiene standards (Tzanakakis et al. 2014).

6.2.1 International Scenario

Wastewater reuse for agricultural irrigation is more important in water-scarce countries. For irrigational purposes, quality and nature of wastewater reuse vary among different countries. Antwi-Agyei et al. (2016) interviewed 490 respondents of Accra, Ghana during two cropping seasons and found that awareness about the sources of wastewater in consumers and street food vendors was low as compared to market vendors. In terms of health risk awareness, it was generally low among the farmers and high among consumers and salespersons. This study promoted the intervention that directly benefits farmers, vendors, and consumers along with improvement in the knowledge of food safety and hygiene. Thebo et al. (2017) reported that at global level, approximately 11% of urban and peri-urban agriculture field area was irrigated with untreated wastewater. They concluded that the consumers, who depend on wastewater irrigated crops, must be assured for food safety by estimating the level of contaminants.

Moussaoui et al. (2017) reported that urban untreated wastewater is reused in agro-forestry sectors of Marrakesh city in a sustainable way. Local climatic condition of this city is an ultimate challenge for conserving the quality of wastewater resources, due to the degradation of organic matter, i.e., foul smell and availability of pathogenic microbes which had negative impact on soil quality in terms of fertility as well as its productivity. To face this problem treated wastewater irrigation becomes reliable and sustainable strategy, which enhanced soil fertility and combating water scarcity. Rezapour et al. (2019) studied Cd accumulation in wheat grain cultivated in the wastewater irrigated agriculture field of Western Azerbaijan Province, northwestern Iran. The accumulation of Cd and its translocation and carcinogenic health risk were estimated through food chain contamination that ultimately affect the human life. Cadmium (Cd) is non-biodegradable and highly persistent in nature. Translocation factor for Cd through root to grain ranged from 0.18–0.24 and its carcinogenic health risk was found to be low to moderate risk category (Rezapour et al. 2019). Wastewater irrigation from outlet of electroplating factory threatened the agriculture quality (Xiao et al. 2019). Soil pH as well as concentrations of total and secondary phase fraction, i.e., bioavailable fraction (acid-soluble, reducible, oxidizable, and residual fraction) of heavy metals lowered down as the distance increased from electroplating factories. Redundancy analysis and stepwise regression analysis showed that soil pH, silt content, amorphous Fe oxides, and Mn oxides affected the Secondary Phase Fraction. Concentration of heavy metal concentration in soil and varied up to 68.8% and 43.5%, respectively.

6.2.2 National Scenario

Wastewater reuse for irrigation in Indian agriculture field is in practice from several decades. The treated or untreated forms of wastewater are widely used for irrigation in urban and peri-urban areas. Although there are no certain comprehensive data for total wastewater used for irrigating agriculture lands, some studies reported its considerable use. One of the studies from the International Water Management Institute by Amerasinghe et al. (2013) has reported that about 50,000 ha land was irrigated by urban wastewater. In peri-urban areas, a variety of crops were grown under wastewater irrigation, most commonly being vegetables for local urban markets. Vegetables are important dietary constituents of human food, which fulfil various needs of nutritional components. Sharma et al. (2007) found that farmers of suburban areas of Varanasi irrigated their vegetables like palak (*Beta vulgaris* L. var. All green H1) with wastewater. This vegetable was grown due to its short growth period, number of harvested times in single cropping, required low agriculture area, low labor cost, etc. Analyses of heavy metal in irrigational water, soil and vegetables in both summer and winter seasons had shown that concentrations of Cu, Zn, Pb, Cr, Mn, and Ni in irrigational water and soil were below the recommended limit, except Cd which was higher in winter than summer. However, in edible part of palak Cd concentration was higher during summer; but, Pb and Ni contamination were higher in both the seasons.

Municipalities at Bhavnagar, Rajkot, Gujarat put some charges for irrigating the agricultural lands with wastewater like municipal corporation of Bhavnagar and Rajkot charges Rs. 750/ha and Rs. 2500–3000/ha, respectively (Palrecha et al. 2012). Farmers of peri-urban area of Hyderabad use wastewater for irrigation and prefer to pay charges which help the industries to treat their wastewater and this information is also valuable for planning small on-site wastewater treatment systems which help in improving livelihood in risky environment (Saldías et al. 2017).

Sahay et al. (2019) analyzed the effect of wastewater irrigation on growth, physiological characteristics, and productivity of different *Brassica* species. In this experiment, wastewater irrigation was done along with the amendment of NPK at two different doses (N_{60} P_{30} K_{30} and N_{80} P_{45} K_{45}). The amendment of NPK with 80:45:45 did not show significant effect; however with 60:30:30 dose, the crop showed more production due to the reduction of Ni, Cd, Cr, Pb, and Cu accumulation in plant parts. Radhika and Kulkarni (2019) studied in Hubli-Dharwad, twin cities (Karnataka), second largest populated city after Bangalore, where sewage flow rate rapidly increased by 12 times with 1.07% per annum growth rate. This trend of sewage flow with growth rate creates a challenging task for government to manage sanitation problem and its utilization.

Heavy metals content in soil from Dankaur and Kasna villages, Greater Noida, UP (India) irrigated with treated wastewater released from 137 MLD sewage treatment plant was analyzed by Hussain et al. (2019). A pot experiment performed by Sharma and Agrawal (2006) has reported that the vegetables such as spinach (*Spinacea oleracea*), radish (*Raphanus sativus*), and carrot (*Daucus carota* Sub sp. *sativus*) showed more accumulation of Cd, Cu, Zn, Co, and Mn in their leaves. Treated wastewater irrigated soil showed variations in enrichment factors with decreasing trend like Zn > Ni > Pb > Cr > Cu > Co > Mn > Cd (Hussain et al. 2019).

6.3 Composition of Wastewater

On the basis of sources, wastewater can be categorized into three types, i.e., (i) storm runoff, mainly natural causes, (ii) industrial wastewater, released through several industries, and (iii) domestic wastewater, released through households. Domestic wastewater is categorized into two parts: (i) black wastewater and (ii) grey wastewater. Black wastewater contains higher amount of organic materials than the greywater. Storm runoff water dilutes the wastewater and diluted wastewater has also higher concentrations of organic material than storm runoff water. Detailed classification of wastewater is given in Fig. 6.2.

Concentration of wastewater can be estimated by the addition of pollution load in a particular amount of water or vice versa. Pollution load analysis on a daily, monthly, or yearly basis is a good indicator of the composition of wastewater. Composition of wastewater is influenced by both the location and time as all places do not have similar types of pollution sources, i.e., either domestic or industrial sources (Mohan et al. 2014). Seasonal variation in wastewater contaminants is also ascribed to dilution process, e.g., in rainy season more dilution of wastewater leads to lower concentration of contaminants. Wastewater contains different types of contaminants such as different chemical compounds (inorganic, organic, and different emerging contaminants) and harmful microbial populations.

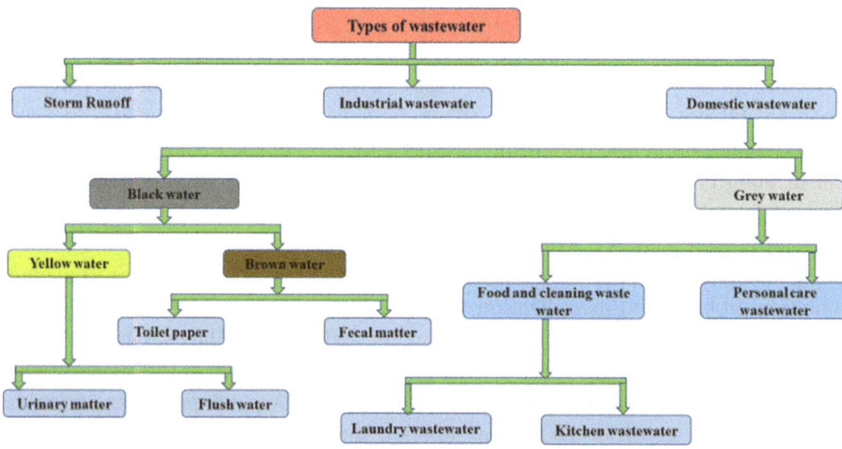

Fig. 6.2 Classification of wastewater. (Modified from Boutin and Eme 2016)

6.3.1 *Chemical Compounds in Wastewater*

6.3.1.1 Inorganic and Organic Chemicals

In wastewater, different types of inorganic constituents occur such as heavy metals, different salts, oxyhalides (bromate, chlorate, and perchlorate), and types of nanomaterials whose concentration depends upon their source and degree of treatments received. National (Bureau of Indian Standard) and International agencies (FAO/WHO) set guidelines for industrial discharge of wastewater to reduce health risk. These guidelines set limit for the standards for metals and metalloids in the sewage discharge. Heavy metals and oxyhalides are bioaccumulative in nature and highly persistent in nature. Due to long-term wastewater irrigation in agriculture field, the contaminants get rapidly accumulated in soil and cause an adverse effect on living organisms through food chain contamination. Macronutrients such as nitrogen and phosphorous also present in excess amount in wastewater stream. These excess occurrences of macronutrients cause eutrophication in water body due to more algal growth.

Sources of organic components in wastewater include household wastes (liquid waste, humic substances, fecal materials through toilets), types of industrial waste, fats, greases, and oils, etc. Wastewater irrigation of soil is low-cost resource of macronutrients and organic matter content. These strategies can be successfully applied in restoration of degraded croplands, eroded soil in hilly areas, or arid and semi-arid areas (Chatzistathis and Koutsos 2017). Organic materials act as a source of carbon for microbial growth and also sometimes lead to clogging which promote growth of microorganism in wastewater. In wastewater treatment process, secondary treatment is done for organic matter removal through microbial population. Organic materials present in the wastewater are beneficial for agricultural purpose as it contain not only organic carbon but also macronutrients, which help in enhancing the fertility of irrigated soil.

6.3.1.2 Emerging Contaminants

The emerging contaminants (EC) are newly identified compounds, organic in nature, and released into the environment by anthropogenic activities (Yankui et al. 2019). The pharmaceutical companies of personal care products, pesticides synthesizer, household chemicals, transformation products, etc., are mostly responsible for their generation. ECs include different natural or synthetic hormones, pharmaceutical products, endocrine disrupting compounds, artificial sweetener, etc. These contaminants are new in the origin, have alternate route to human exposure. A sequence of biological treatment coupled with advanced tertiary treatment such as activated carbon adsorption or chemical oxidation is used to remove such trace chemicals from wastewater.

Tran et al. (2019) reported 31 ECs in untreated wastewater, treated wastewater, urban storm runoff, agricultural runoff, and in freshwater which frequently influence the quality of urban surface water. For identifying and characterizing the origin of diffuse source in urban wastewater, selected ECs are developed as a marker. Twenty-one target ECs were detected 100% in the collected raw wastewater, samples with median concentrations ranging from 49.6 to 77,721 ng/L, while only 13 compounds were found with detection frequency > 50% in freshwater bodies. The median concentration of the majority of detected ECs was below 100 ng/L in freshwater samples. Thus, analysis of ECs such as acesulfame, acetaminophen, cyclamate, and saccharine may serve as a suitable marker of diffuse source in surface water.

6.4 Impacts of Wastewater on Natural Ecosystem

Water is necessary for survival of living organisms as it is susceptible in quality and limited in quantity or availability (Khan et al. 2018). Due to the increasing demand of freshwater or potable water for different purposes in different sectors like agriculture, industries, domestic, etc., threaten its sustainable use. Wastewater irrigation appears to be a good alternative source to enhance the availability of potable water for various purposes. However, wastewater also contains several chemical toxicants and harmful microbes. Long-term uses of untreated or treated wastewater in agriculture field considerably enhance the potential toxic metal content in soil and growing plants. These toxicants may further threaten food chain because of several human health hazardous effects (Dogan et al. 2014). Suitable strategies for proper foretelling of heavy metal uptake by food crops will help in appropriate risk assessment of wastewater irrigated soils. Soil, groundwater, or surface water and growing plants are impacted by long-term wastewater irrigation which is summarized as below.

6.4.1 Soil Characteristics

6.4.1.1 pH and Electrical Conductivity

Wastewater irrigation affects the physicochemical properties of soil such as pH, electrical conductivity, soil temperature, organic matter, cation exchange capacity, bulk density, soil porosity, soil hydraulic conductivity, and infiltration rate. Several studies have shown that soil pH lowered down by wastewater irrigation due to oxidation of participating cations (Khalid et al. 2018). Solubility of heavy metals increases at lower pH and causes an increase in heavy metal bioavailability in the soil which can be easily taken up by the growing plants (Zhao et al. 2015). Net negative charge, i.e., cation exchange capacity increases at higher pH and positive charge, i.e., anion exchange capacity increases at lower pH. Adsorption of metals is also affected by soil pH due to the change in surface charge (Bhargava et al. 2012). Electrical conductivity of soil increases more due to wastewater irrigation than groundwater irrigation. Ashraf et al. (2013) reported that electrical conductivity of wastewater reduced the yield of tomato by enhancing the salinity of soil.

6.4.1.2 Soil Organic Matter

Organic matter is the major component to maintain the productivity of soil. The organic matter content in soil increases due to wastewater irrigation that enhanced the fertility of soil. Organic matter in soil is mainly in the form of humic substances or humus and non-humic substances. Major components of humus are humic acids and fulvic acids. At lower pH, high molecular weight organic acid, i.e., humic acids are not soluble in the soil solution and removed via precipitation. In case of fulvic acids, a low molecular weight organic acid is easily soluble at all pH and has more active sites than humic acid (Gupta et al. 2019a).

6.4.1.3 Soil Temperature and Redox Potential

Temperatures of soil play an important role in the mobility and availability of metals in the soil and also affect the soil microbial health which helps in regulation of soil fertility. Soil temperature may be elevated through wastewater irrigation due to decomposition of soil organic matter which increases the availability of metals in soil. Cornu et al. (2016) reported that transfer of Cd and Zn from soil to plant increases at high soil temperature. Several metals are present in soil solution in oxidized forms or reduced forms and their mobility depends on its redox state, i.e., acceptance or removal of electrons from soil solution determined by its redox potential (Sheoran et al. 2016; Khalid et al. 2018).

6.4.1.4 Some Other Characteristics

Soil bulk density is determined by the mass of undisturbed soil per unit volume and is represented as g cm^{-3}. Soil bulk density is always lower than the particle density due to occurrence of pore space. If, the bulk density of soil increases its porosity always decreases. Long-term wastewater irrigation changes the soil bulk density and porosity which depends on the quality of wastewater, i.e., concentrations of dissolved and particulate constituents in the wastewater. Shariot-Ullah (2019) studied that the effluents from the North Bengal Sugar Mill in Bangladesh affected agriculture soil properties, e.g., bulk density reduced from 1.44 to 1.42 g/cm^3 and porosity of soil increased approximately 2.17%.

The hydraulic conductivity and infiltration rate of soil is altered by wastewater irrigation. The factors such as types of soil, clay content, $CaCO_3$ amount, soil humidity, and types of wastewater are responsible for the alteration of hydraulic conductivity and infiltration rate of soil (Lado et al. 2005). Infiltration rate of wastewater irrigated soil is mainly affected by duration of the wastewater irrigation. Gharaibeh et al. (2007) observed that wastewater irrigation of soil up to 5 years significantly lowered down the infiltration rate, although irrigation period extends to 15 years, increased the infiltration rate due to the large cracks formation.

6.4.2 Water Resources

The long-term wastewater irrigation not only affect the soil properties, i.e., alkalinity, salinity, nitrates, presence of potential contaminants, pathogenic threats, etc., but also lead to deterioration of groundwater quality due to the presence of excess salt, nitrates, and other toxic pollutants. Groundwater shares the largest portion of freshwater on earth, which is continuously overexploited by the various anthropogenic activities such as agriculture, industries, and also in domestic sectors, etc. Any substance which is soluble in nature, easily penetrate to groundwater and deteriorate its quality, e.g., wastewater contain water-soluble complexes with organic ligands which are easily leached down into different soil layer and ultimately mixed with the groundwater. Freshwater resources like surface water vary with different climatic conditions but groundwater is not directly affected by such conditions. Groundwater contamination is increasing day to day, unfortunately it is contaminated with several potential toxic pollutants which become extremely dangerous for the living stocks. Treatment of groundwater is very costly because it requires expensive operation systems or other facilities. According to Agrawal et al. (2016), there is an urgent need for consistent monitoring of the risk assessment and proper management of groundwater to reduce the pollutant loads.

Madhav et al. (2018) reported the groundwater pollution in Bhadohi district of Uttar Pradesh due to massive industrialization of carpet sectors, 40% groundwater samples of this area have nitrate content above the permissible limit (>50 mg/l), which cause harmful effect on human health, especially in children cause methemoglobinemia (blue baby syndrome) which leads to create hypoxic condition. Gupta et al. (2019b) studied different villages area of Jajmau, Kanpur, Uttar Pradesh to assess the groundwater quality which is badly affected by rapid industrialization. Kanpur city is hub of various industries mainly for tannery industries, approximately more than 800 in number. Industrial discharge containing different types of contaminants caused serious health problems to local people. These contaminants not only affect the surface water but groundwater also gets contaminated due to leaching processes. Results of this study showed that groundwater contains higher amounts of TDS ranged between 2835 mg/l – 2581 mg/l, and heavy metal, mainly chromium concentration (0.004–0.13 mg/l), exceeded its permissible limit. Cr is carcinogenic in nature and also its hexavalent form is more dangerous than the trivalent form.

6.4.3 *Food Chain Contamination*

Wastewater irrigation affects human health in both aspects either positively or negatively. Positive aspect of wastewater irrigation is related to food security of the water-scarce zone and negative aspect of wastewater is based upon the presence of different toxic potential contaminants and pathogens, etc., above their permissible limits. Among all the wastewater irrigated food plants, vegetables are the major food crops, consumed at a global level. Vegetables accumulated the toxicants in their edible parts. Approximately 90% of total metals enter to the human body via intake by vegetables and other edible crops as it is the major component of the human diet and the remaining 10% metal through inhalation of dust and dermal contacts (Martorell et al. 2011; Khan et al. 2014; Gupta et al. 2019a). Heavy metals negatively affect the nutritional value of the foods. The plants irrigated with wastewater were found nutrient deficient due to antagonistic interaction between metal and nutrient such as Cd suppress the uptake of Zn due to same carrier transporter protein (Salgare and Acharekar 1992; Sharma and Agrawal 2006).

For security of human health and food safety, it is important to characterize the sources and concentration of heavy metals in wastewater, irrigated soil, and growing plants in order to establish quality standards (Sun et al. 2013). Because of higher demands of food in the recent decades, food safety has become one of the burning issues with respect to human health. This context provokes researchers and scientist to work on food chain contamination by heavy metals and its associated health risks. The concentration of heavy metal (mg/kg) in wheat grain irrigated with wastewater varied between 0.1–0.9, 0.3–0.5, 0.7–1.4, 0.8–1.6, 0.6–0.9, 1.2–1.6, and 0.06–0.2 for Zn, Cr, Cu, Mn, Ni, Cd, and Pb, respectively. Cd concentration exceeded permissible limit (0.2 mg/kg) set by FAO/WHO. Similarly, vegetables and other crops as well as milk produced under wastewater irrigation regimes were also found contaminated with potentially toxic heavy metals (Sharma et al. 2007; Gupta et al. 2019a).

6.5 Control Measures for Wastewater Reuse

Due to the rapid growth in industrialization, the level of contaminants is increasing day by day in industrial discharge, so, for healthy and safe environment their removal from wastewater is very important. Generally, conventional wastewater treatment approach is the combined form of different physical, chemical, and biological processes for the removal of contaminants including different hazardous metals, organic materials, colloids, and types of emerging contaminants from wastewater. Each treatment process has its own advantages and disadvantages in terms of cost, labor requirement, efficiency, reliability, practicability, feasibility, ecofriendly nature, operation process, energy cost, and quality of byproducts, etc. Treatment approaches also depend on the source and types of wastewater. Various physical and chemical processes applied for the removal of contaminants suffer from several limitations due to higher cost, energy requirements, basic properties alteration, and its residual byproducts (Crini and Lichtfouse 2019). For resolving such challenges, several types of green and sustainable approaches are developed by perceiving them as an ecofriendly approach. These techniques require less energy cost-effective, reliable, efficient, and highly accepted by the public. The different kinds of approaches used for removing the contaminants from wastewater are discussed as below:

6.5.1 Physicochemical Approach

Researchers are using different types of the physicochemical methods such as gravity sedimentation process, coagulation, ion exchange, chemical precipitation, adsorption through abiotic materials, catalytic removal and nanotechnology, etc., to remove the toxic pollutants from the wastewater since many years.

6.5.1.1 Gravity Sedimentation Process

Sedimentation is a physical wastewater treatment process as well as also a part of primary wastewater treatment process, in which gravity plays an important role in the removal of suspended solids from the wastewater. Settlement ponds are used for the removal of higher weight suspended particles from wastewater with the help of sedimentation process. Clarifiers are built with mechanical means for consistent removal of suspended solids from the wastewater. In sedimentation process, removal of suspended solids is affected by its size and specific gravity (Demirbas et al. 2017).

6.5.1.2 Coagulation

In wastewater treatment, suspended materials of wastewater can be removed simply through settling process, but for the removal of smaller particles especially for colloidal particles, coagulation process plays an important role. Coagulation is a chemical process and it is achieved through the use of different coagulants, e.g., aluminum chloride ($AlCl_3.6H_2O$), sodium aluminates ($NaAlO_2$), alum, ferrous sulfate, etc. Due to the use of coagulants, aggregation of destabilize particles, and sedimentation substances form a larger particle called “Floc” and the process is called flocculation. The advantage of these flocs due to higher weight can easily be removed by sedimentation and filtration (Ebeling et al. 2003). Farhan (2019) analyzed the efficiency of different coagulants such as alum, ferrous sulfate and ferric chloride, etc., with the addition of different organic coagulants, e.g., polyethylene glycol, polyacrylamide, and inorganic, e.g., bentonite for removing higher and lower turbidity of the wastewater. Optimal dose of coagulants is also determined for higher removal of suspended solids and leaves less turbidity. Removal capacity of turbidity from the wastewater by alum, ferrous sulfate, ferric chloride, and calcium oxides was found to be 99.59%, 98.26%, 98.66%, and 92.18%, respectively (Farhan 2019).

6.5.1.3 Ion Exchange

Ion exchange process is a physical approach for removing heavy metals from the wastewater. In this process, ions are reversibly interchanged between the solid, i.e., natural or synthetic solid resin and liquid phase. Resins uptake ions from electrolytic solution and release ions which are similar in charge or chemically equivalent (Cavaco et al. 2007).

6.5.1.4 Through Adsorption Process

Adsorption process is a widely accepted physicochemical technique. It is based on the exchange of mass between liquid phase, i.e., wastewater and solid phase, i.e., adsorbent. Adsorption process is completed in several steps, first of all, contaminants move from wastewater to the surface of adsorbent, in the second step contaminants lock or adsorbed on adsorbent surface, and in the final step penetration of contaminants to the adsorbent structure. Adsorption process has been done in both, natural byproducts, e.g., rubber, cotton, rice husk, jute fiber, etc., and synthetic byproducts, e.g., zeolite, activated carbon, fly ash, iron slag, etc. (Salam et al. 2011). Huang et al. (2018) reported the removal of heavy metal (Pb, Cu, etc.) from wastewater through two types of zeolite, i.e., ZIF-8 and ZIF-67. ZIF-8 synthesized by addition of 2-methylimidazole and zinc nitrate hexahydrate ($Zn\ (NO_3)_2{\cdot}6H_2O$) in solvent of N, N-dimethylformamide (DMF). ZIF-67 synthesized by addition of 2-methylimidazole and cobalt nitrate hexahydrate ($Co(NO_3)_2{\cdot}6H_2O$) in a DMF solvent. Zeolite is highly porous material, which helps in improving the adsorption capacity. Adsorption capacity of ZIF-8 and ZIF-67 was 1119.80 and 1348.42 mg/g for Pb, and 454.72 and 617.51 mg/g for Cu, respectively. Adsorption capacity of such types of zeolite was very higher, take few minutes to reach adsorption equilibrium. These characteristics proved that ZIF-8 and ZIF-67 have good adsorption capacity for the removal of heavy metals from wastewater (Huang et al. 2018). De Abreu Domingos and da Fonseca (2018) performed a comparative study for the treatment of petroleum refinery wastewater by an adsorption process using activated carbon and ion exchange process through resins (styrene–divinylbenzene). The study verified the feasibility and reliability of both the processes and showed that the removal of contaminants from petroleum refinery wastewater through adsorption by activated carbon was 55% higher than the ion exchange process.

6.5.1.5 Catalytic Removal

Now a day's catalytic removal of toxicants from the wastewater is a promising technology. Removal of non-biodegradable contaminants from the wastewater is achieved in presence of a catalyst through process of oxidation, rate of enhanced degradation. In the wastewater treatment process, oxidation of dissolved organic carbon in liquid phase is achieved in the presence of catalysts at lower temperature and pressure. Catalysts should be cost-effective, reliable, feasible, and have numerous active sites in its surface for improving its remediation capacity (Levec and Pintar 2007). For catalytic activity, noble metals, e.g., platinum, ruthenium, rhodium, palladium, etc., are used for abatement of contaminants from wastewater. Catalyst activities also depend on several factors such as wastewater pH, types and quality of wastewater, concentration of contaminants, etc. In corrosive environment, catalysts should be stable and mechanically strong after several run for higher oxidation and complete mineralization of organic components (Matatov-Meytal and Sheintuch 1998; Monteoliva et al. 2020). Yin and Zhu (2019) suggested use of solar energy in photocatalytic oxidation technology for the treatment of the wastewater from printing and dyeing chemical industries. Wastewater became more hazardous due to the presence of methylene blue dye effluents. For degradation of methylene blue dye effluents, a highly effective catalysts, Ag^+ nodes with 1,3,5-trimesic acid linker by the process of hydrothermal reaction were constructed. During the process of photocatalytic degradation, Ag^+ oxidized and become Ag^{2+} and degradation mechanism completed in multistep phenomenon. At pH 3 and 20 min contact time, Ag^+ catalyst photocatalytically degrade 99.8% methylene blue of wastewater from the printing and dying chemical industries.

6.5.1.6 Use of Nanotechnology

The availability of clean and affordable water for human need is a great challenge during the present time. Therefore, there is need of some innovational technologies for wastewater treatment. Nanotechnologies provide good potential for removing the toxicants from the wastewater and create an opportunity for advanced wastewater treatment technology. There are several properties of nanomaterials which improve its contaminants removal capacity such as large surface area, strong sorption capacity, higher fast dissolution rate, highly reactive, super magnetism, etc. There are some limitations of nanotechnology, e.g., higher cost, high quality of nanomaterials, technical proficiency, etc. Collaboration between industries, research centers, government, etc., is needed for avoiding these challenges and to improve the efficiency, reliability, and feasibility of nanotechnology (Qu et al. 2013). Hassan and Mahmood (2019) analyzed the efficiency of iron oxide (Fe_2O_3) nanoparticles for abatement of cadmium from sewage water of Medical City, Baghdad. The best conditions for maximum Cd removal, i.e., 96.9% are at pH 5.5, for 95.8 min contact time and with 20.77 mg/L of iron oxide nanoparticles (Hassan and Mahmood 2019).

6.5.2 Biological Approaches

Physicochemical approaches have certain drawbacks like higher cost, high energy input, changes in water quality, and disturbance in native flora communities. Therefore, recently biological or green approaches come into practices that are cost-effective, require less energy, more reliable, feasible, and eco-friendly. In this, different microorganisms such as bacteria, fungus, algae, etc., are used to degrade wastewater contaminants. Microorganisms are used in different engineering set up such as trickling filter, activated sludge process, membrane bioreactors, rotating biological contactors, etc., for wastewater treatment. Several sustainable and green approaches used for wastewater treatment are given in Table 6.1 and discussed as below.

6.5.2.1 Bacterial Biodegradation

In wastewater treatment technology, the use of bacterial community in different engineered systems such as trickling filter, rotating biological contactor, activated sludge process, etc., or as a biosorbent is highly considerable. The metabolic rate of bacteria determines its effectiveness in wastewater treatment. Microbial communities rich in species degrade a wider range of substrates than its pure culture. Biosynthesis of extra cellular polymeric substances by microbial aggregates through generated by cell lysis, secretion, released materials from cell surfaces are necessary for maintaining biomass structure and protects bacterial cells against various contaminants. Biological wastewater treatment system has been designed by considering three aspects, such as engineering, ecological, and microbial. With design and operation of wastewater treatment plants, integration of theoretical ecology allow better prediction of microbial population, variations in communities structure and function with changes in environmental conditions (Cydzik-Kwiatkowska and Zielińska 2016). Investigation of microbial communities is important for degradation of wastewater released from pharmaceutical, petroleum refineries, textiles, paper pulp industries, etc. (Ma et al. 2015).

Ajaz et al. (2019) reported that *Alcaligenes aquatilis* was used to decolorize 82% synazol red 6HBN from wastewater at pH 7 after incubation of 4 days at 37 °C. Under static conditions, maximum decolorization of wastewater by *A. aquatilis* was achieved in the presence of sawdust and yeast extract as a source of carbon and nitrogen, respectively. Results further showed that the maximum wastewater decolorization, i.e., approximately 86% of multiple dyes was found in 5 days incubation. *A. aquatilis* not only potentially decolorize wastewater but also transformed azo dyes into different useful end products applied in the environmental biotechnology.

Table 6.1 List of sustainable and green approaches for removal of contaminants from wastewater

S. N.	Sustainable process	Description	References
1.	Application of aerobic upflow submerged attached growth reactor	Clear away ammonia, thiocyanate, and cyanate from gold mine wastewater	Di Biase et al. (2020)
2.	Application of zooplankton (*Daphnia* sp.)-based reactor	Removed organic matter and heavy metal Cu from wastewater	Pous et al. (2020)
3.	Use of microalgal consortia for treatment of municipal wastewater	Removed organic carbon and Cu, Cr, Pb, and Cd from sewage wastewater	Sharma et al. (2020b)
4.	Use of Enhanced Biological Phosphorus Removal (EBPR)	Chemical precipitation with evidence of phosphate accumulating bacterial population	Meza et al. (2020)
5.	Use of microbial community structure	Removed COD, terephthalic acid, para-toluic acid, benzoic acid, and acetic acid from wastewater	Ma et al. (2020)
6.	Application of biological aerated filter	Removed ammonia, nitrate and phosphorus from wastewater	Li et al. (2020)
7.	Use of aerobic granular sludge reactor	Clear away naphthalene, acenaphthylene, and acenaphthene from wastewater	Ofman et al. (2020)
8.	Fenton's treatment followed by subsequent biological treatment	Reduced BOD and COD of pharmaceuticals wastewater	Changotra et al. (2019)
9.	Integrated bio-oxidation and adsorptive filtration reactor	Removed arsenic from different industrial effluents	Kamde et al. (2019)
10.	Integrated biofilter, i.e., biofilm, clam (*Tegillarca granosa*), and macrophytes (*Spartina anglica*)	Bio-filters significantly influenced the biodegradability and resolvability of particulate organic matter and control dissolve oxygen, water temperature, and nitrogen	Lukwambe et al. (2019)
11.	Use of hybrid membrane photobioreactor and membrane photobioreactor	Removed atrazine (2-chloro-4-ethylamino-6-isopropylamino-s-triazine) and CNP from wastewater	Derakhshan et al. (2019)
12.	Use of indigenous microalgae (*Dictyosphaerium* sp. strain MM-IR2)	Improved sulfate removal from the power plant wastewater	Mohammadi et al. (2019)
13.	Application of rotating biological contactors	Abatement of pharmaceuticals and personal care products contaminants from domestic wastewater	Delgado et al. (2019)
14.	Biodegradation through halophilic *Halomonas* strain Gb bacterium	Toluidine red biodegradation in a synthetic wastewater	Amini et al. (2019)
15.	Used filamentous microalgae *Tribonema* sp. culture	Removed COD, total N, total P and organic contaminants from petrochemical wastewater	Huo et al. (2019)
16.	Use of moving biological bed reactors (MBBR) and sequencing batch reactors (SBR)	Shortening the level of COD, BOD, ammonium, and phosphate ion from fish processing wastewater	Nowak et al. (2019)

(continued)

Table 6.1 (continued)

S. N.	Sustainable process	Description	References
17.	Use of bioaugmented multistage biofilter	Deactivation of pathogenic micro-organisms from municipal wastewater	Ibrahim et al. (2020)
18.	Application of aerobic and anaerobic membrane bioreactors	Abatement of 15 trace organic contaminants from synthetic wastewater	Liu et al. (2020)
19.	Application of anaerobic moving bed biofilm reactors	Removed BOD and COD from oil-contaminated wastewater	Morgan-Sagastume et al. (2019)
20.	Biofilm-membrane bioreactor (BF-MBR)	Use of biomarkers in *Mytilus galloprovincialis* in an integrated bioremediation approach for remediation of oily wastewater	Pirrone et al. (2018)
21.	Use of biofilm	Removed hydrocarbons from polluted industrial wastewater	Rodríguez-Calvo et al. (2018)
22.	Use of cocoa pod husks	Removed Congo red (CR) dye from industrial effluents	Olakunle et al. (2018)
23.	Use of *Marsileaq uadrifolia* as a bioremediating agent	Removed BOD, COD, soluble phosphorous, total N, and heavy metal Zn	Abbasi et al. (2018)
24.	Application of trickling biofilter (TBF) system	Reduction of BOD, COD, TDS, EC, phosphate, sulfate, and total nitrogen from domestic wastewater	Rasool et al. (2018)

6.5.2.2 Fungal Biodegradation

The fungal communities play an important role as a biological agent in degradation of toxicants from wastewater. Fungi can grow at a higher rate, produce much tolerant spores, enzymes, and have capability to degrade toxicants and they are ecologically safe in nature (Williams et al. 2003; Uniyal et al. 2019). Most of the fungal species commonly grow in soil, root, and foliar parts of plants, partially decomposed organic materials. It has high resistance capacity against potential toxic compounds such as heavy metals, emerging contaminants, pathogens as well as antibiotics. Different types of fungi like basidiomycetes and ascomycetes have been used in decolorization of natural and synthetic melanoidins from distillery wastewater and at the same time also generate fungal protein-enriched biomass for animal feed as a valuable product (Pant and Adholeya 2007).

Arikan et al. (2019) reported the contaminant removal capacity of two filamentous fungal species namely, *Aspergillus carbonarius* M333 and *Penicillium glabrum* Pg1 immobilized in macroporous polymeric support from textile wastewater. They studied the removal efficiency of color and chemical oxygen demand (COD) tested in two set up, i.e., batch experiments and continuously operated upflow packed bed bioreactor. The optimized pH was found 5.5 (4.5 to 6.0) for its higher removal potency to color and COD of wastewater. Results further indicated that color and COD removal efficiency from wastewater was found 98.2% and 69.8% for batch experiment and 78.8% and 67.7% for upflow packed bed bioreactor, respectively.

6.5.2.3 Algal Biodegradation

Removal of potentially toxic substances from the wastewater by algal species termed as phycoremediation. Phycoremediation is a promising, efficient, eco-friendly, cost-effective technology. Generally, contaminants removal from wastewater works on the principle of adsorption and absorption; cell surface adsorption is independent on cell metabolism but in case of absorption or intracellular uptake depends on cell metabolism. The capability of algal species for the adsorption and absorption of contaminants depends on large surface area, higher metal binding affinity on their cell surface, feasible in metal uptake and storage systems (Ahmad et al. 2020). Natural ability of alga for uptake of nutrients and heavy metals, and degradation of organic contaminants is possible via symbiotic association with aerobic bacteria. Algae showed higher photosynthetic activity due to the presence of pigments and become greater source of oxygen in an aquatic ecosystem. They can also enhance the degradation of organic compounds in aerobic condition and helpful in biofuel production (Majumder et al. 2014; Gupta et al. 2020).

Ali et al. (2018) used *Scenedesmus obliquus* isolated from phytoplankton community of Nile River for the removal of pharmaceutical compounds from industrial effluents. Pharmaceutical residues are also considered as an emerging pollutant and have potential negative impacts on environment. Algal biomass was modified using alkaline solution and used for biosorption of different pharmaceuticals compounds namely cefadroxil, paracetamol, ibuprofen, ciprofloxacin, etc. At natural pH, 0.5 g/L of modified algal biomass has adsorption capacities for cefadroxil, paracetamol, ibuprofen, and ciprofloxacin as 68 mg/g, 58 mg/g, 42 mg/g, and 39 mg/g, respectively. Results further showed reusability of modified algal biomass for biosorption of different pharmaceutical compounds; its efficiency is decreased by 4.5% after three uses. However, there is a need to explore various algal remediation techniques as an eco-friendly alternatives for a better environmental condition.

6.5.2.4 Integrated Aerobic-Anaerobic Biodegradation

Assessment of some important factors like composition and concentration of contaminants in wastewater, volume generation, cost-effectiveness, treatment susceptibility, environmental impacts, etc., is necessary for the selection of the best wastewater treatment technology. Wastewater treatments by a combination of aerobic and anaerobic degradation in an individual bioreactor have higher removal efficiency. Different engineering setups such as rotating biological contactor, membrane bioreactor, trickling filter, sequential batch reactor, etc., are used for aerobic degradation of contaminant. Upflow anaerobic film, membrane anaerobic reactor system, etc., are used for anaerobic degradation of contaminants in wastewater. Generally, these integrated approaches have been applied for removing higher level of organic matters from the wastewater. In the present time, awareness about these integrated approaches rapidly improved due to number of advantages, e.g., low chemicals requirement, less sludge production, higher resource recovery potential, less equipment requirement, and easy handling. Some disadvantages are also found related to this integrated approach such as membrane fouling, equipment malfunctions, and need of higher proficiency for its operation, but overall performance of these integrated approach is impressive (Goli et al. 2019).

Fazal et al. (2019) investigated the simultaneous removal of COD, nitrate, and sulfate from the wastewater using aerobic and anaerobic biodegradation approach. Medium properties, i.e., pH was the main determining factor for biodegradation of COD, nitrate, and sulfate. In aerobic treatment process, COD was degraded completely in CO_2 and H_2O, nitrate removal efficiency ranged from 83% to 90% in cellular metabolism. However, anaerobic process removes from 68% to 80% COD via methanogenesis process, nitrate removal efficiency ranged between 93% and 98%, sulfate changed into elemental sulfur from 85% to 97%. In anaerobic process, COD conversion rate was also found lower than those of the aerobic process. More studies are still required for integrated approach to enhance its contaminants removal efficiency, energy production as well as for reducing its operational cost and negative environmental impacts.

6.5.3 Molecular Approach

Conventional microbial techniques for wastewater treatment are mainly based on pure culture isolation. The morphological, physiological, and biochemical assay of microbes extensively provided information of microbial biodiversity in natural and engineered systems. However, improving the efficiency of conventional wastewater treatment techniques using molecular approaches is very important. Molecular docking is important bioinformatics modelings in which contaminants bind with other substances and stabilized it (Liu et al. 2018). Molecular techniques such as cloning and gene library generation, fluorescent *in-situ* hybridization with DNA probes (FISH), and denaturant gradient gel electrophoresis (DGGE), etc., are also used in wastewater treatment. Although cloning and gene library generation are time-consuming processes they provide very significant taxonomical information. For identification of microbes at any desired taxonomic level, FISH techniques are used depending on the used probe specificity. DGGE is a simple and quick method for characterizing band patterns of different samples and sample profiling rapidly but genetic analysis is analyzed by particular band sequencing (Sanz and Köchling 2007).

Zahedi et al. (2019) assessed the eukaryotic waterborne pathogens in wastewater treatment plants (WWTPs) responsible for public health threat. In this study, they identified fecal pathogenic eukaryotes including *Cryptosporidium* sp., at different stages of treatment (initial or influents, intermediate and discharge, or effluents) in samples collected from 4 WWTPs in Western Australia, in which 3 WWTPs were used as stabilization ponds and one WWTP was used as activated sludge treatment technology. For next-generation sequencing (NGS), 18s rRNA of eukaryotic pathogen from wastewater as well as the mammalian blocking primer were used for reducing mammalian DNA amplification in wastewater. With the help of bioinformatics, 49 eukaryotic phyla were detected in wastewater samples, in which 3 phyla of human intestinal parasites, which were primarily detected in raw wastewater. Six subtypes of *Blastocystis* sp. and 4 *Entamoeba* sp. were identified by 18s NGS, but *Giardia* sp. and *Cryptosporodium* sp. were not detected. Real-time polymerase chain reaction was failed to detect *Giardia* sp. but detected *Cryptosporidium* sp. in WWTPs with the help of specific NGS. NGS is a more specific and reliable approach for detection of pathogen in wastewater and could be used in assessment of wastewater pathogen in future.

6.6 Conclusions

Rapid growth in population, urbanization, industrialization, and socio-economic development enlarge the gap between water availability and demand, particularly in semi-arid and arid area. Such problem can be avoided by enhancing water use efficiency or use of wastewater, i.e., domestic wastewater, storm runoff, and industrial effluents as an alternative water source. Long-term wastewater irrigation negatively affects the soil quality, e.g., salinization, reduced porosity, infiltration and hydraulic conductivity of soil, water repellency, reduced high bulk density, heavy metal contamination, and groundwater pollution. Agriculture sector is the major consumer of wastewater and considered not only as a valuable water resource but also has nutritive value. However, in the current scenario, safe use of wastewater is a matter of major concern as it is contaminated with potentially toxic elements. There should be development of suitable strategies for efficient and reliable use of wastewater and for reducing its negative impact on ecosystems, especially for water-scarce areas. In this chapter, wastewater availability and reuse at global and regional levels are briefly summarized and composition and characterization of wastewater are also highlighted. Long-term wastewater irrigation can adversely affect the physicochemical and biological properties of soil and pose risk to safety and quality of food through food chain contamination. Different control measures, i.e., physical, biological, and molecular for removing contaminants from wastewater are also discussed in brief. Abatement of contaminants in the wastewater irrigated soil using inorganic and organic amendments, bacterial inoculants, and cycling with qualitative water can provide a future research direction for developing eco-friendly, economically, and efficient management strategies for sustainable use of wastewater in agriculture. Success of sustainable wastewater reuse in different sectors of society mostly depends on public acceptance, public-private partnership model, and awareness spread by government and non-government organizations.

References

Abbasi SA, Ponni G, Tauseef SM (2018) *Marsilea quadrifolia*: a new bioagent for treating wastewater. Water Air Soil Pollut 229(4):133

Agrawal N, Singh R, Chaubey PM (2016) Assessment of ground water toxicity due to tannery industry at Jajmau, Kanpur (UP). Int J Innov Res Sci Eng Technol 5(3):4055–4405

Ahmad S, Pandey A, Pathak VV, Tyagi VV, Kothari R (2020) Phycoremediation: algae as eco-friendly tools for the removal of heavy metals from wastewaters. In: Bioremediation of industrial waste for environmental safety. Springer, Singapore, pp 53–76

Ajaz M, Rehman A, Khan Z, Nisar MA, Hussain S (2019) Degradation of azo dyes by *Alcaligenes aquatilis* 3c and its potential use in the wastewater treatment. AMB Express 9(1):64. https://doi.org/10.1186/s13568-019-0788-3

Ali ME, El-Aty AMA, Badawy MI, Ali RK (2018) Removal of pharmaceutical pollutants from synthetic wastewater using chemically modified biomass of green alga *Scenedesmus obliquus*. Ecotoxicol Environ Saf 151:144–152

Amerasinghe P, Bhardwaj RM, Scott C, Jella K, Marshall F (2013) Urban wastewater and agricultural reuse challenges in India. Colombo, Sri Lanka: International Water Management Institute (IWMI). 36p. IWMI research report 147. https://doi.org/10.5337/2013.200

Amini B, Otadi M, Partovinia A (2019) Statistical modeling and optimization of Toluidine Red biodegradation in a synthetic wastewater using *Halomonas strain* Gb. J Environ Health Sci Eng:1–12

Antwi-Agyei P, Peasey A, Biran A, Bruce J, Ensink J (2016) Risk perceptions of wastewater use for urban agriculture in Accra, Ghana. PLoS One 11(3):e0150603. https://doi.org/10.1371/journal.pone.0150603

Arikan EB, Isik Z, Bouras HD, Dizge N (2019) Investigation of immobilized filamentous fungi for treatment of real textile industry wastewater using up flow packed bed bioreactor. Bioresour Technol Rep 7:100197

Ashraf M, Imtiaz M, Abid M, Afzal M, Shahzad SM (2013) Reuse of wastewater for irrigating tomato plants (*Lycopersicon esculentum* L.) through silicon supplementation. J Water Reuse Desal 3(2):128–139

Aziz F, Farissi M (2014) Reuse of treated wastewater in agriculture: solving water deficit problems in arid areas. Ann West Univ Timisoara Ser Biol 17(2):95

Bhargava A, Carmona FF, Bhargava M, Srivastava S (2012) Approaches for enhanced phytoextraction of heavy metals. J Environ Manag 105:103–120

Boutin, C, Eme C (2016) Domestic wastewater characterization by emission source. 13eme congress specialise IWA on Small water and wastewater systems, Sept 2016, Athènes, Greece. 8 p. hal01469077

Cavaco SA, Fernandes S, Quina MM, Ferreira LM (2007) Removal of chromium from electroplating industry effluents by ion exchange resins. J Hazard Mater 144(3):634–638

Central Pollution Control Board (2013) Performance evaluation of sewage treatment plants under NRCD. http://164.100.107.13/upload/NewItems/NewItem_195_STP_REPORT.pdf

Changotra R, Rajput H, Dhir A (2019) Treatment of real pharmaceutical wastewater using combined approach of Fenton applications and aerobic biological treatment. J Photochem Photobiol A 376:175–184

Chatzistathis T, Koutsos T (2017) Olive mill wastewater as a source of organic matter, water and nutrients for restoration of degraded soils and for crops managed with sustainable systems. Agric Water Manag 190:55–64

Cornu JY, Denaix L, Lacoste J, Sappin-Didier V, Nguyen C, Schneider A (2016) Impact of temperature on the dynamics of organic matter and on the soil-to-plant transfer of Cd, Zn and Pb in a contaminated agricultural soil. Environ Sci Pollut Res 23(4):2997–3007

Crini G, Lichtfouse E (2019) Advantages and disadvantages of techniques used for wastewater treatment. Environ Chem Lett 17(1):145–155

CWC (2000) Water and related statistics. Information Systems Organization, Water Planning and Organization Wing, Central Water Commission, RK Puram, New Delhi 66: 451
Cydzik-Kwiatkowska A, Zielińska M (2016) Bacterial communities in full-scale wastewater treatment systems. World J Microbiol Biotechnol 32(4):66. https://doi.org/10.1007/s11274-016-2012-9
De Abreu Domingos R, da Fonseca FV (2018) Evaluation of adsorbent and ion exchange resins for removal of organic matter from petroleum refinery wastewaters aiming to increase water reuse. J Environ Manag 214:362–369
Delgado N, Navarro A, Marino D, Peñuela GA, Ronco A (2019) Removal of pharmaceuticals and personal care products from domestic wastewater using rotating biological contactors. Int J Environ Sci Technol 16(1):1–10
Demirbas A, Bamufleh HS, Edris G, Alalayah WM (2017) Treatment of contaminated wastewater. Petrol Sci Technol 35(9):883–889
Derakhshan Z, Ehrampoush MH, Mahvi AH, Dehghani M, Faramarzian M, Eslami H (2019) A comparative study of hybrid membrane photobioreactor and membrane photobioreactor for simultaneous biological removal of atrazine and CNP from wastewater: a performance analysis and modeling. Chem Eng J 355:428–438
Di Biase A, Wei V, Kowalski MS, Bratty M, Hildebrand M, Jabari P, Devlin TR, Oleszkiewicz JA (2020) Ammonia, thiocyanate, and cyanate removal in an aerobic up-flow submerged attached growth reactor treating gold mine wastewater. Chemosphere:125395. https://doi.org/10.1016/j.chemosphere.2019.125395
Dogan Y, Baslar S, Ugulu I (2014) A study on detecting heavy metal accumulation through biomonitoring: content of trace elements in plants at Mount Kazdagi in Turkey. Appl Ecol Environ Res 12(3):627–636
Ebeling JM, Sibrell PL, Ogden SR, Summerfelt ST (2003) Evaluation of chemical coagulation–flocculation aids for the removal of suspended solids and phosphorus from intensive recirculating aquaculture effluent discharge. Aquac Eng 29(1–2):23–42
Farhan AAR (2019) Evaluation of a number of coagulants and coagulation aids in removing water turbidity. In: IOP conference series: materials science and engineering, vol 518, no 6. IOP Publishing, p 062016
Fazal S, Xu H, Xu X, Huang S, Zhang Y, Ali A, Hayat W, You Y, Du X, Ullah Z (2019) Simultaneous removal of carbon, sulfur, and nitrogen from biological synthetic wastewater: aerobic and anaerobic approach. Arab J Geosci 12(5):168. https://doi.org/10.1007/s12517-019-4346-8
Gharaibeh MA, Eltaif NI, Al-Abdullah B (2007) Impact of field application of treated wastewater on hydraulic properties of vertisols. Water Air Soil Pollut 184(1–4):347–353
Goli A, Shamiri A, Khosroyar S, Talaiekhozani A, Sanaye R, Azizi K (2019) A review on different aerobic and anaerobic treatment methods in dairy industry wastewater. J Environ Treat Tech 6 (1):113–141
Gupta N, Yadav KK, Kumar V, Kumar S, Chadd RP, Kumar A (2019a) Trace elements in soil-vegetables interface: translocation, bioaccumulation, toxicity and amelioration – a review. Sci Total Environ 651:2927–2942
Gupta R, Jindal T, Khan AS, Srivastava P, Kanaujia A (2019b) Assessment of ground water quality and its suitability for drinking in industrial area Jajmau, Kanpur, India. Plant Arch 19 (1):1569–1571
Gupta GS, Yadav G, Tiwari S (2020) Bioremediation of heavy metals: a new approach to sustainable agriculture. In: Restoration of wetland ecosystem: a trajectory towards a sustainable environment. Springer, Singapore, pp 195–226
Hassan DF, Mahmood MB (2019) Using of iron oxide nanoparticles and application in the removing of heavy metals from sewage water. Iraqi J Sci 60(4):732–738
Huang Y, Zeng X, Guo L, Lan J, Zhang L, Cao D (2018) Heavy metal ion removal of wastewater by zeolite-imidazolate frameworks. Sep Purif Technol 194:462–469

Huo S, Chen J, Zhu F, Zou B, Chen X, Basheer S, Cui F, Qian J (2019) Filamentous microalgae *Tribonema sp.* cultivation in the anaerobic/oxic effluents of petrochemical wastewater for evaluating the efficiency of recycling and treatment. Biochem Eng J 145:27–32

Hussain A, Priyadarshi M, Dubey S (2019) Experimental study on accumulation of heavy metals in vegetables irrigated with treated wastewater. Appl Water Sci 9(5):122. https://doi.org/10.1007/s13201-019-0999-4

Ibrahim S, El-Liethy MA, Abia ALK, Hasan MAEG, Al Zanaty AM, Kamel MM (2020) Design of a bioaugmented multistage biofilter for accelerated municipal wastewater treatment and deactivation of pathogenic microorganisms. Sci Total Environ:134786. https://doi.org/10.1016/j.scitotenv.2019.134786

Jaramillo MF, Restrepo I (2017) Wastewater reuse in agriculture: a review about its limitations and benefits. Sustainability 9:1734–1753

Jeong H, Kim H, Jang T (2016) Wastewater reuse in agriculture: a contribution toward sustainable wastewater reuse in South Korea. Water 8:169–187. https://doi.org/10.3390/w8040169

Kamde K, Dahake R, Pandey RA, Bansiwal A (2019) Integrated bio-oxidation and adsorptive filtration reactor for removal of arsenic from wastewater. Environ Technol 40(10):1337–1134

Khalid S, Shahid M, Bibi I, Sarwar T, Shah AH, Niazi NK (2018) A review of environmental contamination and health risk assessment of wastewater use for crop irrigation with a focus on low and high-income countries. Int J Environ Res Pub He 15(5):895. https://doi.org/10.3390/ijerph15050895

Khan S, Reid BJ, Li G, Zhu YG (2014) Application of biochar to soil reduces cancer risk via rice consumption: a case study in Miaoqian village, Longyan. China Environ Int 68:154–161

Khan ZI, Ugulu I, Umar S, Ahmad K, Mehmood N, Ashfaq A, Bashir H, Sohail M (2018) Potential toxic metal accumulation in soil, forage and blood plasma of buffaloes sampled from Jhang, Pakistan. Bull Environ Contam Toxicol 101(2):235–242

Kibuye FA, Gall HE, Elkin KR, Ayers B, Veith TL, Miller M, Jacob S, Hayden KR, Watson JE, Elliott HA (2019) Fate of pharmaceuticals in a spray-irrigation system: from wastewater to groundwater. Sci Total Environ 65(4):197–208

Kumar MD, Tortajada C (2020) Assessing wastewater management in India. SpringerBriefs in Water Science and Technology. https://doi.org/10.1007/978-981-15-2396-0_6

Lado M, Ben-Hur M, Assouline S (2005) Effects of effluent irrigation on seal formation, infiltration, and soil loss during rainfall. Soil Sci Soc Am J 69(5):1432–1439

Levec J, Pintar A (2007) Catalytic wet-air oxidation processes: a review. Catal Today 124 (3–4):172–184

Li Y, Guo J, Li H, Song Y, Chen Z, Lu C, Han Y, Hou Y (2020) Effect of dissolved oxygen on simultaneous removal of ammonia, nitrate and phosphorus via biological aerated filter with sulfur and pyrite as composite fillers. Bioresour Technol 296:122340. https://doi.org/10.1016/j.biortech.2019.122340

Liu WX, Shen LF, Liu JW, Wang YW, Li SR (2007) Uptake of toxic heavy metals by rice (*Oryza sativa* L.) cultivated in the agricultural soil near Zhengzhou City, People's Republic of China. Bull Environ Contam Toxicol 79(2):209–213

Liu Z, Liu Y, Zeng G, Shao B, Chen M, Li Z, Jiang Y, Liu Y, Zhang Y, Zhong H (2018) Application of molecular docking for the degradation of organic pollutants in the environmental remediation: a review. Chemosphere 203:139–150

Liu W, Song X, Huda N, Xie M, Li G, Luo W (2020) Comparison between aerobic and anaerobic membrane bioreactors for trace organic contaminant removal in wastewater treatment. Environ Technol Innov 17:1–10. [100564]. https://doi.org/10.1016/j.eti.2019.100564

Lukwambe B, Zhao L, Nicholaus R, Yang W, Zhu J, Zheng Z (2019) Bacterioplankton community in response to biological filters (clam, biofilm, and macrophytes) in an integrated aquaculture wastewater bioremediation system. Environ Pollut 254:113035. https://doi.org/10.1016/j.envpol.2019.113035

Ma Q, Qu Y, Shen W, Zhang Z, Wang J, Liu Z, Li D, Li H, Zhou J (2015) Bacterial community compositions of coking wastewater treatment plants in steel industry revealed by Illumina high-throughput sequencing. Bioresour Technol 179:436–443

Ma XC, Li XK, Wang XW, Liu GG, Zuo JL, Wang ST, Wang K (2020) Impact of salinity on anaerobic microbial community structure in high organic loading purified terephthalic acid wastewater treatment system. J Hazard Mater 383:121132. https://doi.org/10.1016/j.jhazmat.2019.121132

Madhav S, Ahamad A, Kumar A, Kushawaha J, Singh P, Mishra PK (2018) Geochemical assessment of groundwater quality for its suitability for drinking and irrigation purpose in rural areas of Sant Ravidas Nagar (Bhadohi), Uttar Pradesh. Geol Ecol Landsc 2(2):127–136

Majumder S, Gupta S, Raghuvanshi S (2014) Removal of dissolved metals by bioremediation. In: Heavy metals in water: presence, removal and safety, pp 44–56

Martorell I, Perelló G, Martí-Cid R, Llobet JM, Castell V, Domingo JL (2011) Human exposure to arsenic, cadmium, mercury, and lead from foods in Catalonia, Spain: temporal trend. Biol Trace Elem Res 142(3):309–322

Matatov-Meytal YI, Sheintuch M (1998) Catalytic abatement of water pollutants. Ind Eng Chem Res 37(2):309–326

Meza LC, Piotrowski P, Farnan J, Tasker TL, Xiong B, Weggler B, Murrell K, Dorman FL, Heuvel JPV, Burgos WD (2020) Detection and removal of biologically active organic micropollutants from hospital wastewater. Sci Total Environ 700:134469. https://doi.org/10.1016/j.scitotenv.2019.134469

Mohammadi M, Mowla D, Esmaeilzadeh F, Ghasemi Y (2019) Enhancement of sulfate removal from the power plant wastewater using cultivation of indigenous microalgae: stage-wise operation. J Environ Chem Eng 7(1):102870. https://doi.org/10.1016/j.jece.2018.102870

Mohan GR, Speth TF, Murray D, Garland JL (2014) Municipal wastewater: a rediscovered resource for sustainable wáter reuse. In: Potable water, pp 153–179

Monteoliva GA, Martín-Pascual J, Muñío MM, M-Poyatos J (2020) Effects of carrier addition on water quality and pharmaceutical removal capacity of a membrane bioreactor–Advanced oxidation process combined treatment. Sci Total Environ:135104. https://doi.org/10.1016/j.scitotenv.2019.135104

Morgan-Sagastume F, Jacobsson S, Olsson LE, Carlsson M, Gyllenhammar M, Horváth IS (2019) Anaerobic treatment of oil-contaminated wastewater with methane production using anaerobic moving bed biofilm reactors. Water Res 163:114851. https://doi.org/10.1016/j.watres.2019.07.018

Moussaoui T, Wahbi S, Mandi L, Masi S, Ouazzani N (2017) Reuse study of sustainable wastewater in agroforestry domain of Marrakesh city. J Saudi Soc Agric Sci 18(3):288–293

Muchuweti M, Birkett JW, Chinyanga E, Zvauya R, Scrimshaw MD, Lester JN (2006) Heavy metal content of vegetables irrigated with mixtures of wastewater and sewage sludge in Zimbabwe: implications for human health. Agric Ecosyst Environ 112(1):41–48

Nowak A, Mazur R, Panek E, Dacewicz E, Chmielowski K (2019) Treatment efficiency of fish processing wastewater in different types of biological reactors. Phys Chem Earth 109:40–48

Ofman P, Struk-Sokołowska J, Skoczko I, Wiater J (2020) Alternated biodegradation of naphthalene (NAP), acenaphthylene (ACY) and acenaphthene (ACE) in an aerobic granular sludge reactor (GSBR). J Hazard Mater 383:121184. https://doi.org/10.1016/j.jhazmat.2019.121184

Olakunle MO, Inyinbor AA, Dada AO, Bello OS (2018) Combating dye pollution using cocoa pod husks: a sustainable approach. Int J Sustain Eng 11(1):4–15

Palrecha A, Kapoor D, Malladi T (2012) Wastewater irrigation in gujarat an exploratory study. http://www.iwmi.cgiar.org/iwmi-tata/PDFs/2012_Highlight-30.pdf

Pant D, Adholeya A (2007) Biological approaches for treatment of distillery wastewater: a review. Bioresour Technol 98(12):2321–2334

Pirrone C, Rossi F, Cappello S, Borgese M, Mancini G, Bernardini G, Gornati R (2018) Evaluation of biomarkers in *Mytilus galloprovincialis* as an integrated measure of biofilm-membrane bioreactor (BF-MBR) system efficiency in mitigating the impact of oily wastewater discharge to marine environment: a microcosm approach. Aquat Toxicol 198:49–62

Pous N, Hidalgo M, Serra T, Colomer J, Colprim J, Salvadó V (2020) Assessment of zooplankton-based eco-sustainable wastewater treatment at laboratory scale. Chemosphere 238:124683. https://doi.org/10.1016/j.chemosphere.2019.124683

Qu X, Alvarez PJ, Li Q (2013) Applications of nanotechnology in water and wastewater treatment. Water Res 47(12):3931–3946

Radhika VS, Kulkarni GN (2019) Status of sewage water generation in Karnataka, India. Asian J Agric Exten Econ Sociol:1–5

Rarasati AD, Nabila RA, Sumabrata J, Rianto S (2017) Comparative study of wastewater knowledge towards sustainable infrastructure: a case study of Germany and Indonesian university students. Symposium II UNIID 2017. 2(1):554–559

Rasool T, Rehman A, Naz I, Ullah R, Ahmed S (2018) Efficiency of a locally designed pilot-scale trickling biofilter (TBF) system in natural environment for the treatment of domestic wastewater. Environ Technol 39(10):1295–1306

Rezapour S, Atashpaz B, Moghaddam SS, Kalavrouziotis IK, Damalas CA (2019) Cadmium accumulation, translocation factor, and health risk potential in a wastewater-irrigated soil-wheat (*Triticum aestivum* L.) system. Chemosphere 231:579–587

Rodríguez-Calvo A, Silva-Castro GA, Robledo-Mahón T, González-López J, Calvo C (2018) Capacity of hydrophobic carriers to form biofilm for removing hydrocarbons from polluted industrial wastewater: assay in microcosms. Water Air Soil Pollut 229(6):175–188

Rutkowski T, Raschid-Sally L, Buechler S (2007) Wastewater irrigation in the developing world—two case studies from the Kathmandu Valley in Nepal. Agric Water Manag 88(1–3):83–91

Sahay S, Iqbal S, Inam A, Gupta M, Inam A (2019) Waste water irrigation in the regulation of soil properties, growth determinants, and heavy metal accumulation in different Brassica species. Environ Monit Assess 191(2):107. https://doi.org/10.1007/s10661-019-7228-3

Salam OEA, Reiad NA, ElShafei MM (2011) A study of the removal characteristics of heavy metals from wastewater by low-cost adsorbents. J Adv Res 2(4):297–303

Saldías C, Speelman S, Drechsel P, Van Huylenbroeck G (2017) A livelihood in a risky environment: farmers' preferences for irrigation with wastewater in Hyderabad, India. Ambio 46 (3):347–360

Salgare SA, Acharekar C (1992) Effect of industrial pollution on growth and content of certain weeds. J. Nat Conserve 4:1–6

Sanz JL, Köchling T (2007) Molecular biology techniques used in wastewater treatment: an overview. Process Biochem 42(2):119–133

Shariot-Ullah M (2019) Impacts of sugar mill's wastewater use on physical and hydraulic properties of soil: evidence from North Bengal Sugar Mill, Natore, Bangladesh. Sustain Water Res Manag 5(2):863–871

Sharma RK, Agrawal M (2006) Single and combined effects of cadmium and zinc on carrots: uptake and bioaccumulation. J Plant Nutr 29(10):1791–1804

Sharma RK, Agrawal M, Marshall F (2007) Heavy metal contamination of soil and vegetables in suburban areas of Varanasi, India. Ecotoxicol Environ Safe 66(2):258–266

Sharma N, Sharma RK, Samant SS, Pande V, Khatri S (2020a) Exposure of mountain ecosystem to heavy metal assessed by soil and vegetable analysis in Kullu Valley, Northern India. J Toxicol He 109:560–572

Sharma J, Kumar V, Kumar SS, Malyan SK, Mathimani T, Bishnoi NR, Pugazhendhi A (2020b) Microalgal consortia for municipal wastewater treatment–lipid augmentation and fatty acid profiling for biodiesel production. J Photochem Photobiol B Biol 202:111638. https://doi.org/10.1016/j.jphotobiol.2019.111638

Sheoran V, Sheoran AS, Poonia P (2016) Factors affecting phytoextraction: a review. Pedosphere 26(2):148–166

Singh KP, Mohan D, Sinha S, Dalwani R (2004) Impact assessment of treated/untreated wastewater toxicants discharged by sewage treatment plants on health, agricultural, and environmental quality in the wastewater disposal area. Chemosphere 55(2):227–255

Sugam R, Ghosh A (2013) Urban water and sanitation in India: multi-stakeholder dialogues for systemic solutions. CEEW-Veolia report, November, pp i–147

Sun C, Liu J, Wang Y, Sun L, Yu H (2013) Multivariate and geostatistical analyses of the spatial distribution and sources of heavy metals in agricultural soil in Dehui, Northeast China. Chemosphere 92(5):517–523

The World Bank Group (2016) Annual freshwater withdrawals, agriculture of total freshwater withdrawal (WWW Document). http://data.worldbank.org/indicator/ER.H2O.FWAG.ZS?

Thebo AL, Drechsel P, Lambin EF, Nelson KL (2017) A global, spatially-explicit assessment of irrigated croplands influenced by urban wastewater flows. Environ Res Lett 12(7):074008. https://doi.org/10.1088/1748-9326/aa75d1

Tran NH, Reinhard M, Khan E, Chen H, Nguyen VT, Li Y, Goh SG, Nguyen QB, Saeidi N, Gin KYH (2019) Emerging contaminants in wastewater, stormwater runoff, and surface water: application as chemical markers for diffuse sources. Sci Total Environ 676:252–267

Tzanakakis VE, Koo-Oshima S, Haddad M, Apostolidis N, Angelakis AN, Angelakis A, Rose J (2014) The history of land application and hydroponic systems for wastewater treatment and reuse. In: Evolution of sanitation and wastewater technologies through the centuries. IWA Publishing, London, p 457

Uniyal S, Kumar I, Singh PK, Sharma RK (2019) Chapter 8: Genetic engineering: an emerging approach for fostering bioremediation, current status, challenges and outlook. In: Sinha R, Pandey S, Ghoshal N (eds) Innovations in life science research. NOVA Science Publishers, New York. ISBN:978-1-53615-868-7

Urie DH (1986) The status of wastewater irrigation of forest. In: Cole DW, Henry CL, Nutter WL (eds) The forest alternative for treatment and utilization of municipal and industrial wastes. University of Washington Press, Seattle, pp 26–40

WHO/UNICEF Joint Water Supply, & Sanitation Monitoring Programme (2014) Progress on drinking water and sanitation: 2014 update. World Health Organization. https://www.who.int/water_sanitation_health/publications/2014/jmp-report/en/

Williams J, Clarkson JM, Mills PR, Cooper RM (2003) A selective medium for quantitative reisolation of *Trichoderma harzianum* from *Agaricusbisporus* compost. Appl Environ Microbiol 69(7):4190–4191

Xia J, Duan QY, Luo Y, Xie ZH, Liu ZY, Mo XG (2017) Climate change and water resources: case study of eastern monsoon region of China. Adv Clim Chang Res 8(2):63–67

Xiao L, Guan D, Chen Y, Dai J, Ding W, Peart MR, Zhang C (2019) Distribution and availability of heavy metals in soils near electroplating factories. Environ Sci Pollut Res 26(22):22596–22610. https://doi.org/10.1007/s11356-019-04706-0

Yankui T, Yin M, Yang W, Li H, Zhong Y, Mo Liyang Y, Ma X, Sun X (2019) Emerging pollutants in water environment: occurrence, monitoring, fate, and risk assessment. Water Environ Res 91:984–991. https://doi.org/10.1002/wer.1163

Yin XJ, Zhu LG (2019) High-efficiency photocatalytic performance and mechanism of silver-based metal–organic framework. J Mater Res 34(6):991–998

Zahedi A, Greay TL, Paparini A, Linge KL, Joll CA, Ryan UM (2019) Identification of eukaryotic microorganisms with 18S rRNA next-generation sequencing in wastewater treatment plants, with a more targeted NGS approach required for Cryptosporidium detection. Water Res 158:301–312

Zhao K, Fu W, Ye Z, Zhang C (2015) Contamination and spatial variation of heavy metals in the soil-rice system in Nanxun Country, Southeastern China. Int J Environ Res Pub He 12 (2):1577–1594

Chapter 7
Nitrogenous Wastes and Its Efficient Treatment in Wastewater

Parmita Chawley, Krishna Yadav, and Sheeja Jagadevan

7.1 Introduction

Overgrowing population, luxurious lifestyle, and rapid industrialization demand huge supply of both potable and non-potable water. The above mentioned development and lifestyle lead to a scarcity of freshwater supply. Growing imbalance between freshwater availability and consumption by all living organisms necessitates recycling and reuse of wastewater. The need for adequate treatment of wastewater which enters the aquatic ecosystem due to anthropogenic activities from rural and urban sectors is therefore essential. Among various pollutants such as polyaromatic hydrocarbons, polychlorobiphenyls, sulfonamides, pyridine, ammonia, peptides, nitrogenous wastes and phosphates which destabilize the quality of natural water bodies, nitrogenous wastes are of main concern. Inorganic and bioavailable forms of nitrogen such as NO_3^-, NH_4^+, and NO_2^- and the presence of organic nitrogen (urea, protein hydrolysates) originating from decomposition of nitrogenous wastes serve as nutrients in the environment. These are some of the major causes of eutrophication of lakes, ponds, streams, rivers, and other surface water bodies (Kumari et al. 2019). Advances in wastewater treatment (WWT) technologies are crucial for the improvement of environmental quality. Moreover, it is imperative to limit the entry of point sources of chemical contaminants such as industrial discharges into surface waters.

Several kinds of physical, chemical, and biological treatment of wastewater for removal of nitrogenous wastes are in practice. Physical treatment processes, such as membrane filtration, electrodialysis, etc., are the conventionally employed wastewater treatment technologies. Chemical treatment technologies including conventional

P. Chawley · K. Yadav · S. Jagadevan (✉)
Department of Environmental Science and Engineering, Indian Institute of Technology (Indian School of Mines), Dhanbad, Jharkhand, India
e-mail: sheejaj@iitism.ac.in

A. Singh et al. (eds.), *Water Pollution and Management Practices*,
https://doi.org/10.1007/978-981-15-8358-2_7

oxidation processes such as Fenton oxidation, ozonation, photolysis, and advanced oxidation processes (AOPs) are commonly applied. Additionally, biological nitrogen removal techniques involving nitrification, denitrification, anammox, and comammox processes have been found to be suitable (Li et al. 2018; Wang et al. 2018; Capodici et al. 2019; Yan et al. 2019; Cotto et al. 2020). However, there are advantages and disadvantages for every treatment technology, as enlisted in Table 7.1.

Until recently, not only nutrient (nitrogen, phosphorous, etc.) removal processes but also removal of micropollutants, polyaromatic hydrocarbons, etc. have gained much attention. Micropollutants are synthetic organic chemicals found in concentration range from few nanograms per liter to a few hundred micrograms per liter and include pharmaceutical organic contaminants, personal care products, and endocrine-disrupting compounds (Men et al. 2017). Biological treatment processes used for removal of nitrogenous pollutants have a good correlation with removal of these micropollutants and aromatic hydrocarbons (Fernandez-Fontaina et al. 2012; Buitrón et al. 2015; Wang et al. 2019a; Zhou et al. 2019). Simultaneous removal of $NH_4^+ - N$ with micropollutants and organic aromatic pollutants has been reported by ammonia-oxidizing bacteria (AOB), which are responsible for nitrification. AOB are involved in the first step of nitrification wherein it catalyzes the oxidation of NH_4^+ into NO_2^- which is further oxidized into NO_3^- by nitrite-oxidizing bacteria. The first step performed by AOB is regarded as the rate-limiting step of nitrification and, thus, grabs considerable attention in treatment of nitrogenous waste.

This chapter henceforth summarizes various physical, chemical, and biological treatment processes for removal of nitrogenous waste from wastewater. In addition, AOB have been given emphasis on biological removal of nitrogen along with removal of micropollutants and aromatics from wastewater. Apart from the multifunctional role of AOB in wastewater treatment, they have been further recognized with their bio-valorization potential in biofuel production.

7.2 Sources of Nitrogenous Waste in Water

7.2.1 Organic Wastes

Wastewater from food processing industries primarily contributes toward organic nitrogen wastes into wastewater. Fish processing industries mainly liberate proteins, peptides, and volatile amines as nitrogenous wastes (Chowdhury et al. 2010). Mushroom production industries release proteinaceous wastes, carbohydrates, and fats and generate spent mushroom substrate as a byproduct, which is also rich in protein, carbohydrate, and fat (Lou et al. 2017; Meng et al. 2017). In addition, organic nitrogenous wastes are also generated from beverage industries such as dairy processing units and corn steep liquor manufacturing industries. Corn steep liquor contains only 2% protein, but corn gluten water obtained from separation of corn

Table 7.1 Various methods for removal of nitrogenous wastes

Treatment process	Advantages	Challenges	References
Physical			
Reverse osmosis (wind driven)	Capable of removing organic chemicals, protein, and ions from brackish water and seawater	Reduced permeability due to precipitation of salts such as calcium sulfate and calcium chloride, which are responsible for fouling	Afonso et al. (2004) and Tang et al. (2020)
Ion exchange	Removal of other anions along with nitrogenous waste such as nitrate	Extra operational cost to this process	Shrimali and Singh (2001) and Karanasios et al. (2010)
Electrodialysis	Does not need osmotic pressure and has low investment for the certain feed and energy	Only removes ions, not harmful microorganisms and organic compounds	Strathmann (2010)
Chemical			
Fenton oxidation	Degrade the contaminates	Unintended consumption of produced hydroxyl radical	de Luna et al. (2012)
Photocatalysts (TiO_2)	Aqueous medium utilization, mostly as amine neutralization	Low technology readiness level, currently not viable for commercial application	Ghayur et al. (2019)
Ozonation	Showed strong affinity toward emerging contaminates in hydrogen peroxide	Needs 40–50% extra energy demand over conventional wastewater treatment plant	Deegan et al. (2011)
Electrochemical conversion	Aqueous medium utilization, mostly as amine neutralization	High catalyst cost, low technology readiness level	Ghayur et al. (2019) and Meng et al. (2020)
Conventional			
Biological activated carbon	Removal of contaminants from wastewater with less toxic products	Regeneration and disposal of sludge, high operational cost	Rivera-Utrilla et al. (2013)
Microalgae reactor	No acute toxicity, recovery of algal biomass as resource	Cold season has negative impact and reduced removal efficiency	Matamoros et al. (2015)
Activated sludge	Environmentally friendly and have low operational cost	Low removal efficiency for contaminants derived from pharmaceutical waste	Sreekanth et al. (2009)
Nonconventional			
Membrane bioreactor	Suitable for removal of bio-recalcitrant	Consumption of energy and aeration cost were found high, problems with fouling	Deegan et al. (2011)
Constructed wetland	Low operational and maintenance cost, effective for removal of pathogens and pesticides	Dependency on the seasons, clogging, solid entrapment, growth of biofilms, required huge land area	Töre et al. (2012)

starch milk is predominantly proteinaceous (Wang et al. 2016). Moreover, nitrogenous sources exist in the form of hazardous heterocyclic aromatic nitrogen compounds, such as pyridine, quinoline, etc. (Padoley et al. 2008). Coal processing industries generate appreciable fraction of pyridine and quinoline as its derivatives (Shi et al. 2019).

7.2.2 *Inorganic Wastes*

Manure originating from poultry farming and cattle rearing contributes ammonia in wastewater (Wang et al. 2014). Similarly, aquaculture wastewater also carries nitrogen in the form of NH_3, NO_2^-, and NO_3^- coming out from the excreta of fisheries and from decomposition of dead fishes (Yin et al. 2018;Li et al. 2019). Some fish processing industries also reported the presence of ammonia in fishery wastewater (Technical Report Series FREMP 1994). Agro-based industries and olive mill wastewater contribute to significant amount of total nitrogen in wastewater (van der Wiel et al. 2019; Achak et al. 2019). Coke oven wastewater generated from steel industries is also rich in ammonia, in addition to toxic pollutants such as phenol and cyanide (Tyagi et al. 2018).

7.3 Conventional Processes for Removal of Nitrogenous Waste

Finding an appropriate technique for removal of nitrogenous waste produced from wastewater has become a key concern across the world. When these compounds reach water bodies, it leads to eutrophication and inhibits growth of microorganisms. Therefore, there is need to explore cost-effective removal techniques for these nitrogenous wastes.

7.3.1 *Physical Processes*

Nitrogenous waste present in water and wastewater can be removed by physical, chemical, and biological processes. Physical processes include filtration techniques such as reverse osmosis (Afonso et al. 2004), ion exchange (Karanasios et al. 2010), and electrodialysis (Strathmann 2010; Abdel-Shafy et al. 2016; Quist-Jensen et al. 2017). The aforementioned processes involve transfer of nitrogenous contaminants from one phase to another without undergoing any conversion or decomposition of the parent material. Physical processes are applicable to low molecular weight nitrogenous wastes, and allows the pollutant or material to be recovered (Sohni et al. 2018).

7.3.1.1 Reverse Osmosis

To meet the increasing demands of water supply, people all around the world are looking for utilization of wastewater by improving the quality of water through reverse osmosis (RO). Reverse osmosis offers a great capacity to remove organic nitrogen (Merlo et al. 2012), proteins, and ions from brackish water and seawater (Afonso et al. 2004). However, this process produces RO concentrate (ROC), containing several contaminants and nutrients. Nearly, 30–50% of RO input is left as ROC in municipal reclamation system (Dialynas et al. 2008), since ROC derived from municipal wastewater contains nitrogen as the primary nutrient. In spite of direct disposal of ROC, zero liquid discharge (recovery of feed water up to 95–98%) could be one suitable approach to reduce volume of wastewater. A typical ROC with high amount of total nitrogen (TN) and low C/N ratio was utilized in membrane-aerated reactor to check the nitrogen removal efficiency. Results showed that TN removal percentage was 79.2 with 5.8 C/N ratio, 24-h retention time, and 0.02 MPa aeration pressure (Quan et al. 2018).

RO was performed for removal of nitrogen (TN) from domestic wastewater and in combination with industrial wastewater by tubular and spiral-wound membrane elements. The separation efficiency of total nitrogen from domestic wastewater through RO approach (tubular membrane) was found to be 95% (Bilstad 1995). Tubular membrane did not require any pretreatment, while spiral membrane element necessitated an additional 25–200 micrometer cartridge filter as pretreatment (Bilstad 1995). In another study, higher concentration of nitrate-nitrogen (35–75 mg/L) in water was reduced to level of potable water having concentration less than 10 mg/L with the help of reverse osmosis. In this study, two sets of initial concentration of nitrate-nitrogen were investigated: (i) low to medium (35–43 mg/L) and (ii) medium to high (54–72 mg/L). Better removal was reported in the first condition with initial concentration of nitrate-nitrogen reduced to 1.4–5.5 mg/L, while in the second case, it was found to be 12–17 mg/L in treated water (Schoeman 2009).

The intrinsic properties such as high efficiency of membrane in selective mineral rejection, permeability to water, and ease of operation at room temperature make it a suitable choice for WWT (López-Ramírez et al. 2006). However, the operational cost for reverse osmosis is found to be high owing to its high energy consumption for huge project. This problem can be resolved by introducing renewable energy sources such as solar, wind, and wave energy (Liu et al. 2002, 2007). Studies have reported wind-driven reverse osmosis system to be capable of removing nitrogen from aquaculture wastewater (Liu et al. 2007). The nitrate removed by such process however gets accumulated in the brine system, thus defeating the purpose for installation of such a hybrid process (Matos et al. 2009). In addition, permeability of the membrane is compromised due to precipitation of salts (calcium sulfate and calcium chloride), thereby leading to fouling (Hasson et al. 2001).

7.3.1.2 Ion Exchange

During ion-exchange process, reversible interchange of ions between two phases, i.e., solid and liquid, takes place in which moving ions of ion-exchange membrane get exchanged with the adjacent medium (Bashir et al. 2010). Ion-exchange resins have been used for treatment of wastewater consisting of different functional groups and ions. When nitrogenous waste derived from landfill leachate was passed through natural zeolite, Dowex 50W-X8, and Purolite MN 500 resin beads, enhanced ammonium ion uptake was observed on addition of Na^+ ion and citric acid (Jorgensen and Weatherley 2003). Results indicated that the uptake of ammonium was highest with Dowex 50W-X8 followed by Purolite MN 500 and lowest in case of natural zeolite (Jorgensen and Weatherley 2003). This process can also remove sulfate along with nitrate ions, prevalently found in fertilizer wastewater (Leaković et al. 2000). Furthermore, exhausted resin can be regenerated with NaCl by displacing negatively charged ions to chloride ions (Roquebert et al. 2000). However, the discharge of toxic ions such as calcium, phosphate, magnesium, and fluoride should be taken into account, as it contributes to extra operational cost in this process (Shrimali and Singh 2001; Karanasios et al. 2010).

7.3.1.3 Electrodialysis

Electrodialysis (ED) is a membrane technique used to separate charged ions from aqueous solution in the presence of electric potential gradient, which provides driving force to separate the ions. Electrodialysis has been employed in large scale (>20,000 $m^3\ d^{-1}$) and found to be the most suitable technique for desalination and demineralization of brackish and industrial process water (Strathmann 2010). Ammonia-nitrogen-containing wastewater was applied to electrodialysis system, and the performance of the system was evaluated through different factorially designed experiments (Song et al. 2012). The experimental results showed that optimum condition for ED, i.e., operation voltage and concentration ratio of two cells, were obtained as 12 V and 1:3, respectively. In addition, removal of ammonia-nitrogen was also influenced by higher feed concentration of stream (Song et al. 2012). Electrodialysis can be a suitable option for low to high concentration of nitrate-nitrogen in rural area, but maintenance and operation of electrodialysis are difficult tasks (Schoeman 2009).

Electrodialysis can also be used in nutrient recovery from municipal wastewater. The total concentration of ammonia-nitrogen was found to be 7100 ± 300 mg/L from concentrated product derived from wastewater dilute feed stream to product stream (Ward et al. 2018). In another study, electrodialysis was employed to treat university sewage water. The study was conducted to monitor the behavior of membrane with respect to ammoniacal nitrogen. Results regarding removal of ammoniacal nitrogen and total nitrogen were found to be 95–98% and 80–99%, respectively (Albornoz et al. 2019). Additionally, electrodialysis requires very few pretreatments to reduce fouling of membrane as compared to reverse osmosis and thus owing to reverse polarity operation (Strathmann 2010). This technique does not need osmotic pressure and has low investment in terms of feed and energy. The main drawback of this method is that it removes only ions and is not effective against microorganisms and organic compounds which have been produced during this process. In addition, when the salt concentration of effluent is increased, it involves high energy investment.

7.3.2 Chemical Processes

Chemical methods involve complete mineralization and degradation of the nitrogenous waste and hence are deemed to be robust, reliable, and effective. Methods such as advanced oxidation processes (AOPs) are increasingly becoming popular these days. Chemical species such as ozone, chlorine, and hydrogen peroxide are commonly employed as an oxidant, and AOPs such as Fenton oxidation and electrochemical oxidation involve a combination of one or more of these oxidants with transition metals or metal oxide. The primary goal of chemical oxidation process is to mineralize the contaminants and convert it into water, nitrogen, and other minerals. To run the oxidation processes, energy source such as electric current, gamma radiation, ultrasonic, and UV-Vis radiation is needed (Ikehata et al. 2008). The aforementioned processes are commonly used for the treatment of wastewater.

7.3.2.1 Fenton Oxidation

Contaminants derived from pharmaceutical, pesticide, tannery, and textile industries are having ample amount of ammoniacal nitrogen, which can be effectively transformed to nitrogen and nitrate with the help of the Fenton process (Dantas et al. 2003). Ammoniacal nitrogen is one of the primary pollutants present in tannery wastewater and its concentration in untreated effluent is higher than 100 mg/L (Wang et al. 2012). Fenton oxidation is an oxidation process involving the formation of hydroxyl radical from a reaction between iron and hydrogen peroxide, in which ferrous can be regenerated as depicted by Eqs. 7.1 and 7.2 (Shemer et al. 2006). Reaction (7.2) being very slow results in the accumulation of ferric (Fe^{3+}) in the solution, which further precipitates as $Fe(OH)_3$ (de Luna et al. 2012). This inherently leads to reduction in the treatment efficiency due to the removal of iron from the solution. Moreover, it entails significant amount of addition of the reagent, resulting in enhanced operational cost.

$$Fe^{2+} + H_2O_2 \rightarrow Fe^{3+}OH^{\bullet} + OH^{-} \quad (7.1)$$

$$Fe^{3+} + H_2O_2 \rightarrow Fe^{2+} + O_2H^{\bullet} + H^{+} \quad (7.2)$$

Unintended consumption of hydroxyl radicals by ferrous ions and hydrogen peroxide (Eqs. 7.3 and 7.4) could be another limitation of the Fenton process.

$$OH^{\bullet} + Fe^{2+} \rightarrow Fe^{3+} + OH^{-} \quad (7.3)$$

$$OH^{\bullet} + H_2O_2 \rightarrow H_2O + O_2H^{\bullet} \quad (7.4)$$

Additionally, the efficiency of the Fenton process could also be hindered at higher reagent concentrations due to the presence of weaker $O_2H^{\bullet}$ as compared to $OH^{\bullet}$ radical.

7.3.2.2 Ozonation

Ozonation is a form of advanced oxidation process, which includes direct reaction of the contaminants with ozone and a secondary oxidant (hydroxy radical) in aqueous medium (Maldonado et al. 2006; Esplugas et al. 2007). Ozonation has been a major tool in enhancing biodegradability and efficiency of oxidation process. This process can remove all kinds of emerging contaminants up to 90–100%. Production of ozone requires huge energy, sometimes 40–50% extra energy over conventional wastewater treatment methods, thus making it an economically nonviable method (Deegan et al. 2011). In this method, low concentrated ammonia can be converted to nitrate with the help of ozonation; the reaction can be represented as:

$$NH_3 + 4O_3 \rightarrow H^{+} + NO_3 + H_2O + 4O_2 \quad (7.5)$$

Since the lifetime of ozone in water is very short, its rate of reaction with ammonia is also very slow (Capodaglio et al. 2015). An advanced oxidation process consisting of two oxidants, i.e., ozone and hydrogen peroxide, was found suitable for removal of ammonia from wastewaters, and it was observed that ozone molecules were involved in ammonia oxidation through direct oxidation (Kuo et al. 1997). Furthermore, the kinetics for the hybrid treatment (ozone and hydrogen peroxide) in alkaline condition was observed to be of second order.

Another study showed enhanced removal efficiency of ammonia during ozonation process (Khuntia et al. 2013). In this process, the ozonation unit is assembled with microbubbles which provide high gas–liquid interfacial area and longer persistence, and lesser amount of ozone is required because hydroxyl radical is also engaged in pollutant oxidation (Khuntia et al. 2013). Generation of hydroxyl radical in acidic condition was also reported by Li et al. 2009, which facilitated enhanced oxidation of ammonia owing to the presence of these radicals. Moreover, ammonia could be effectively oxidized even at low concentration in alkaline conditions too.

Hence, microbubble-added ozonation process can be employed to treat wastewater at large scale because this process is fast and suitable and can be used for a wide range of pH. Lastly, after completion of ozonation, ozone self-decomposes to oxygen without adding any other pollutant to water.

7.3.2.3 Electrochemical Oxidation

Electrochemical oxidation, also called as anodic oxidation, is a versatile technique to remove contaminants from industrial wastewater due to its ease in operation, lack of secondary contaminant generation, and better treatment efficiency (Song et al. 2019). Prime benefits of this technique include the following: (i) no requirement of extraneous addition of chemicals, (ii) can take place in a simple electrochemical cell, and (iii) at ambient temperature and pressure conditions. Several recent studies have reported electrochemical oxidation of model pollutants from synthetic wastewater (Panizza and Cerisola 2008; Lizhang et al. 2016; Cotillas et al. 2018). Additionally, this technique has been employed to treat real wastewater which consists of multiple contaminants and complex compounds.

Electrochemical oxidation degrades pollutants either through direct or indirect oxidation (Garcia-Segura et al. 2018). In case of direct oxidation process, pollutants oxidize at the surface of electrode through electron transfer mode, while indirect oxidation involves oxidizing agents such as hydroxyl radicals, produced during electrolysis. Through electrochemical oxidation process, ammonia can be converted into other forms of nitrogen either by direct or indirect oxidation as presented in Eqs. 7.6, 7.7, and 7.8.

$$2NH_{3(aq)} + 6OH^- \rightarrow N_2 + 6H_2O + 6e^- \text{ (Anode reaction)} \tag{7.6}$$

$$6H_2O + 6e^- \rightarrow 3H_2 + 6OH^- \text{ (Cathode reaction)} \tag{7.7}$$

$$2NH_{3(aq)} \rightarrow N_2 + 3H_2 \text{ (Overall reaction)} \tag{7.8}$$

From the above equations, it was found that direct oxidation has many advantages but application of this process at large scale can reduce the performance of the system. On the other hand, indirect oxidation process involves production of strong oxidizing agents such as HClO, H_2O_2, and $OH^{\bullet}$ during electrochemical reaction (Liu et al. 2019), which act as scavenger of ammonia (Xiao et al. 2009; Capodaglio et al. 2015).

Another study has reported enhanced removal of NH_4^+-N ions from lead smelting wastewater via electrochemical oxidation coupled with coagulation-flocculation technique by employing a low-cost graphite anode (Meng et al. 2020). The results indicated reduced concentration of NH_4^+-N ions by converting ammoniacal nitrogen to nitrogen and organic carbon to carbon dioxide, thereby meeting the emission standard of wastewater. Furthermore, removal of organic and ammonium-nitrogen from swine wastewater was carried out using the aforementioned technique. Results suggested that removal of ammonium-nitrogen was favored in the presence of sodium chloride and retarded in the absence of sodium chloride (Huang et al. 2018).

7.4 Biological Treatment Processes

7.4.1 Anammox

Anammox is a chemoautotrophic biological process which involves conversion of ammonium (NH_4^+) into dinitrogen (N_2) gas by anaerobic ammonia-oxidizing bacteria by utilizing nitrite (NO_2^-) as the electron acceptor, as indicated in Eq. (7.9) (Qi et al. 2018).

$$NH_4^+ + 1.32NO_2 + 0.066HCO_3 + 0.13H^+ \rightarrow 1.02N_2 + 0.26NO_3 + 0.066H_2CO_{0.5}N_{0.15} + 2.03H_2O \quad (7.9)$$

Anammox chemolithotrophy is performed within specialized organelle called as anammoxosome. Anammox process is becoming an attractive technique in wastewater treatment, as it simultaneously removes ammonia and nitrite with no or limited generation of greenhouse gas such as N_2O and exerts lower oxygen demand and negligible sludge generation (Li et al. 2018). Anammox bacteria show nitrogen removal activity at varied temperature range 10–35 °C (Wang et al. 2018). For the enriched anammox biomass, a specific activity of 30–44 mg N/gram of volatile solids of sludge/day was obtained when the reactor was fed with 61 mg (NH_4^+ + NO_2^-)-N/L, at 10 °C containing *Candidatus Brocadia fulgida* as the dominant species (Hendrickx et al. 2014). It has been mostly studied in treatment of nitrogen-rich saline wastewater. Saline wastewater generated from sea food processing, underground brine wastewater from gas field, mustard pickling industry, etc. is rich in nitrogen (Shinohara et al. 2009; Chen et al. 2015; Picos-Benítez et al. 2019). Anammox bacteria are enriched from two sources: brackish or marine sediments called marine anammox bacteria (MAB) and freshwater-derived anammox bacteria (FAB). MAB could survive under a salinity significantly greater than 30 g/L and able to produce N_2 from nitrogen removal up to 70% (Li et al. 2018). FAB was found to be inefficient in nitrogen removal at lower temperatures and high saline conditions than MAB (Kawagoshi et al. 2012; Zhu et al. 2017). High nitrogen removal rate (NRR, 10.7 kg N/(m^3.d)) was acquired with MAB (Yokota et al. 2018). Several reactors such as sequencing batch reactor (SBR), upflow anaerobic sludge blanket reactor (UASB), and fixed bed biofilm reactor (FBBR) have been used for enrichment of FAB and MAB to achieve nitrogen removal from saline wastewater (Li et al. 2018). Several limitations such as high sensitivity to external environment, weak tolerance capacity to substrate concentration, long doubling time, lesser cell yield, and difficulty to maintain anoxic environment pose hindrance toward further application of anammox process (Li et al. 2018; Qi et al. 2018).

7.4.2 Nitrification

Biological nitrification is a two-step oxidation process. The first step is the oxidation of ammonia into nitrite by ammonia-oxidizing bacteria (AOB), and the second step involves oxidation of nitrite into nitrate by nitrite-oxidizing bacteria (NOB).

$$NH_3 + 1.5O_2 \rightarrow NO_2^- + H_2O + H^+ \text{ (AOB)} \tag{7.10}$$

$$NO_2 + 0.5O_2 \rightarrow NO_3^- \text{ (NOB)} \tag{7.11}$$

Oxidation of ammonia into nitrite by AOB is also called as partial nitrification (PN). Ammonia removal from wastewater through PN process has been successfully achieved for wastewater treatment containing high nitrogen concentration or low carbon/nitrogen (C/N) ratio, such as municipal wastewater, landfill leachate, anaerobic sludge digester liquor, etc. (Wang et al. 2010; Ge et al. 2014; Zhang et al. 2016; Capodici et al. 2019). Several factors such as dissolved oxygen (DO), pH value, free ammonia, and free nitrous acid affect the PN process (Wei et al. 2015). Up to 1400 mg/L ammonium was efficiently removed by an air lift reactor through PN system (Chai et al. 2015). Ammonia removal has been also achieved at temperature as low as 14 °C when 65 mg/L ammonia was present in activated sludge (Zhang et al. 2016). However, removal of nitrogen compound in conventional WWT process is limited by slow growth of nitrifying bacteria on suspended systems. As compared to suspended growth of AOB, biofilm system is more stable and protects slow-growing nitrifying bacteria from washout in its competition with heterotrophic bacteria. Therefore, various biofilm systems such as sequencing batch reactor (SBBR), moving bed biofilm reactor (MBBR), and fixed bed biofilm reactor (FBBR) have been applied for treatment of nitrogen-rich wastewater (Cruvellier et al. 2017; Wei et al. 2017; Zheng et al. 2019; Ashkanani et al. 2019). An average ammonia removal efficiency of 98.2% was achieved in SBBR through PN process when 600 mg/L ammonia was present (Wei et al. 2017). MBBRs are also effective in treatment of ammonia-rich wastewater under different working conditions and a wide range of temperatures (1–21 °C) (Ashkanani et al. 2019). Ammonia concentration as high as 676 ± 195 mg/L was treated in an MBBR pilot-scale plant for 211 days, accounting for 38.6 ± 14.8% ammonia removal (Abzazou et al. 2016). A combination of SBR and MBBR has also been employed to remove total nitrogen from wastewater, with 50–93% removal efficiency (Ferrentino et al. 2018). Pure culture of AOB such as *Nitrosomonas europaea* and *Nitrobacter winogradskyi* has been employed for removal of ammonia load up to 2.5 kgN/m^3.d in a fixed bed bioreactor (Cruvellier et al. 2017).

A hybrid treatment system of a combination of partial nitrification with anammox process has been investigated for the removal of nitrogenous wastes (Ma et al. 2019; Zhang et al. 2019). This combined system was able to remove NH_3-N without organic carbon consumption and consumes less oxygen and causes less production of CO_2, N_2O and sludge than the traditional process.

7.4.3 Denitrification

Biological denitrification removes NO_3^--N from polluted water. It is performed by heterotrophic facultative microorganisms that follow sequential reduction of NO_3^- to N_2, with NO_2^-, NO and N_2O being the most common intermediates, in an anaerobic environment (Eqs. 7.12, 7.13, 7.14, and 7.15).

$$NO_3^- + 2e^- + 2H^+ \xrightarrow{\text{nitrate reductase}} NO_2^- + H_2O \tag{7.12}$$

$$2NO_2^- + 2e^- + 4H^+ \xrightarrow{\text{nitrite reductase}} 2NO + 2H_2O \tag{7.13}$$

$$2NO + 2e^- + 2H^+ \xrightarrow{\text{nitric oxide reductase}} N_2O + H_2O \tag{7.14}$$

$$N_2O + 2e^- + 2H^+ \xrightarrow{\text{nitrous oxide reductase}} N_2 + H_2O \tag{7.15}$$

Denitrification is dependent on three basic factors – dissolved organic carbon, pH, and temperature. This biochemical process also requires availability of electron donors which is supplied by oxidation of carbon source. Dissolved organic carbon sources provide electron donors that get consumed during sequential reduction of nitrate to N_2 gas. The optimal pH and temperature for biological denitrification are 7–9 and 20–30 °C, respectively (Lu et al. 2014). Complete denitrification with reduced nitrite formation and reduction in emission of N_2O gas presents a challenging task to be addressed in the future. For efficient removal of nitrogen from wastewater, a combination of nitrification, denitrification, and anammox process is employed in WWT plants.

Simultaneous nitrification and denitrification (SND) is an efficient technique for treatment of wastewater with low C/N ratio. It allows both processes to occur in the same compartment which is facilitated by growth of both nitrifying and denitrifying bacteria along an oxygen gradient in biofilm structures or dissolved oxygen levels. This process is able to maintain neutral pH as alkalinity is consumed during nitrification and produced during denitrification. It eliminates sludge recycling and also abolishes the need to add carbon supply for denitrification process to occur. It is therefore an economical option for removal of NH_4^+-N from wastewater. SND systems have been successfully used for treatment of ammonia-rich wastewater, though the exact mechanism of this system is not fully understood. The removal efficiencies of NH_4^+-N and total nitrogen in an SBR have been reported to be 97.91 ± 2.04% and 72.28 ± 2.23%, respectively, by maintaining low DO (dissolved oxygen) levels (0.7 ± 0.1 mg/L) (Yan et al. 2019). Pure culture of novel isolated bacterium *Ochrobactrum anthropic* LJ81 has been found to be capable of converting 80% of NH_4^+-N into N_2 gas through SND process (Lei et al. 2019).

Nowadays, partial nitrification is being coupled with anammox process to enhance nitrogen removal efficiency (Ma et al. 2019). This combined system is particularly useful for nitrogen removal at places where denitrification is not possible due to the presence of lower biodegradable organic matters. In order to allow denitrification to occur at such places, external carbon supply would be needed which is not economically feasible. Coupled partial nitrification and anammox process have been successfully used for nitrogen removal from mature landfill leachates as it contains less biodegradable organic matters (Sun et al. 2016; Wen et al. 2016). This system also helps to promote enrichment of anammox microorganisms by the activity of AOB. Nitrite produced by AOB during partial nitrification is used up by anammox microbial communities and thus further helps to enhance anammox bacterial activity (Miao et al. 2020).

Hybrid treatment strategies, where advanced oxidation processes have been coupled with anammox system, have also been studied to enhance nitrogen removal efficiency. In a mature landfill leachate site, ozonation and photo-Fenton oxidation process coupled with anammox system was found to be capable of removing nitrogen with 87% efficiency (Anfruns et al. 2013). In another study, nano zero-valent iron (nZVI) and ferroferric oxide (Fe_3O_4) have been found to enhance formation of anammox granulation and facilitate anammox biofilm formation (Gao et al. 2014). Low dose of nZVI promotes anammox activity by dissolved oxygen depletion, but nZVI at 75 mg/L concentration has been found to be inhibitory to anammox activity (Yan et al. 2019). Free Fe^{2+}/Fe^{3+} released from corrosion of nZVI regulate intracellular Fenton oxidation reaction. Furthermore, they are important components of heme-containing enzymes such as cytochrome c proteins, which play a key role in anammox catabolism.

7.5 Ammonia-Oxidizing Bacteria: The First-Stage Nitrifying Microorganism

In nature, ammonia-oxidizing bacteria (AOB), ammonia-oxidizing archaea (AOA), anammox (anaerobic ammonia oxidizer), and comammox (complete ammonia oxidizer) microorganisms are known to oxidize NH_3. Among these microorganisms, several strains of AOB have been isolated and cultured as pure cultures, for example, *Nitrosomonas europaea*, *Nitrosomonas eutropha*, *Nitrosospira multiformis*, etc. Only recently, a pure culture of comammox has been isolated, namely, *Nitrospira inopinata* (Kits et al. 2019). Though several ammonia oxidizers occur in nature, an understanding of biochemical reactions of ammonia oxidation is known only for AOB. Out of several known pure culture of AOB, *N. europaea* is the most widely studied. They are predominantly chemolithotrophic bacteria, obtaining energy from inorganic source (NH_3) and assimilate carbon from CO_2. However, growth on organic carbon has been found in *N. europaea* in the presence of fructose and pyruvate (Hommes et al. 2003). AOB are very slow growing, with doubling time of 7–24 h (Thandar et al. 2016). Morphologically, they bear extensively thick intracytoplasmic membrane (ICM) which is stacked in layers (Fig. 7.1). These thick ICM layers harbor the enzyme responsible for ammonia oxidation, i.e.,

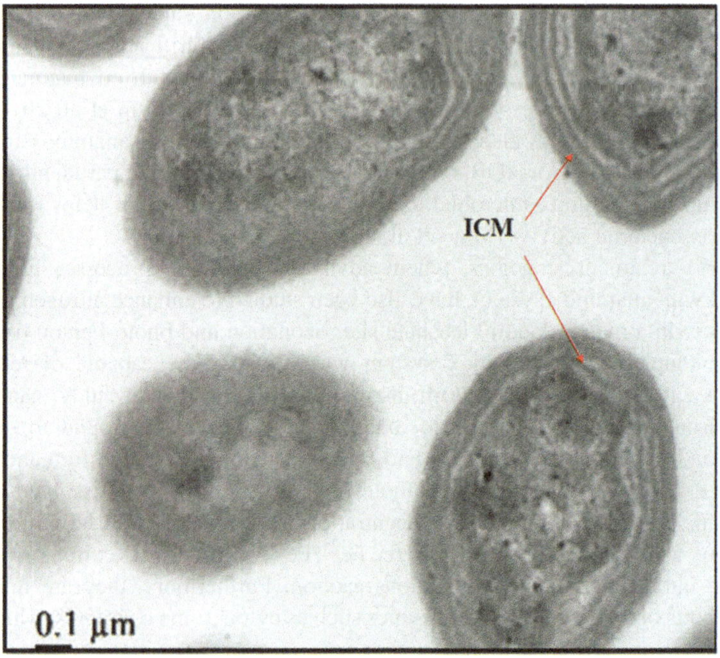

Fig. 7.1 Transmission electron microscope image of *Nitrosomonas europaea* (Wu et al. 2018)

ammonia monooxygenase (AMO) (Suzuki et al. 1974). The basic catabolic reaction carried out by AOB is represented in Eqs. 7.16, 7.17, 7.18, and 7.19.

$$NH_3 + O_2 + 2H^+ + 2e^- \overset{AMO}{\rightarrow} NH_2OH + H_2O \quad (7.16)$$

$$NH_2OH + H_2O \overset{HAO}{\rightarrow} NO_2^- + 5H^+ + 4e^- \quad (7.17)$$

$$HCO_3^- + xe^- \rightarrow \text{Biomass} \quad (7.18)$$

$$0.5O_2 + 2H^+ + (2 - xe^-) \rightarrow H_2O \quad (7.19)$$

The first step in nitrification is catalyzed by AMO enzyme which oxidizes ammonia into hydroxylamine (NH_2OH) with the requirement of two reducing equivalents and concomitant reduction of one mole of O_2 into H_2O. The second step is catalyzed by hydroxylamine oxidoreductase (HAO) enzyme which oxidizes hydroxylamine formed in the first step to nitrite, with the release of four electrons. This reaction catalyzed by HAO is the only known source of generation of electrons in AOB cellular biochemical reactions. HAO is a homotrimeric multi-subunit periplasmic enzyme which requires eight c-type hemes (Arp et al. 2002). Out of four electrons released by HAO, two electrons flow back to AMO, x electrons are utilized during autotrophic biomass accumulation, and ($2-xe^-$) electrons further reduce 0.5 mole of O_2 into H_2O. Thus, 1 mole of oxidation of NH_3 consumes 1.5 moles of O_2 to produce 1 mole of NO_2^- (Eq. 7.20).

$$NH_3 + 1.5O_2 \rightarrow NO_2^- + H_2O + H^+ \tag{7.20}$$

AOB have been found to be dominant ammonia-oxidizing microorganisms in agricultural soil (Ouyang et al. 2016; Norman and Barrett 2016). There exists a strong relationship between soil pH and availability of NH_3 (Eq. 7.21). pH strongly affects the ionization of NH_4^+/NH_3, thereby affecting the availability of NH_3 as described by Eq. 7.21 (a modification of the Henderson-Hasselbalch equation that assumes a pKa value of 9.25 for the ionization of NH_4^+/NH_3) (Norman and Barrett 2016):

$$[NH_3] = [NH_4^+]\left[10^{(\text{soil pH}-9.25)}\right] \tag{7.21}$$

AOB find importance in the treatment of nitrogen-rich wastewater because they are the only group of culturable microorganisms that performs the first and rate-limiting step in nitrification. AOB are abundantly found in domestic wastewater at high dissolved oxygen and influent ammonium concentration of 36.1–422.3 mg/L nitrogen (Kim et al. 2013). Though several studies have reported existence of both AOB and AOA in wastewater treatment plants, AOB have been found to be the major contributor to ammonia oxidation in highly aerated activated sludge system (Liu et al. 2016; Islam et al. 2019). They are also valuable in the treatment of swine wastewater which is rich in NH_4^+ (Zhang et al. 2017). Ninety-eight percent NH_4^+ removal and 96% total nitrogen removal efficiency has been achieved from swine wastewater in integrated constructed wetland on account of microbial action of both AOB and AOA.

Biological WWT system further allows energy generation and resource recovery from wastewater. BES (bioelectrochemical systems) associated with oxidation-reduction reactions in microorganisms are an effective way to produce energy while simultaneously consuming organic carbon in wastewater. Microbial fuel cell is one such technology which helps to treat wastewater and simultaneously generate electricity out of organic materials present in it (Das et al. 2018; Das and Ghangrekar 2019). Some research evidences have shown nitrogen removal coupled with electricity production using AOB as well as anammox microorganisms (He et al. 2009; Di Domenico et al. 2015). Nutrient recovery from wastewater is another feasible option available along with biological WWT. Approximately, 20% phosphorous is present in domestic wastewater (Batstone et al. 2015). Swine wastewater is rich in both nitrogen and phosphorous, the later can be found in the range of 100–1400 mg/L phosphorous, which has been recovered with 97.69% efficiency in the form of struvite (Wang et al. 2019b).

7.6 Role of Ammonia-Oxidizing Bacteria (AOB) in Wastewater Treatment

7.6.1 Removal of Micropollutants

Micropollutants (MPs) are toxic compounds that are present in nanogram to microgram per liter range in an engineered system. AOB are capable of removal of MP by the action of ammonia monooxygenase enzyme via co-metabolism. Co-metabolism involves simultaneous transformation of a nongrowth substrate as well as growth substrate by the same enzyme, wherein the growth substrate further enters in the metabolic pathway to produce energy. The removal of MP (nongrowth substrate) by AOB is thus dependent on energy produced by the growth substrate (–NH_3). Low substrate specificity and broad substrate spectrum allow co-metabolism of MPs by AMO (Rasche et al. 1991). Simultaneous oxidation of both growth and nongrowth substrate by AMO results in competitive inhibition of the primary growth substrate, i.e., NH_3. Biotransformation of MPs by AMO causes toxicity to cells, resulting in loss of AMO activity. The removal of MPs therefore depends on the extent of competitive inhibition to ammonia oxidation and the ability of cells to overcome toxic effects of MPs (Yang et al. 1999).

The biotransformation of MPs follows either first-order or pseudo first-order kinetics, and the transformation was found to increase with addition of nitrogen load to nitrifying sludge (Fernandez-Fontaina et al. 2012; Dawas-Massalha et al. 2014; Xu et al. 2016; Zhou et al. 2019). However, extremely higher ammonia concentrations competitively inhibit biotransformation until all ammonia gets depleted (Dawas-Massalha et al. 2014). A particular ratio of NH_3/MP is therefore necessary to be maintained to bring about the removal of MP. Table 7.2 lists micropollutants which have been biologically transformed or removed by the activity of AOB.

7.6.2 Removal of Aromatic Pollutants

Removal of aromatic pollutants through AOB has been achieved due to the broad-spectrum specificity nature of AMO enzyme, although co-oxidation of aromatic compounds and ammonia by AMO inhibits ammonia oxidation by varying extent. Inhibition in ammonia oxidation arises possibly because of competition between ammonia and other compounds for reactive oxygen species, availability of reducing power, and other growth medium constituents. However, ammonia oxidation activity has been found to be restored to normal levels when pure culture of AOB such as *N. europaea* cells were resuspended in fresh culture medium without aromatics (Keener and Arp 1994). *N. europaea* has been extensively studied for its ability to transform benzene and substituted benzenes such as ethylbenzene, styrene, halobenzenes, pyridine, phenol, p-cresol, o-cresol, etc. (Keener and Arp 1994).

Table 7.2 Lists of micropollutants biologically degraded by activity of ammonia-oxidizing bacteria

Micropollutant	Degraded by	References
Sulfonamides: sulfadiazine, sulfamethazine, sulfamethoxazole, sulfadoxine, sulfamerazine, sulfamonomethoxine, sulfathiazole	*Nitrosomonas nitrosa* Nm90, nitrifying sludge	Zhou et al. (2019)
Sulfamethoxazole	Nitrifying activated sludge, enriched AOB	Men et al. (2017) and Kassotaki et al. (2016)
Cephalexin	Enriched nitrifying sludge	Wang et al. (2019a)
Diclofenac	Enriched nitrifying sludge	Wu et al. (2019)
Atenolol	Enriched nitrifying sludge	Xu et al. (2017)
17α-Ethinylestradiol, estriol, 17-estradiol	*Nitrosomonas europaea*	Khunjar et al. (2011) and Shi et al. (2004)
Carbamazepine, iopromide, trimethoprim	Nitrifying biomass	Khunjar and Love (2011)
Tetracyclines	Nitrosifying sludge	Sheng et al. (2018)
Quinoline	Ammonia-oxidizing bacteria	Fu and Zhao (2015)
Triclosan, bisphenol A	*Nitrosomonas europaea*	Roh et al. (2009)

Nitrosomonas eutropha C91 was shown to degrade p-cresol (Kjeldal et al. 2014), and *N. europaea* cells have been shown to transform a wide range of polycyclic aromatic hydrocarbons such as naphthalene, 2-methylnaphthalene, acenaphthalene, acenaphthene, fluorene, anthracene, and phenanthrene (Chang et al. 2002).

In WWT process, a mixed nitrifying culture was reported to remove p-cresol along with ammonium oxidation (Beristain-Cardoso et al. 2011). Efficient consumption of toluene at concentration of 20 mg/L has been reported upon long-term exposure to nitrifying sludge (Youssef et al. 2013). Biodegradation of p-nonylphenol has been achieved through nitrifying sludge in a bioreactor which showed that p-nonylphenol biodegradation is related to ammonium oxidation activity (Buitrón et al. 2015). Consumption of 2-chlorophenol by nitrifying consortium was also reported in WWT process (Martínez-Jardines et al. 2018).

7.6.3 Production of Alternative Liquid Fuel

Apart from utility in wastewater treatment, AOB find application in the production of alternative liquid fuel (methanol) by virtue of its bio-valorization potential. In nature, two groups of bacteria are capable of producing methanol: (i) aerobic methanotrophic bacteria, which uses CH_4 as their source of carbon and energy, thereby oxidizing methane into methanol by methane monooxygenase (MMO) enzyme (Jagadevan and Semrau 2013), and (ii) ammonia-oxidizing bacteria, which can partially oxidize CH_4 to methanol by using ammonia as an energy source (Taher and Chandran 2013). Besides these, some methane-oxidizing archaea are also found to carry out anaerobic oxidation of methane (AOM) (Hanson and Hanson 1996; Boetius et al. 2000; Beal et al. 2009; Ge et al. 2014).

Ammonia monooxygenase (AMO) encoded by ammonia-oxidizing bacteria (AOB) is another enzyme that can oxidize methane to methanol. Both MMO and AMO enzyme are evolutionary related (Tavormina et al. 2011; Lawton et al. 2014). AMO enzyme possesses a broad range specificity of substrates such as methane, methanol, ethylene, methyl bromide, etc. (Taher and Chandran 2013). Hence, AOB are capable to produce alternative liquid fuel (methanol) from methane oxidation, while using NH_3 as energy source. The requirement of reducing power is essential for this conversion. The pioneering work for methane as an alternative substrate for AOB has been reported by Hyman and Wood (1983). They reported that CH_4 inhibits ammonia consumption by *Nitrosomonas europaea* with inhibition constant of 2 mM. In addition, methane oxidation by pure culture of AOB such as *Nitrosomonas europaea* and *Nitrosomonas oceanus* have been studied previously (Jones and Morita 1983; Hyman and Wood 1983). *N. europaea* resists the inhibitory effect of CH_4 up to 1 mM concentration, whereas *N. oceanus* was inhibited by less than 0.1 mM CH_4 (Jones and Morita 1983).

N. europaea was investigated in the production of methanol in biofilm system with maximum yield of 0.09 mg methanol/mg biomass (VS)/d (Thorn 2006). There have been some evidences of oxidation of methane by nitrifying bacteria consisting of *Nitrosomonas* spp. in a mixed consortium. For example, *Nitrosomonas* species have been found in plastic waste biodegradation site where synthetic biogas has been artificially pumped to promote methane oxidation (Muenmee et al. 2015). *Nitrosomonas* and *Nitrosospira* species have been found in oxidation of dissolved methane which is produced from anaerobic digestion of wastewater (Hatamoto et al. 2011). In a recent work, a maximum of 1.6 ± 0.15 mg-COD CH_3OH L^{-1} was reported to be produced in a continuous stirred tank reactor (CSTR) at hydraulic retention time of 7 h by AOB (Su et al. 2019). However, the maximum yield of methanol of 59.89 ± 1.12 mg-COD CH_3OH L^{-1} was obtained from a mixed nitrifying enrichment culture (Taher and Chandran 2013).

7.7 Inhibitors of AOB

Despite several functional capabilities of AOB, nitrification process in industrial wastewater treatment plants is inhibited due to the presence of various toxic chemicals. Heavy metals (e.g., zinc, copper, arsenic, etc.) are one of the major contributors toward nitrification inhibition. Inhibitors of nitrification are classified as chemical and biological inhibitors which are discussed in the following sections.

7.7.1 Chemical Inhibitors

In the past decade, ionic liquids (ILs) have gained considerable attention as alternative to commonly used volatile organic compounds (VOCs). These are organic salts consisting of inorganic anions such as Cl^-, BF_4^-, Br^-, PF_6^-, $(CN)_2N^-$, CNS^-, $AlCl_4^-$, etc. which make weak coordination bond with organic cations (Thamke et al. 2019). They are employed in organic synthesis as solvents and catalysts, metallic nanoparticle synthesis, biocatalysis, separation of azeotropic mixtures, CO_2 capture, extractions, etc. (Tzani et al. 2019). Discharge from industries that manufacture or utilize ILs creates environmental risks. Imidazolium bromide ILs such as $(C_4mim)Br$, $(C_6mim)Br$, and $(C_{10}mim)Br$ have been found to reduce microbial abundance in soil including AOB and AOA (Cheng et al. 2019).

7.7.2 Biological Inhibitors

Some plant root extracts have been found to act as potential inhibitors of nitrification.

Brachiaria humidicola grasses are found to possess highest biological nitrification inhibition (BNI) capacity (Subbarao et al. 2017). Root exudates from rice and *Leymus racemosus*, which is a family of wheat, also show BNI potential (Sun et al. 2016; Lu et al. 2019). Biological nitrification inhibitors such as 1,9-decanediol, methyl 3-(4-hydroxyphenyl) propionate (MHPP), and α-linolenic acid (LN) have been found to inhibit both AOB and AOA population (Lu et al. 2019). However, different chemical compounds from root extract of same plant produce differing action on nitrification. For example, sorgoleone, a hydrophobic BNI from sorghum root extract, causes significant suppression in nitrification, but sakuranetin (a hydrophilic BNI) also extracted from sorghum root has no nitrification inhibition potential (Subbarao et al. 2013). Although these compounds play inhibitory role in nitrification, they find importance in preservation of NH_4^+-N in agricultural soil. AOB and AOA population are also naturally present in terrestrial environments. Application of nitrogenous fertilizers for crop improvement promotes growth of AOB, but increase in AOB population alleviates nitrogen from soil and makes it less available for plants.

7.8 Conclusions

Conventional wastewater treatment processes such as reverse osmosis, electrodialysis, ion exchange, etc. were used way back in the 1990s. Considering the loopholes in such physical methods, there came a paradigm shift in WWT technologies toward advanced oxidation processes such as ozonation, electrochemical oxidation, etc. To expand this technology, more endeavors are required in the development of novel catalysts and regulatory mechanism in the amount of hydroxyl radical produced in advanced oxidation processes. Presently, biological treatment processes are commonly applied for removal of nitrogen wastes. These processes do not require expensive chemical catalysts and thus come across as a cost-effective technique. Furthermore, incorporation of hybrid systems in conjunction with anammox process or partial nitrification-denitrification can be an effective way for removal of nitrogen from wastewater. However, good regulatory control over dissolved oxygen and biodegradable organic matter is necessary for easy running of biological treatment processes. This approach also allows energy and resource recovery from wastewater and thus conceptualizes existence of circular economy.

Acknowledgment The authors are thankful to the University Grants Commission (UGC), New Delhi, and Indian Institute of Technology (Indian School of Mines) Dhanbad for research fellowship to P.C. and K.Y respectively. The authors thank DST-FIST for the support to the Department of Environmental Science and Engineering, Indian Institute of Technology (Indian School of Mines) Dhanbad.

References

Abdel-Shafy HI, El-Khateeb MA, Mansour MSM (2016) Treatment of leather industrial wastewater via combined advanced oxidation and membrane filtration. Water SciTechnol 74(3):586–594. https://doi.org/10.2166/wst.2016.234

Abzazou T, Araujo RM, Auset M, Salvadó H (2016) Tracking and quantification of nitrifying bacteria in biofilm and mixed liquor of a partial nitrification MBBR pilot plant using fluorescence in situ hybridization. Sci Total Environ 541:1115–1123. https://doi.org/10.1016/j.scitotenv.2015.10.007

Achak M, Boumya W, Ouazzani N, Mandi L (2019) Preliminary evaluation of constructed wetlands for nutrients removal from olive mill wastewater (OMW) after passing through a sand filter. Ecol Eng 136:141–151. https://doi.org/10.1016/j.ecoleng.2019.06.007

Afonso MD, Jaber JO, Mohsen MS (2004) Brackish groundwater treatment by reverse osmosis in Jordan. Desalination 164(2):157–171. https://doi.org/10.1016/S0011-9164(04)00175-4

Albornoz LL, Marder L, Benvenuti T, Bernardes AM (2019) Electrodialysis applied to the treatment of an university sewage for water recovery. J Environ Chem Eng 7(2):102982. https://doi.org/10.1016/j.jece.2019.102982

Anfruns A, Gabarró J, Gonzalez-Olmos R, Puig S, Balaguer MD, Colprim J (2013) Coupling anammox and advanced oxidation-based technologies for mature landfill leachate treatment. J Hazard Mater 258–259:27–34. https://doi.org/10.1016/j.jhazmat.2013.04.027

Arp DJ, Sayavedra-Soto LA, Hommes NG (2002) Molecular biology and biochemistry of ammonia oxidation by *Nitrosomonas europaea*. Arch Microbiol 178(4):250–255. https://doi.org/10.1007/s00203-002-0452-0

Ashkanani A, Almomani F, Khraisheh M, Bhosale R, Tawalbeh M, AlJaml K (2019) Bio-carrier and operating temperature effect on ammonia removal from secondary wastewater effluents using moving bed biofilm reactor (MBBR). Sci Total Environ 693:133425. https://doi.org/10.1016/j.scitotenv.2019.07.231

Bashir MJK, Aziz HA, Yusoff MS, Huqe AA, Mohajeri S (2010) Effects of ion exchange resins in different mobile ion forms on semi-aerobic landfill leachate treatment. Water Sci Technol 61(3):641–649. https://doi.org/10.2166/wst.2010.867

Batstone DJ, Hülsen T, Mehta CM, Keller J (2015) Platforms for energy and nutrient recovery from domestic wastewater: a review. Chemosphere 140:2–11. https://doi.org/10.1016/j.chemosphere.2014.10.021

Beal EJ, House CH, Orphan VJ (2009) Manganese- and iron-dependent marine methane oxidation. Science (New York, NY) 325(5937):184–187. https://doi.org/10.1126/science.1169984

Beristain-Cardoso R, Pérez-González DN, González-Blanco G, Gómez J (2011) Simultaneous oxidation of ammonium, p-cresol and sulfide using a nitrifying sludge in a multipurpose bioreactor: a novel alternative. Bioresour Technol 102(3):3623–3625. https://doi.org/10.1016/j.biortech.2010.10.127

Bilstad T (1995) Nitrogen separation from domestic wastewater by reverse osmosis. J Membrane Sci 102:93–102. https://doi.org/10.1016/0376-7388(94)00279-8

Boetius A, Ravenschlag K, Schubert CJ, Rickert D, Widdel F, Gieseke A et al (2000) A marine microbial consortium apparently mediating anaerobic oxidation of methane. Nature 407 (6804):623–626. https://doi.org/10.1038/35036572

Buitrón G, Torres-Bojorges AX, Cea-Barcia G (2015) Removal of p-nonylphenol isomers using nitrifying sludge in a membrane sequencing batch reactor. Chem Eng 281:860–868. https://doi.org/10.1016/j.cej.2015.07.018

Capodaglio AG, Hlavínek P, Raboni M (2015) Physico-chemical technologies for nitrogen removal from wastewaters: a review. Ambiente e Agua 10(3):481–498

Capodici M, Corsino SF, Di Trapani D, Viviani G (2019) Achievement of partial nitrification under different carbon-to-nitrogen ratio and ammonia loading rate for the co-treatment of landfill leachate with municipal wastewater. Biochem Eng 149:107229. https://doi.org/10.1016/j.bej.2019.05.006

Chai L-Y, Ali M, Min X-B, Song Y-X, Tang C-J, Wang H-Y et al (2015) Partial nitrification in an air-lift reactor with long-term feeding of increasing ammonium concentrations. Bioresour Technol 185:134–142. https://doi.org/10.1016/j.biortech.2015.02.091

Chang SW, Hyman MR, Williamson KJ (2002) Cooxidation of naphthalene and other polycyclic aromatic hydrocarbons by the nitrifying bacterium, *Nitrosomonas europaea*. Biodegradation 13 (6):373–381. https://doi.org/10.1023/A:1022811430030

Chen Y-P, Ma T-F, Hu X, Fang F, Shen Y, Yang J-X et al (2015) Start-up of a combined anaerobic/partial nitritation/ANAMMOX process for high-salt mustard wastewater treatment. Appl Biochem Biotechnol 175(1):119–134. https://doi.org/10.1007/s12010-014-1247-x

Cheng C, Ma J, Wang J, Du Z, Li B, Wang J et al (2019) Toxicity comparison of three imidazolium bromide ionic liquids to soil microorganisms. Environ Pollut 255:113321. https://doi.org/10.1016/j.envpol.2019.113321

Chowdhury P, Viraraghavan T, Srinivasan A (2010) Biological treatment processes for fish processing wastewater – a review. Bioresour Technol 101(2):439–449. https://doi.org/10.1016/j.biortech.2009.08.065

Cotillas S, Llanos J, Cañizares P, Clematis D, Cerisola G, Rodrigo MA et al (2018) Removal of Procion Red MX-5B dye from wastewater by conductive-diamond electrochemical oxidation. Electrochim Acta 263:1–7. https://doi.org/10.1016/j.electacta.2018.01.052

Cotto I, Dai Z, Huo L, Anderson CL, Vilardi KJ, Ijaz U et al (2020) Long solids retention times and attached growth phase favor prevalence of comammox bacteria in nitrogen removal systems. Water Res 169:115268. https://doi.org/10.1016/j.watres.2019.115268

Cruvellier N, Poughon L, Creuly C, Dussap C-G, Lasseur C (2017) High ammonium loading and nitrification modelling in a fixed-bed bioreactor. J Water Process Eng 20:90–96. https://doi.org/10.1016/j.jwpe.2017.10.006

Dantas TLP, José HJ, Moreira RDFPM (2003) Oxidaçãoemprocessos Fenton e Foto-Fenton emefluentes de curtumes. Acta Scientiarum Technol 25(1):91–95. https://doi.org/10.4025/actascitechnol.v25i1.2254

Das S, Ghangrekar MM (2019) Tungsten oxide as electrocatalyst for improved power generation and wastewater treatment in microbial fuel cell. Environ Technol. https://doi.org/10.1080/09593330.2019.1575477

Das I, NooriMd T, Bhowmick GD, Ghangrekar MM (2018) Synthesis of bimetallic iron ferrite $CoO.5ZnO.5Fe_2O_4$ as a superior catalyst for oxygen reduction reaction to replace noble metal catalysts in microbial fuel cell. Int J Hydrog Energy 43(41):19196–19205. https://doi.org/10.1016/j.ijhydene.2018.08.113

Dawas-Massalha A, Gur-Reznik S, Lerman S, Sabbah I, Dosoretz CG (2014) Co-metabolic oxidation of pharmaceutical compounds by a nitrifying bacterial enrichment. Bioresour Technol 167:336–342. https://doi.org/10.1016/j.biortech.2014.06.003

de Luna MDG, Veciana ML, Su C-C, Lu M-C (2012) Acetaminophen degradation by electro-Fenton and photoelectro-Fenton using a double cathode electrochemical cell. J Hazard Mater 217–218:200–207. https://doi.org/10.1016/j.jhazmat.2012.03.018

Deegan AM, Shaik B, Nolan K, Urell K, Oelgemöller M, Tobin J, Morrissey A (2011) Treatment options for wastewater effluents from pharmaceutical companies. Int J Environ Sci Technol 8 (3):649–666. https://doi.org/10.1007/BF03326250

Di Domenico EG, Petroni G, Mancini D, Geri A, Palma LD, Ascenzioni F (2015) Development of electroactive and anaerobic ammonium-oxidizing (anammox) biofilms from digestate in microbial fuel cells. Biomed Res Int 2015:351014. https://doi.org/10.1155/2015/351014

Dialynas E, Mantzavinos D, Diamadopoulos E (2008) Advanced treatment of the reverse osmosis concentrate produced during reclamation of municipal wastewater. Water Res 42 (18):4603–4608. https://doi.org/10.1016/j.watres.2008.08.008

Esplugas S, Bila DM, Krause LGT, Dezotti M (2007) Ozonation and advanced oxidation technologies to remove endocrine disrupting chemicals (EDCs) and pharmaceuticals and personal care products (PPCPs) in water effluents. J Hazard Mater 149(3):631–642. https://doi.org/10.1016/j.jhazmat.2007.07.073

Fernandez-Fontaina E, Omil F, Lema JM, Carballa M (2012) Influence of nitrifying conditions on the biodegradation and sorption of emerging micropollutants. Water Res 46(16):5434–5444. https://doi.org/10.1016/j.watres.2012.07.037

Ferrentino R, Ferraro A, Mattei MR, Esposito G, Andreottola G (2018) Process performance optimization and mathematical modelling of a SBR-MBBR treatment at low oxygen concentration. Process Biochem 75:230–239. https://doi.org/10.1016/j.procbio.2018.08.023

Fu Z, Zhao J (2015) Impact of quinoline on activity and microbial culture of partial nitrification process. Bioresour Technol 197:113–119

Gao F, Zhang H, Yang F, Li H, Zhang R (2014) The effects of zero-valent iron (ZVI) and ferroferric oxide (Fe_3O_4) on anammox activity and granulation in anaerobic continuously stirred tank reactors (CSTR). Process Biochem 49(11):1970–1978. https://doi.org/10.1016/j.procbio.2014.07.019

Garcia-Segura S, Ocon JD, Chong MN (2018) Electrochemical oxidation remediation of real wastewater effluents – a review. Process Saf Environ 113:48–67. https://doi.org/10.1016/j.psep.2017.09.014

Ge X, Yang L, Sheets JP, Yu Z, Li Y (2014) Biological conversion of methane to liquid fuels: status and opportunities. Biotechnol Adv 32(8):1460–1475. https://doi.org/10.1016/j.biotechadv.2014.09.004

Ghayur A, Verheyen TV, Meuleman E (2019) Biological and chemical treatment technologies for waste amines from CO_2 capture plants. J Environ Manag 241:514–524. https://doi.org/10.1016/j.jenvman.2018.07.033

Hanson RS, Hanson TE (1996) Methanotrophic bacteria. Microbiol Rev 60(2):439–471

Hasson D, Drak A, Semiat R (2001) Inception of $CaSO_4$ scaling on RO membranes at various water recovery levels. Desalination 139(1):73–81. https://doi.org/10.1016/S0011-9164(01)00296-X

Hatamoto M, Miyauchi T, Kindaichi T, Ozaki N, Ohashi A (2011) Dissolved methane oxidation and competition for oxygen in down-flow hanging sponge reactor for post-treatment of anaerobic wastewater treatment. Bioresour Technol 102(22):10299–10304. https://doi.org/10.1016/j.biortech.2011.08.099

He Z, Kan J, Wang Y, Huang Y, Mansfeld F, Nealson KH (2009) Electricity production coupled to ammonium in a microbial fuel cell. Environ Sci Technol 43(9):3391–3397. https://doi.org/10.1021/es803492c

Hendrickx TLG, Kampman C, Zeeman G, Temmink H, Hu Z, Kartal B, Buisman CJN (2014) High specific activity for anammox bacteria enriched from activated sludge at 10°C. Bioresour Technol 163:214–221. https://doi.org/10.1016/j.biortech.2014.04.025

Hommes NG, Sayavedra-Soto LA, Arp DJ (2003) Chemolithoorganotrophic growth of *Nitrosomonas europaea* on fructose. J Bacteriol 185(23):6809–6814. https://doi.org/10.1128/JB.185.23.6809-6814.2003

Huang KL, Wei KC, Chen MH, Ma CY (2018) Removal of organic and ammonium nitrogen pollutants in swine wastewater using electrochemical advanced oxidation. Int J Electrochem Sci 13(12):11418–11431. https://doi.org/10.20964/2018.12.32

Hyman MR, Wood PM (1983) Methane oxidation by *Nitrosomonas europaea*. J Biochem 212 (1):31–37. https://doi.org/10.1042/bj2120031

Ikehata K, El-Din MG, Snyder SA (2008) Ozonation and advanced oxidation treatment of emerging organic pollutants in water and wastewater. Ozone Sci Eng 30(1):21–26. https://doi.org/10.1080/01919510701728970

Islam GM, Vi P, Gilbride KA (2019) Functional relationship between ammonia-oxidizing bacteria and ammonia-oxidizing archaea populations in the secondary treatment system of a full-scale municipal wastewater treatment plant. J Environ Sci 86:120–130. https://doi.org/10.1016/j.jes.2019.04.031

Jagadevan S, Semrau JD (2013) Priority pollutant degradation by the facultative methanotroph, *Methylocystis strain* SB2. Appl Microbiol Biotechnol 97(11):5089–5096. https://doi.org/10.1007/s00253-012-4310-y

Jones RD, Morita RY (1983) Methane oxidation by *Nitrosococcus oceanus* and *Nitrosomonas europaea*. Appl Environ Microbiol 45(2):401–410

Jorgensen TC, Weatherley LR (2003) Ammonia removal from wastewater by ion exchange in the presence of organic contaminants. Water Res 37(8):1723–1728. https://doi.org/10.1016/S0043-1354(02)00571-7

Karanasios KA, Vasiliadou IA, Pavlou S, Vayenas DV (2010) Hydrogenotrophic denitrification of potable water: a review. J Hazard Mater 180(1):20–37. https://doi.org/10.1016/j.jhazmat.2010.04.090

Kassotaki E, Buttiglieri G, Ferrando-Climent L, Rodriguez-Roda I, Pijuan M (2016) Enhanced sulfamethoxazole degradation through ammonia oxidizing bacteria co-metabolism and fate of transformation products. Water Res 94:111–119. https://doi.org/10.1016/j.watres.2016.02.022

Kawagoshi Y, Fujisaki K, Tomoshige Y, Yamashiro K, Qiao Y (2012) Temperature effect on nitrogen removal performance and bacterial community in culture of marine anammox bacteria derived from sea-based waste disposal site. J Biosci Bioeng 113(4):515–520. https://doi.org/10.1016/j.jbiosc.2011.11.024

Keener WK, Arp DJ (1994) Transformations of aromatic compounds by *Nitrosomonas europaea*. Appl Environ Microbiol 60(6):1914–1920

Khunjar WO, Love NG (2011) Sorption of carbamazepine, 17α-ethinylestradiol, iopromide and trimethoprim to biomass involves interactions with exocellular polymeric substances. Chemosphere 82(6):917–922. https://doi.org/10.1016/j.chemosphere.2010.10.046

Khunjar WO, Mackintosh SA, Skotnicka-Pitak J, Baik S, Aga DS, Love NG (2011) Elucidating the relative roles of ammonia oxidizing and heterotrophic bacteria during the biotransformation of 17α-ethinylestradiol and trimethoprim. Environ Sci Technol 45:3605–3612. https://doi.org/10.1021/es1037035

Khuntia S, Majumder SK, Ghosh P (2013) Removal of ammonia from water by ozone microbubbles. Ind Eng Chem Res 52(1):318–326. https://doi.org/10.1021/ie302212p

Kim YM, Park H, Cho KH, Park JM (2013) Long term assessment of factors affecting nitrifying bacteria communities and N-removal in a full-scale biological process treating high strength hazardous wastewater. Bioresour Technol 134:180–189. https://doi.org/10.1016/j.biortech.2013.02.036

Kits KD, Jung M-Y, Vierheilig J, Pjevac P, Sedlacek CJ, Liu S et al (2019) Low yield and abiotic origin of N_2O formed by the complete nitrifier *Nitrospira inopinata*. Nat Commun 10(1):1–12. https://doi.org/10.1038/s41467-019-09790-x

Kjeldal H, Pell L, Pommerening-Röser A, Nielsen JL (2014) Influence of p-cresol on the proteome of the autotrophic nitrifying bacterium *Nitrosomonas eutropha* C91. Arch Microbiol 196 (7):497–511. https://doi.org/10.1007/s00203-014-0985-z

Kumari S, Jose S, Jagadevan S (2019) Optimization of phosphate recovery as struvite from synthetic distillery wastewater using a chemical equilibrium model. Environ Sci Pollut Res 26 (29):30452–30462. https://doi.org/10.1007/s11356-019-06152-4

Kuo C-H, Yuan F, Hill DO (1997) Kinetics of oxidation of ammonia in solutions containing ozone with or without hydrogen peroxide. Ind Eng Chem Res 36(10):4108–4113. https://doi.org/10.1021/ie9702082

Lawton TJ, Ham J, Sun T, Rosenzweig AC (2014) Structural conservation of the B subunit in the ammonia monooxygenase/particulate methane monooxygenase superfamily. Proteins 82 (9):2263–2267. https://doi.org/10.1002/prot.24535

Leaković S, Mijatović I, Cerjan-Stefanović Š, Hodžić E (2000) Nitrogen removal from fertilizer wastewater by ion exchange. Water Res 34(1):185–190. https://doi.org/10.1016/S0043-1354(99)00122-0

Lei X, Jia Y, Chen Y, Hu Y (2019) Simultaneous nitrification and denitrification without nitrite accumulation by a novel isolated *Ochrobactrum anthropic* LJ81. Bioresour Technol 272:442–450. https://doi.org/10.1016/j.biortech.2018.10.060

Li P, Takahashi M, Chiba K (2009) Enhanced free-radical generation by shrinking microbubbles using a copper catalyst. Chemosphere 77(8):1157–1160. https://doi.org/10.1016/j.chemosphere.2009.07.062

Li J, Qi P, Qiang Z, Dong H, Gao D, Wang D (2018) Is anammox a promising treatment process for nitrogen removal from nitrogen-rich saline wastewater? Bioresour Technol 270:722–731. https://doi.org/10.1016/j.biortech.2018.08.115

Li C, Li J, Liu G, Deng Y, Zhu S, Ye Z et al (2019) Performance and microbial community analysis of Combined Denitrification and Biofloc Technology (CDBFT) system treating nitrogen-rich aquaculture wastewater. Bioresour Technol 288:121582. https://doi.org/10.1016/j.biortech.2019.121582

Liu CCK, Jae-Woo P, Migita R, Gang Q (2002) Experiments of a prototype wind-driven reverse osmosis desalination system with feedback control. Desalination 150(3):277–287. https://doi.org/10.1016/S0011-9164(02)00984-0

Liu CCK, Xia W, Park JW (2007) A wind-driven reverse osmosis system for aquaculture wastewater reuse and nutrient recovery. Desalination 202(1):24–30. https://doi.org/10.1016/j.desal.2005.12.034

Liu Y, Ngo HH, Guo W, Peng L, Pan Y, Guo J et al (2016) Autotrophic nitrogen removal in membrane-aerated biofilms: archaeal ammonia oxidation versus bacterial ammonia oxidation. Chem Eng J 302:535–544. https://doi.org/10.1016/j.cej.2016.05.078

Liu Y-J, Hu C-Y, Lo S-L (2019) Direct and indirect electrochemical oxidation of amine-containing pharmaceuticals using graphite electrodes. J Hazard Mater 366:592–605. https://doi.org/10.1016/j.jhazmat.2018.12.037

Lizhang W, Shengxiang Y, Bo W, Peng L, Zhe'nan L, Yuemin Z (2016) The influence of anode materials on the kinetics toward electrochemical oxidation of phenol. Electrochim Acta 206:270–277. https://doi.org/10.1016/j.electacta.2016.04.168

López-Ramírez JA, Oviedo MDC, Alonso JMQ (2006) Comparative studies of reverse osmosis membranes for wastewater reclamation. Desalination 191(1):137–147. https://doi.org/10.1016/j.desal.2005.08.013

Lou Z, Sun Y, Zhou X, Baig SA, Hu B, Xu X (2017) Composition variability of spent mushroom substrates during continuous cultivation, composting process and their effects on mineral nitrogen transformation in soil. Geoderma 307:30–37. https://doi.org/10.1016/j.geoderma.2017.07.033

Lu H, Chandran K, Stensel D (2014) Microbial ecology of denitrification in biological wastewater treatment. Water Res 64:237–254. https://doi.org/10.1016/j.watres.2014.06.042

Lu Y, Zhang X, Jiang J, Kronzucker HJ, Shen W, Shi W (2019) Effects of the biological nitrification inhibitor 1,9-decanediol on nitrification and ammonia oxidizers in three agricultural soils. Soil Biol Biochem 129:48–59. https://doi.org/10.1016/j.soilbio.2018.11.008

Ma Y, Wei D, Zhang X, Fu H, Chen T, Jia J (2019) An innovative strategy for inducing anammox from partial nitrification process in a membrane bioreactor. J Hazard Mater 379:120809. https://doi.org/10.1016/j.jhazmat.2019.120809

Maldonado MI, Malato S, Pérez-Estrada LA, Gernjak W, Oller I, Doménech X, Peral J (2006) Partial degradation of five pesticides and an industrial pollutant by ozonation in a pilot-plant scale reactor. J Hazard Mater 138(2):363–369. https://doi.org/10.1016/j.jhazmat.2006.05.058

Martínez-Jardines M, Martínez-Hernández S, Texier A-C, Cuervo-López F (2018) 2-Chlorophenol consumption by cometabolism in nitrifying SBR reactors. Chemosphere 212:41–49. https://doi.org/10.1016/j.chemosphere.2018.08.064

Matamoros V, Gutiérrez R, Ferrer I, García J, Bayona JM (2015) Capability of microalgae-based wastewater treatment systems to remove emerging organic contaminants: a pilot-scale study. J Hazard Mater 288:34–42. https://doi.org/10.1016/j.jhazmat.2015.02.002

Matos CT, Sequeira AM, Velizarov S, Crespo JG, Reis MAM (2009) Nitrate removal in a closed marine system through the ion exchange membrane bioreactor. J Hazard Mater 166(1):428–434. https://doi.org/10.1016/j.jhazmat.2008.11.038

Men Y, Achermann S, Helbling DE, Johnson DR, Fenner K (2017) Relative contribution of ammonia oxidizing bacteria and other members of nitrifying activated sludge communities to micropollutant biotransformation. Water Res 109:217–226. https://doi.org/10.1016/j.watres.2016.11.048

Meng L, Li W, Zhang S, Wu C, Lv L (2017) Feasibility of co-composting of sewage sludge, spent mushroom substrate and wheat straw. Bioresour Technol 226:39–45. https://doi.org/10.1016/j.biortech.2016.11.054

Meng X, Khoso SA, Jiang F, Zhang Y, Yue T, Gao J et al (2020) Removal of chemical oxygen demand and ammonia nitrogen from lead smelting wastewater with high salts content using electrochemical oxidation combined with coagulation–flocculation treatment. Sep Purif Technol 235:116233. https://doi.org/10.1016/j.seppur.2019.116233

Merlo R, Wong J, Occiano V, Sandera K, Pai A, Sen S et al (2012) Analysis of organic nitrogen removal in municipal wastewater by reverse osmosis. Water Environ Res 84(7):588–595. https://doi.org/10.2175/106143012x13373550426797

Miao Y, Zhang J, Peng Y, Wang S (2020) An improved start-up strategy for mainstream anammox process through inoculating ordinary nitrification sludge and a small amount of anammox sludge. J Hazard Mater 384:121325. https://doi.org/10.1016/j.jhazmat.2019.121325

Muenmee S, Chiemchaisri W, Chiemchaisri C (2015) Microbial consortium involving biological methane oxidation in relation to the biodegradation of waste plastics in a solid waste disposal open dump site. Int Biodeter Biodeg 102:172–181. https://doi.org/10.1016/j.ibiod.2015.03.015

Norman JS, Barrett JE (2016) Substrate availability drives spatial patterns in richness of ammonia-oxidizing bacteria and archaea in temperate forest soils. Soil Biol Biochem 94:169–172. https://doi.org/10.1016/j.soilbio.2015.11.015

Ouyang Y, Norton JM, Stark JM, Reeve JR, Habteselassie MY (2016) Ammonia-oxidizing bacteria are more responsive than archaea to nitrogen source in an agricultural soil. Soil Biol Biochem 96:4–15. https://doi.org/10.1016/j.soilbio.2016.01.012

Padoley KV, Mudliar SN, Pandey RA (2008) Heterocyclic nitrogenous pollutants in the environment and their treatment options – an overview. Bioresour Technol 99(10):4029–4043. https://doi.org/10.1016/j.biortech.2007.01.047

Panizza M, Cerisola G (2008) Removal of colour and COD from wastewater containing acid blue 22 by electrochemical oxidation. J Hazard Mater 153(1):83–88. https://doi.org/10.1016/j.jhazmat.2007.08.023

Picos-Benítez AR, Peralta-Hernández JM, López-Hincapié JD, Rodríguez-García A (2019) Biogas production from saline wastewater of the evisceration process of the fish processing industry. J Water Process Eng 32:100933. https://doi.org/10.1016/j.jwpe.2019.100933

Qi P, Li J, Dong H, Wang D, Bo Y (2018) Performance of anammox process treating nitrogen-rich saline wastewater: kinetics and nitrite inhibition. J Clean Prod 199:493–502. https://doi.org/10.1016/j.jclepro.2018.07.201

Quan X, Huang K, Li M, Lan M, Li B (2018) Nitrogen removal performance of municipal reverse osmosis concentrate with low C/N ratio by membrane-aerated biofilm reactor. Front Environ Sci Eng 12(6):5. https://doi.org/10.1007/s11783-018-1047-6

Quist-Jensen CA, Macedonio F, Horbez D, Drioli E (2017) Reclamation of sodium sulfate from industrial wastewater by using membrane distillation and membrane crystallization. Desalination 401:112–119. https://doi.org/10.1016/j.desal.2016.05.007

Rasche ME, Hyman MR, Arp DJ (1991) Factors limiting aliphatic chlorocarbon degradation by *Nitrosomonas europaea*: cometabolic inactivation of Ammonia monooxygenase and substrate specificity. Appl Environ Microbiol 57(10):2986–2994

Rivera-Utrilla J, Sánchez-Polo M, Ferro-García MÁ, Prados-Joya G, Ocampo-Pérez R (2013) Pharmaceuticals as emerging contaminants and their removal from water. A review. Chemosphere 93(7):1268–1287. https://doi.org/10.1016/j.chemosphere.2013.07.059

Roh H, Subramanya N, Zhao F, Yu CP, Sandt J, Chu KH (2009) Biodegradation potential of wastewater micropollutants by ammonia-oxidizing bacteria. Chemosphere 77(8):1084–1089

Roquebert V, Booth S, Cushing RS, Crozes G, Hansen E (2000) Electrodialysis reversal (EDR) and ion exchange as polishing treatment for perchlorate treatment. Desalination 131(1):285–291. https://doi.org/10.1016/S0011-9164(00)90026-2

Schoeman JJ (2009) Nitrate-nitrogen removal with small-scale reverse osmosis, electrodialysis and ion-exchange units in rural areas. Water SA 35(5):721–728

Shemer H, Kunukcu YK, Linden KG (2006) Degradation of the pharmaceutical metronidazole via UV, Fenton and photo-Fenton processes. Chemosphere 63(2):269–276. https://doi.org/10.1016/j.chemosphere.2005.07.029

Sheng B, Cong H, Zhang S, Meng F (2018) Interactive effects between tetracycline and nitrosifying sludge microbiota in a nitritation membrane bioreactor. Chem Eng J 341:556–564

Shi J, Fujisawa S, Nakai S, Hosomi M (2004) Biodegradation of natural and synthetic estrogens by nitrifying activated sludge and ammonia-oxidizing bacterium *Nitrosomonas europaea*. Water Res 39(9):2322–2329

Shi J, Xu C, Han Y, Han H (2019) Enhanced anaerobic biodegradation efficiency and mechanism of quinoline, pyridine, and indole in coal gasification wastewater. Chem Eng J 361:1019–1029. https://doi.org/10.1016/j.cej.2018.12.162

Shinohara T, Qiao S, Yamamoto T, Nishiyama T, Fujii T, Kaiho T et al (2009) Partial nitritation treatment of underground brine waste with high ammonium and salt content. J Biosci Bioeng 108(4):330–335. https://doi.org/10.1016/j.jbiosc.2009.04.014

Shrimali M, Singh KP (2001) New methods of nitrate removal from water. Environ Pollut 112 (3):351–359. https://doi.org/10.1016/S0269-7491(00)00147-0

Sohni S, Gul K, Ahmad F, Ahmad I, Khan A, Khan N, Khan SB (2018) Highly efficient removal of acid red-17 and bromophenol blue dyes from industrial wastewater using graphene oxide functionalized magnetic chitosan composite. Polym Compos 39(9):3317–3328. https://doi.org/10.1002/pc.24349

Song HP, Jiang YY, Cheng FQ (2012) Treatment of ammonia-nitrogen wastewater by electrodialysis. https://doi.org/10.4028/www.scientific.net/AMR.391-392.1346

Song Q, Li M, Wang L, Ma X, Liu F, Liu X (2019) Mechanism and optimization of electrochemical system for simultaneous removal of nitrate and ammonia. J Hazard Mater 363:119–126. https://doi.org/10.1016/j.jhazmat.2018.09.046

Sreekanth D, Sivaramakrishna D, Himabindu V, Anjaneyulu Y (2009) Thermophilic degradation of phenolic compounds in lab scale hybrid up flow anaerobic sludge blanket reactors. J Hazard Mater 164(2):532–1539. https://doi.org/10.1016/j.jhazmat.2008.09.070

Strathmann H (2010) Electrodialysis, a mature technology with a multitude of new applications. Desalination 264(3):268–288. https://doi.org/10.1016/j.desal.2010.04.069

Su YC, Sathyamoorthy S, Chandran K (2019) Bioaugmented methanol production using ammonia oxidizing bacteria in a continuous flow process. Bioresour Technol 279:101–107. https://doi.org/10.1016/j.biortech.2019.01.092

Subbarao GV, Nakahara K, Ishikawa T, Ono H, Yoshida M, Yoshihashi T et al (2013) Biological nitrification inhibition (BNI) activity in sorghum and its characterization. Plant Soil 366 (1):243–259. https://doi.org/10.1007/s11104-012-1419-9

Subbarao GV, Arango J, Masahiro K, Hooper AM, Yoshihashi T, Ando Y et al (2017) Genetic mitigation strategies to tackle agricultural GHG emissions: the case for biological nitrification inhibition technology. Plant Sci 262:165–168. https://doi.org/10.1016/j.plantsci.2017.05.004

Sun L, Lu Y, Yu F, Kronzucker HJ, Shi W (2016) Biological nitrification inhibition by rice root exudates and its relationship with nitrogen-use efficiency. New Phytol 212(3):646–656. https://doi.org/10.1111/nph.14057

Suzuki I, Dular U, Kwok SC (1974) Ammonia or ammonium ion as substrate for oxidation by *Nitrosomonas europaea* cells and extracts. J Bacteriol 120(1):556–558

Taher E, Chandran K (2013) High-rate, high-yield production of methanol by ammonia-oxidizing bacteria. Environ Sci Technol 47(7):3167–3173. https://doi.org/10.1021/es3042912

Tang W, Zhang Y, Bai J, Li J, Wang J, Li L et al (2020) Efficient denitrification and removal of natural organic matter, emerging pollutants simultaneously for RO concentrate based on photo electrocatalytic radical reaction. Sep Purif Technol 234:116032. https://doi.org/10.1016/j.seppur.2019.116032

Tavormina PL, Orphan VJ, Kalyuzhnaya MG, Jetten MSM, Klotz MG (2011) A novel family of functional operons encoding methane/ammonia monooxygenase-related proteins in gammaproteobacterial methanotrophs. Environ Microbiol Rep 3(1):91–100. https://doi.org/10.1111/j.1758-2229.2010.00192.x

Thamke VR, Chaudhari AU, Tapase SR, Paul D, Kodam KM (2019) In vitro toxicological evaluation of ionic liquids and development of effective bioremediation process for their removal. Environ Pollut 250:567–577. https://doi.org/10.1016/j.envpol.2019.04.043

Thandar SM, Ushiki N, Fujitani H, Sekiguchi Y, Tsuneda S (2016) Ecophysiology and comparative genomics of *Nitrosomonas mobilis* Ms1 isolated from autotrophic nitrifying granules of wastewater treatment bioreactor. Front Microbiol 7:1869. https://doi.org/10.3389/fmicb.2016.01869

Thorn GJS (2006) Development of an immobilized *Nitrosomonas europaea* bioreactor for the production of methanol from methane. Retrieved from https://ir.canterbury.ac.nz/handle/10092/1867

Töre GY, Meriç S, Lofrano G, DeFeo G (2012) Removal of trace pollutants from wastewater in constructed wetlands. In: Lofrano G (ed) Emerging compounds removal from wastewater. SpringerBriefs in molecular science. Springer, Dordrecht

Tyagi M, Rana A, Kumari S, Jagadevan S (2018) Adsorptive removal of cyanide from coke oven wastewater onto zero-valent iron: optimization through response surface methodology, isotherm and kinetic studies. J Clean Prod 178:398–407. https://doi.org/10.1016/j.jclepro.2018.01.016

Tzani A, Skarpalezos D, Papadopoulos A, Aravopoulou D, Kleidas I, Ioannou E et al (2019) Synthesis of novel non-toxic naphthenic and benzoic acid ionic liquids. Structure-properties relationship and evaluation of their biodegradability potential. J Mol Liq:111927. https://doi.org/10.1016/j.molliq.2019.111927

van der Wiel BZ, Weijma J, van Middelaar CE, Kleinke M, Buisman CJN, Wichern F (2019) Restoring nutrient circularity: a review of nutrient stock and flow analyses of local agro-food-waste systems. Resour Conserv Recy X(3):100014. https://doi.org/10.1016/j.rcrx.2019.100014

Wang CC, Lee PH, Kumar M, Huang YT, Sung S, Lin JG (2010) Simultaneous partial nitrification, anaerobic ammonium oxidation and denitrification (SNAD) in a full-scale landfill-leachate treatment plant. J Hazard Mater 175(1):622–628. https://doi.org/10.1016/j.jhazmat.2009.10.052

Wang Y, Zeng Y, Chai X, Liao X, He Q, Shi B (2012) Ammonia nitrogen in tannery wastewater: distribution, origin and prevention. J Am Leather Chem Assoc 107(2):40

Wang X, Lu X, Li F, Yang G (2014) Effects of temperature and carbon-nitrogen (C/N) ratio on the performance of anaerobic co-digestion of dairy manure, chicken manure and rice straw: focusing on ammonia inhibition. PLoS One 9(5):e97265. https://doi.org/10.1371/journal.pone.0097265

Wang S, Zhang T, Su H (2016) Enhanced hydrogen production from corn starch wastewater as nitrogen source by mixed cultures. Renew Energy 96:1135–1141. https://doi.org/10.1016/j.renene.2015.11.072

Wang G, Zhang D, Xu Y, Hua Y, Dai X (2018) Comparing two start up strategies and the effect of temperature fluctuations on the performance of mainstream anammox reactors. Chemosphere 209:632–639. https://doi.org/10.1016/j.chemosphere.2018.06.134

Wang B, Ni BJ, Yuan Z, Guo J (2019a) Cometabolic biodegradation of cephalexin by enriched nitrifying sludge: process characteristics, gene expression and product biotoxicity. Sci Total Environ 672:275–282. https://doi.org/10.1016/j.scitotenv.2019.03.473

Wang F, Fu R, Lv H, Zhu G, Lu B, Zhou Z et al (2019b) Phosphate recovery from swine wastewater by a struvite precipitation electrolyzer. Sci Rep 9(1):1–10. https://doi.org/10.1038/s41598-019-45085-3

Ward AJ, Arola K, Thompson Brewster E, Mehta CM, Batstone DJ (2018) Nutrient recovery from wastewater through pilot scale electrodialysis. Water Res 135:57–65. https://doi.org/10.1016/j.watres.2018.02.021

Wei D, Du B, Zhang J, Hu Z, Liang S, Li Y (2015) Composition of extracellular polymeric substances in a partial nitrification reactor treating high ammonia wastewater and nitrous oxide emission. Bioresour Technol 190:474–479. https://doi.org/10.1016/j.biortech.2015.02.104

Wei D, Zhang K, Ngo HH, Guo W, Wang S, Li J et al (2017) Nitrogen removal via nitrite in a partial nitrification sequencing batch biofilm reactor treating high strength ammonia wastewater and its greenhouse gas emission. Bioresour Technol 230:49–55. https://doi.org/10.1016/j.biortech.2017.01.033

Wen X, Zhou J, Wang J, Qing X, He Q (2016) Effects of dissolved oxygen on microbial community of single-stage autotrophic nitrogen removal system treating simulating mature landfill leachate. Bioresour Technol 218:962–968. https://doi.org/10.1016/j.biortech.2016.07.023

Wu J, Zhan M, Chang Y, Su Q, Ran Y (2018) Adaption and recovery of *Nitrosomonas europaea* to chronic TiO_2 nanoparticle exposure. 147:429–439. https://doi.org/10.1016/j.watres.2018.09.043

Wu G, Geng J, Li S, Li J, Fu Y, Xu K, Ren H, Zhang X (2019) Abiotic and biotic processes of diclofenac in enriched nitrifying sludge: kinetics, transformation products and reactions. Sci Total Environ 683:80–88. https://doi.org/10.1016/j.scitotenv.2019.05.216

Xiao SH, Qu JH, Zhao X, Liu HJ, Wan DJ (2009) Electrochemical process combined with UV light irradiation for synergistic degradation of ammonia in chloride-containing solutions. Water Res 43:1432–1440. https://doi.org/10.1016/j.watres.2008.12.023

Xu Y, Yuan Z, Ni BJ (2016) Biotransformation of pharmaceuticals by ammonia oxidizing bacteria in wastewater treatment processes. Sci Total Environ 566–567:796–805. https://doi.org/10.1016/j.scitotenv.2016.05.118

Xu Y, Radjenovic J, Yuan Z, Ni BJ (2017) Biodegradation of atenolol by an enriched nitrifying sludge: products and pathways. Chem Eng J 312:351–359. https://doi.org/10.1016/j.cej.2016.11.153

Yan Y, Wang Y, Wang W, Zhou S, Wang J, Guo J (2019) Comparison of short-term dosing ferrous ion and nanoscale zero-valent iron for rapid recovery of anammox activity from dissolved oxygen inhibition. Water Res 153:284–294. https://doi.org/10.1016/j.watres.2019.01.029

Yang L, Chang Y, Chou M (1999) Feasibility of bioremediation of trichloroethylene contaminated sites by nitrifying bacteria through cometabolism with ammonia. J Hazard Mater 69 (1):111–126. https://doi.org/10.1016/s0304-3894(99)00098-9

Yin H, Yang C, Jia Y, Chen H, Gu X (2018) Dual removal of phosphate and ammonium from high concentrations of aquaculture wastewaters using an efficient two-stage infiltration system. Sci Total Environ 635:936–946. https://doi.org/10.1016/j.scitotenv.2018.04.218

Yokota N, Watanabe Y, Tokutomi T, Kiyokawa T, Hori T, Ikeda D et al (2018) High-rate nitrogen removal from waste brine by marine anammox bacteria in a pilot-scale UASB reactor. Appl Microbiol Biotechnol 102(3):1501–1512. https://doi.org/10.1007/s00253-017-8663-0

Youssef CB, Zepeda A, Gomez J (2013) Modelling the combined inhibition and time-dependent inactivation effects of toluene on ammonia and nitrite oxidation using a nitrifying sludge. Biochem Eng 80:37–44. https://doi.org/10.1016/j.bej.2013.09.005

Zhang X, Liang Y, Ma Y, Du J, Pang L, Zhang H (2016) Ammonia removal and microbial characteristics of partial nitrification in biofilm and activated sludge treating low strength sewage at low temperature. Ecol Eng 93:104–111. https://doi.org/10.1016/j.ecoleng.2016.05.016

Zhang M, Luo P, Liu F, Li H, Zhang S, Xiao R et al (2017) Nitrogen removal and distribution of ammonia-oxidizing and denitrifying genes in an integrated constructed wetland for swine wastewater treatment. Ecol Eng 104:30–38. https://doi.org/10.1016/j.ecoleng.2017.04.022

Zhang J, Miao Y, Zhang Q, Sun Y, Wu L, Peng Y (2019) Mechanism of stable sewage nitrogen removal in a partial nitrification-anammox biofilm system at low temperatures: microbial community and EPS analysis. Bioresour Technol:122459. https://doi.org/10.1016/j.biortech.2019.122459

Zheng M, Zhu H, Han Y, Xu C, Zhang Z, Han H (2019) Comparative investigation on carbon-based moving bed biofilm reactor (MBBR) for synchronous removal of phenols and ammonia in treating coal pyrolysis wastewater at pilot-scale. Bioresour Technol 288:121590. https://doi.org/10.1016/j.biortech.2019.121590

Zhou L-J, Han P, Yu Y, Wang B, Men Y, Wagner M, Wu QL (2019) Cometabolic biotransformation and microbial-mediated abiotic transformation of sulfonamides by three ammonia oxidizers. Water Res 159:444–453. https://doi.org/10.1016/j.watres.2019.05.031

Zhu W, Li J, Dong H, Wang D, Zhang P (2017) Nitrogen removal performance and operation strategy of anammox process under temperature shock. Biodegradation 28(4):261–274. https://doi.org/10.1007/s10532-017-9794-9

Chapter 8
Application of Wastewater in Irrigation and Its Regulation with Special Reference to Agriculture Residues

Monika, Rajeev Pratap Singh, Pooja Singh, Richa Kothari, Kaushik Gautam, and Anita Singh

8.1 Introduction

About 70% of global freshwater is currently used for irrigation and cultivation practices (World Bank Report 2017). The demand for freshwater will go high to feed a projected 9.7 billion population in 2050 (UN 2019). As per the UN-Water Annual Report (2016), population explosion and climate change synergistically affect the availability of freshwater. These situations impose a risk on freshwater for agricultural production and induce societies to switch over another alternative option for continuous supply in agriculture (De Fraiture and Wichelns 2010; Ahmad et al. 2020a, b). Many scientists and their studies recommend wastewater as a viable option in water-scarce nations (Hanjra et al. 2012). According to the International Water Association, Wastewater Report (2018), change is necessary to apprehend the target 6.3 of the Sustainable Development Goals (SDGs), as this commits governments to halve the ratio of untreated wastewater and substantially encourages recycling and safe reuse by 2030.

Wastewater is the constituent of liquid waste discharged from households (excreta, urine, and gray water) and commercial and industrial establishments,

Monika · R. P. Singh (✉)
Department of Environment and Sustainable Development, Institute of Environment and Sustainable Development, Banaras Hindu University, Varanasi, Uttar Pradesh, India

P. Singh
Department of Science Institute of Computer Science and Technology, SHEPA, Varanasi, Uttar Pradesh, India

R. Kothari
Department of Environment Science, Central University of Jammu, Jammu, India

K. Gautam · A. Singh
Centre of Advanced Study in Botany, Institute of Science, Banaras Hindu University, Varanasi, Uttar Pradesh, India

A. Singh et al. (eds.), *Water Pollution and Management Practices*,
https://doi.org/10.1007/978-981-15-8358-2_8

which could also be mixed with stormwater (Aryal et al. 2011). Every day, around 2 million tons of sewage, agricultural, and industrial wastewater are released into the world's water, the equivalent of the weight of the entire human population of 6.8 billion people. The UN estimates that the amount of wastewater produced annually is about 1500 km^3, approximately six times more water than exists in all the rivers of the globe (UN WWAP 2003; World Water Quality Facts And Statistics 2013).

Wastewater utilization for irrigational purposes is one of the emerging perspectives particularly in water-scanty places in the world. Zhang and Shen (2017) reported that wastewater is very advantageous for farming activities because it provides continuous water supply in the duration of the irrigation period and decreases the risk of income losses and crop failure. Wastewater use in agriculture is a reliable supply of water; its supply is irrespective of weather variability and seasonal drought and reduces the dependability on climate or weather conditions for rainwater which is frequently seen in many developing countries (Mateo-Sagasta et al. 2013; Magwaza et al. 2020). In agriculture, crop growth required a mixture of biodegradable organic material (carbon and nitrogen), as well as most of the mineral nutrients (e.g., phosphorous, potassium, magnesium, boron, molybdenum, selenium, and copper); all do present in enough quantity in wastewater (Durán-Álvarez and Jiménez-Cisneros 2015).

Heavy metal is an inevitable element in the wastewater, and they get accumulated in the soil and plants, when irrigated by wastewater. Several techniques and technologies emerged to get rid of trace metals. There are mainly three types of treatment processes: physical, chemical, and biological; these include ion exchange, chemical precipitation, flocculation, membrane filtration, floatation, co-precipitation, electrochemical treatment, and adsorption (Bolisetty et al. 2019). These techniques are not techno-efficient or cost-effective methods. But some techniques like physical adsorption are highly effective and feasible for the elimination of heavy metal from waste stream; it includes activated carbon use as adsorbent in the water and wastewater treatment plants (Ahmad et al. 2020a, b). But activated charcoal is not an economically effective method. After the extensive research, the most cost-effective alternative has come out, i.e., the use of low-cost agricultural and horticultural by-products such as sugarcane bagasse, rice husk, sawdust, coconut husk, oil palm shell, neem bark, etc. for the removal of trace metals from the wastewater, and it has been investigated successfully (Sharma et al. 2013; Castillo-Monroy et al. 2020; Tandon et al. 2020). This quality of wastewater promotes the number of researches and studies globally. The reuse of wastewater in agriculture investigations and their report states that judicious and rationale use will enhance the properties of soil, promote plant growth, and appreciate the bacterial activity. But unjustifiable and irrational use will create plenty of other issues.

With the above context, the present chapter has all the detailed information regarding the status of wastewater generation and their treatment. The consequent effect of wastewater irrigation on soil and plant health is also discussed with a short description on the most cost-effective technique, i.e., the use of agricultural residues to mitigate the problem of metal toxicity up to a certain extent. It will definitely provide crisp information to all the researchers globally, to solve the different issues related to wastewater irrigation as well as how to enhance their beneficial role for irrigational purpose.

8.2 Global Framework on the Utilization of Wastewater

Currently, over 80% of the wastewater generated globally flows back into the ecosystem without being treated or reused (WWAP 2017). Untreated wastewater-irrigated land is ten times larger than the area using treated wastewater (Drechsel and Evans 2010). According to Chopra and Pathak (2015), global wastewater discharge is 400 billion m^3/year, and it pollutes ~5500 billion m^3 of water per year. The wastewater disposal is done by utilizing it in agriculture for centuries in many cities such as Milan, London, Paris, and Berlin (AATSE 2004). However, this type of wastewater disposal is more important in water-scarce region. According to FAO-AQUASTAT (2012), 11.37 km^3/year wastewater was treated in the year 1993 in Japan, whereas according to World Bank (2012) estimation, wastewater treatment in Japan had increased upto14.65 km^3/year. Nyachhyon (2006) found that 0.351 km^3/year volume of wastewater was generated in Nepal but only 2.01 per ha area has been irrigated with untreated water (AQUASTAT 2012). According to UNMIK (2003) report, 167,000 m^3/year of raw sewage was recorded in Kosova. In Spain 3.16 km^3/year was treated, but only 0.368 km^3/year was used (National Statistics Institute, Spain 2012). Out of the total national vegetable production, 26% was irrigated with wastewater in Pakistan (Ensink et al. 2002). In Hanoi, Vietnam, a total of 80% of vegetables is grown by using wastewater from urban and peri-urban areas (Lai 2000a, b). Keraita and Drechsel (2004) estimated 11,500 ha area is irrigated with diluted wastewater in Ghana. In Mexico, about 260,000 ha were irrigated with untreated wastewater (Mexico CAN 2004). The World Health Organization declared that around 20 million hectares of area are irrigated using untreated wastewater throughout the globe (WHO 2006). In Mexico, China, India, and Pakistan, for instance, large areas exist where untreated wastewater recycles in agricultural irrigation from a considerable time (Jimenez and Asano 2008). India is a country with a large population, so it discharges a significant amount of wastewater and stands in the queue of those nations who reuse the water in agriculture.

India is moving toward the deepwater crisis, and it is an alarming situation for the nation (UNICEF FAO 2013). Now time is knocking our door to take a few steps urgently for remediating this miserable situation. The World Bank (2019) predicts a terrific water crisis across many parts of this country by 2050. India would start battling for a withdrawal of 40% of the renewable surface water resources. The rapid expansion of cities and the discharge of gray/wastewater grow up in the same proportion. The Central Public Health and Environmental Engineering Organization estimated that out of the total water supplied for domestic use about 70–80% gets generated as wastewater. As per CPCB (2016), about 61,754 million liters per day

(MLD) of wastewater are generated by the urban population, and 38,791 MLD rest untreated. The per capita wastewater generation by the class I cities and class II towns, representing 72% of the urban population in India, has been estimated to be around 98 liters per capita per day (lpcd). The Central Public Health and Environmental Engineering Organization estimated that out of the total water supplied for domestic use, a major portion (70–80%) is discharged as wastewater. According to the CPCB (2007) report, Maharashtra, Delhi, Uttar Pradesh, West Bengal, and Gujarat are the major contributors of 63% wastewater generation. As per the study, only 3.42 billion cubic meter (BCM) industrial wastewater was utilized for agricultural productivity in India, and this estimation is 1/30 of Japan and Republic of Korea utilization (Van-Rooijen et al. 2008). Bhardwaj (2005) projection for 2050 revealed 48.2 BCM (132 billion per day) wastewater would be generated. Thus, in the coming years, due to increased population and industrialization, there will be major problems of reduced freshwater availability as well as increased wastewater generation (Boretti and Rosa 2019).

The unprocessed and non-treated urban and domestic wastewater loaded with organic and bacterial contamination is responsible for the degraded condition of both surface and groundwater (Durán–Álvarez and Jiménez–Cisneros 2014). Wastewater is a cocktail of both essential and nonessential nutrients, microbes, heavy metals, and many emerging pollutants. Its properties are derived by many factors, place of generation, surroundings, environment, climatic factor, etc. Many countries have their agency to recommend the limits of irrigational water quality criteria and monitor the quality of effluent from sources.

8.3 Physicochemical and Biological Properties of Wastewater

Properties of wastewater are shaped by the result of its inputs. Inputs depend on multiple anthropogenic activities, which produce wastewater in different means either in the form of domestic or commercial and industrial. Several factors extrapolate the quantity, characterization, and composition of urban wastewater such as levels of domestic and industrial effluents and sewer system as well as treatment systems. The lifestyle of inhabitants and standard of living also affect the characteristics of wastewater (Henze and Comeau 2008).

Physicochemical characterization of wastewater evaluation is based on numerous parameters, for instance, pH, electrical conductivity (EC), turbidity, dissolved solids, suspended solids (SS), salinity, sodium absorption rate (SAR), biological oxygen demand, chemical oxygen demand, sodium, potassium, total nitrogen, and total phosphorus. Standard parameters that are not restricted in wastewater reuse for irrigation are as follows: pH 6.0–9.0 (US EPA 2012), suspended solids 30 mg/L (US EPA 2012), maximum sodium absorption ratio is 26 (CPCB 2011), BOD 10 mg/L (US EPA 2012), fecal coliform restricted till 2×10^2 CFU/100 mL

(US EPA 2012), *Escherichia coli* maximum value is 10 CFU/100 mL (WHO 2006), and nematode eggs less than or equal to 1 (WHO 2006). The raw effluent is generally disposed of on agricultural lands within urban and peri-urban areas to grow vegetables even though it contains a range of contaminants such as metal ions (Siebe 1998; Bouwer 2000; Qadir et al. 2000; Elgallal et al. 2016). Wastewater has several benefits, but an excessive amount of heavy metals is the inevitable part of wastewater. India's maximum value of iron and zinc is 7.74 and 3.24, respectively, in wastewater. Wastewater reuse in irrigation has some regulations and standards that have been recommended by many countries and international agencies (CPCB, FAO, WHO, US EPA).

Biological characteristics are an important factor in wastewater. According to Sarkar et al. (2018), the main concerns of the microorganism group in wastewater are bacteria, fungi, algae, protozoa, viruses, and pathogenic microorganisms. Some important bacteria are *Pseudomonas* sp., which reduces NO_3 to NO_2, and *Acinetobacter* sp., which stores large amounts of phosphates under aerobic conditions and releases it under an anaerobic condition (Jin et al. 2015a, b). Wastewater also contains several fungal species, which mainly help in decomposing the complex organic matter to its simple forms. The microbial population of wastewater also includes different types of algal species, which execute eutrophication phenomenon and oxidation of ponds; protozoa, which mainly feed on bacteria and help in the purification of treated wastewater; and viruses (Szymanski and Patterson 2003; Sarkar et al. 2018) (Table 8.1).

8.4 Effect of Wastewater Irrigation

8.4.1 On Physicochemical Properties of Soil

The response of irrigating soils with wastewater has been extensively studied (Kayikcioglu 2012; Durán–Álvarez and Jiménez–Cisneros 2014). Wastewater is a repository of nutrients and improves soil properties. Irrigating soil with wastewater is an attractive option because it can improve the physical, chemical, biological, and biochemical properties of soil (Pomares et al. 1984; Kiziloglu et al. 2008). Physical properties are found to be improved such as electrical conductivity (EC), organic matter, cation exchange capacity, and water-holding capacity (Aydin et al. 2015), and it also maintains the stability of the soil aggregates and soil porosity (Durán–Álvarez and Jiménez–Cisneros 2014). Wastewater-irrigated soils showed slightly lower pH values compared to non-wastewater-irrigated, probably due to the high organic matter content of irrigation water (Nayak et al. 2007; Mosse et al. 2011). Kulandaivelu and Bhat (2012) amended coffee processing wastewater (CPW) at different loading rates such as 25, 50, 75, and 100 L per 1 m^2 at 0–15 cm soil depth. The application of higher rate of CPW, leads to significant increase in the bulk density (BD) and water-holding capacity (WHC), as well reduced the soil

Table 8.1 Comparison between treated wastewater and untreated wastewater

Parameter	Treated wastewater[a]	Untreated wastewater[a]	Recommended standards
pH	8.13	7.50	6.5–8.4
Electrical Conductivity	1225 μScm^{-1}.	1032 μScm^{-1}	0–3000 μScm^{-1}
Total Dissolved solid	624 ppm	560 ppm	0–2000 ppm
Total nitrogen	61.6 ppm	78.4 ppm	0–10 ppm*
BOD	30 mg/ L	650 mg/L	100 mg/l**
COD	145 mg/L	964 mg/L	–
Sodium	60 ppm	48 ppm	0–40 ppm
Potassium	20 ppm	24 ppm	0–2 ppm

[a]Data source: Alghobar and Suresha (2015)
*Laboratory determinations needed to evaluate common irrigation water quality problems (FAO 1985)
**Standard shall be applicable for industries, operations, or processes other than those industries, operations, or processes for which standards have been specified in Schedule of the Environmental Protection Rules, 1989

pH. Consequently, acidic nature in the soil further reduces the soil microbial activities (Herpin et al. 2007; Dhananjaya et al. 2009; Selvamurugan et al. 2010).

Several researches proved that wastewater has fertilizer value; it has all necessary minerals which induce soil health and plant growth. Irrigation with untreated wastewater may increase soil organic matter, nitrogen, and concentrations of Na, K, and Ca (Siebe 1998; Angin et al. 2005; Lucho-Constantino et al. 2005; Mapanda et al. 2005). Singh et al. (2012) reported that adding wastewater to arable land increased organic matter many folds as compared to control treatment (Rutkowski et al. 2007). Microbial species and its population help in monitoring the environment or ecosystem health. Microbial activities are acting as biological indicators. Biological indicators are assigned to represent the living soil and its environment. Microbial activities in soil were escalated under sewage effluent irrigation (Meli et al. 2002; Ramirez-Fuentes et al. 2002; Chaerun et al. 2011). Soil contains consortia of all major groups of microorganisms, i.e., bacteria, actinomycetes, fungi, and algae. These microorganisms convert toxic organic and inorganic compounds to harmless products, often carbon dioxide, water, etc. They also affect the microbiological and biological properties of soil like microbial biomass carbon (MBC) and respiratory rate of the total microbial population. Both short- and long-term effects of wastewater on soil quality can be monitored by the microbial metabolic quotient (qCO2) and enzymatic activities (Gianfreda et al. 2005; Adrover et al. 2012).

However, the long-term application of untreated wastewater in irrigation could lead to inevitable heavy metal accumulation and a resultant loss of soil character (Singh et al. 2005; Mapanda et al. 2005; Sharma and De 2007; Jin et al. 2008; Singh et al. 2012). Mobility and retention of heavy metals are strongly affecting the soil texture; high clay structure of the soil blocks infiltration of pollutants and causes accumulation in the top layer (Christou et al. 2017). Guédron et al. (2014) conveyed via application of untreated wastewater in irrigated land of Mezquital Valley, Mexico City, the presence of high concentrations of metals such as methylmercury and lead. Addition of non-reclaimed wastewater for 10 years in the agricultural land of Harare, Zimbabwe, had exceeded the heavy metals more than acceptable limits (Mapanda et al. 2005). Furthermore, increased soil salinization, alkalinization/acidification, and structural deterioration had observed reduced soil productivity (Klay et al. 2010). Wastewater promotes a continuous supply of organic matter and microorganisms with the help of information and stability of soil aggregates (Durán–Álvarez and Jiménez–Cisneros et al. 2014). Saha et al. (2010) reported that untreated sewage irrigation for 5 years has significantly increased the number of fungi and actinomycete population in the surface soil by, respectively, about 125% and 75%. But ratios of bacteria to fungi and actinomycetes to fungi decreased considerably in both the layers due to sewage irrigation for 5 years. This indicates that an increase in the fungal population was more than the proportional increase in bacteria and actinomycete population, which may be due to the influx of active C through untreated sewage water.

Chaerun et al. (2011) have compared the CO2-C evolution between controlled and industrial wastewater irrigated land. Even after, 31 years of irrigation, the contaminated soil showed higher CO2-C evolution in waste water irrigated soils and has consistent with the exceeded levels of extractable total DNA and higher levels of microbial biomass. The high CO_2 respiration activity in polluted soil was indicative of higher utilization of organic matters by the microbial community, consistent with its high organic C availability (Yang et al. 2019).

To investigate the health of soil microorganisms in response to wastewater application, different soil enzymes are analyzed such as dehydrogenase, urease, alkaline phosphatase, catalase, and glucosidase. Among all the enzymes, dehydrogenase activity depicts the metabolic status of the soil and can act as an important indicator of microbial activity (Nannipieri et al. 2003). The activity of the dehydrogenase enzyme is usually higher in wastewater-irrigated land and strongly depends on concentrations of wastewater (Garcıa-Gil et al. 2000; Arif et al. 2016). Moreover, activities of soil enzymes are very sensitive to the concentrations of toxic substances such as salts and heavy metals. Xian et al. (2015) reported that higher concentrations of heavy metals in wastewater inhibited the activities of soil enzyme and microbial metabolism. Table 8.2 shows numerous studies that have reported positive or negative effects of wastewater on soil properties and ultimately on soil health (Rutkowski et al. 2007; Mosse et al. 2011; Morugán-Coronado et al. 2013).

8.4.2 *On Plant Characteristics*

Agricultural use of wastewater is a sustainable approach, but this technique is followed up from ancient time, or we can say “old wine in a new bottle.” Wastewater is a continuous and good source of nutrients and provides adequate moisture

Table 8.2 Physicochemical and biological responses of soil amended with wastewater

Research area	Soil properties	Effects	References
Island of Mallorca, Spain	*pH*	Decreased	Adrover et al. (2012)
North Coorg, Karnataka, India	*Electrical conductivity*	Increased	Kulandaivelu and Bhat (2012)
Rohtak, Haryana, India	*Organic carbon*	Increased	Rana et al. (2010)
El Gabal El Asfar farm, Cairo governorate, Egypt	*Soil organic matter*	Increased	Mohammed et al. (2018)
North Coorg, Karnataka, India	*Available nitrogen and phosphorus*	Decreased	Kulandaivelu and Bhat (2012)
Island of Mallorca, Spain	*Heavy metals (Zn, Mn, Fe, Cu)*	Increased	Adrover et al. (2012)
Barossa Valley and the McLaren Vale regions, Australia	*Macronutrient (K, Ca, Na)*	Increased	Mosse et al. (2011)
University of California, Riverside, United States	*Microbial communities*	Increased	Ibekwe et al. (2018)
Island of Mallorca, Spain	*Soil enzymes (DHA, urease, FDA)*	Decreased	Adrover et al. (2012)

required for crop growth. A significant number of researchers investigated irrigating plants with wastewater (Saravanamoorthy and Kumari 2007; Gassama et al. 2015). Application of untreated municipal wastewater at different concentrations (0, 2.5, 5, 10, 25, 50, and 100%) on rice variety (MR220) showed that maximum germination percentage and seedling vigor were recorded at 25% concentration with 92% germination rate. Lower concentration of wastewater could be able to improve the seed germination, while higher wastewater concentrations showed reducing effect on the rice variety (Gassama et al. 2015). Wastewater satisfies the need of NPK and reduces the cost of chemical fertilizers. Moreover, those crops, grown in peri-urban agriculture, require specific amounts of NPK for yield; it fulfills the need. Crop productivity may be enhanced due to reducing of chemical fertilization requirement with the supply of nutrient contents in the wastewater (Vergine et al. 2017). Kiziloglu et al. (2008) compared the uptake/enrichment values of various nutrients and trace elements in red cabbage and cauliflower plants irrigated by untreated wastewater (UTWW), treated wastewater, and control. Results depicted that wastewater irrigation positively affected cauliflower and cabbage yields. The highest yields of cauliflower (28,534 kg/ ha) and red cabbage (46,865 kg /ha) were obtained with UTWW. But with numerous benefits, once the nutrient level of NPK exceeded from standard recommended limits, crop growth and yield may negatively be affected. Gassama et al. (2015) investigated the concentration of macronutrient (N, P, K, Ca, Mg) increased at certain concentrations irrigated with untreated wastewater. Untreated municipal wastewater applications might be influencing the physiological process that leads to an increase in growth (Abbas et al. 2007). Brunetti et al. 2007 investigated and found elevation in the yield of maize in olive mill wastewater-amended soil than control soil. It is possibly because of both humified and non-humified soil organic carbon.

The quantity below the prescribed limit with an adequate level of minerals in wastewater improves plant health. Along the plant growth, it has several benefits such as promoting yield, increasing farmer's income, and reducing the cost of chemical fertilizer. Lower concentration of wastewater does not express a toxic response on seed growth; it even has some beneficial effect on the growth and development of the crops in certain concentrations of wastewater (Medhi et al. 2011). There are many experiments and researches done which verify that the minimal concentration of wastewater nourishes the plant properties and improves seed length and root surface and high concentration becomes a risk for the ecosystem (Dash 2012; Fendri et al. 2013; Kaliyamoorthy et al. 2013). Dash (2012) has reported that when 75% of sewage is used then there was decline in seedling length but when treated with 25–50% wastewater concentration then the seedling lengths were increased in both rice and wheat plant. The lower concentrations of effluents provide nitrates and sulfates to the soil that stimulate the protein production and other organic molecules in order to increase the length and growth of plant seedlings (Yousaf et al. 2010).

Although, wastewater irrigation shows positive effects up to a certain concentration, beyond the threshold value, it is perilous for the plant health. Saravanamoorthy and Kumari (2007) applied textile wastewater for seed germination and reported seed germination reduced at 100% concentration. The decrease may be due to the adverse effect of the high toxicity of the wastewater at a higher concentration (Fendri et al. 2013). Daud et al. (2015) clarify in a way that a significant reduction of nutrient uptake under high concentrated effluents might be due to a decrease in water uptake at a higher level of salinity because of the toxicity of high osmotic pressure due to high soluble salts. In some industrial wastewater quality dominated by enormous hazardous chemical pollutants, such contaminants biomagnifies with each upgraded trophic level (Akhtar et al. 2018). Entered elements in the food chain promote various diseases in plants and humans. On the other hand, the dominance of domestic wastewater may result in high salinity levels that may affect the yield of salt-sensitive crops (IWMI 2002). An experiment was conducted in Xinxiang city in China by Ma Shou et al. (2015); this study involves the application of mine wastewater on winter wheat 9023 variety. The study illustrates that mine wastewater has enriched chromium (Cr) and lead (Pb); so it the root physiological system and hampers photosynthetic, biochemical activities in the flowering stage. Heavy metal impedes soil enzymes; it reduces the rate of decomposition and transformation of soil organic matters, synthesis of humus, release, and various redox reactions of soil nutrients, which causes negative effects on plant growth and grain yield.

Vaverková et al. (2019) investigated on seed germination of hemp seed variants in 100% wastewater; the color of the root turned brown after germination resulting in mortality because heavy metal accumulation is known to be a decline for plants affecting ribonuclease, amylase, and protease enzyme activity, thus hindering seed growth and germination (Ahmad and Ashraf 2012). Garg et al. (2006) observed that excessive salt in textile effluent has inhibited root surface area. It might be because of the immense salt concentration in wastewater (Paliwal et al. 1998; Fendri et al. 2013). A study performed by Gupta et al. (2011) in Burdwan, West Bengal, amended mixture of industrial wastewater from Tamla drain water. This mixture integrated with wastewater of steel plants, thermal power plants, alloy steel plants, and also untreated sewage; it is applied to irrigation of mustard (*Brassica nigra*), radish (*Raphanus sativus*), and *Colocasia* (*Colocasia esculenta*). As per experiment plants respond negatively with wastewater. Amino acid content lowers in the shoot and root of *Colocasia*, *Brassica*, and the root of *Raphanus*. Studies have shown that stress induces the decline in insoluble protein and total chlorophyll contents in plants (Hsu and Kao 2003; Rong Guo et al. 2007). The results show a slight increase in phenol content in all three plants. An increase in phenol concentration was observed in either stress condition or a favorable environment (Dučić et al. 2008). The other biochemical parameter, i.e., ascorbic acid, is found to increase marginally in all wastewater-irrigated plants than control. Ascorbic acid as an antioxidant plays an important role in the protection against physiological stress (Guo et al. 2005). Reactive oxygen species (ROS) are produced in plants in response to the damaging effects of environmental stresses, and plants have evolved a variety of antioxidant defense mechanisms in response to stress (Chow et al. 2017, 2018). An experiment was performed at Morocco in a 3 m × 2 m plot amended with 10 L crude olive mill wastewater/m^2 for maize production. The result concluded that the release of phenolic compound and secretion of total peroxidase activity in plants deliver an evidence of their protective role against the physiological stress-induced treatment of olive mill wastewater (Belaqziz et al. 2016). Thus, heavy metal accumulation shows an increase in ascorbic acid content. The effect of toxicants varies from species to species depending upon several factors (Singh et al. 2010) (Fig. 8.1).

8.5 Use of Agricultural Residues to Reduce Metal Contamination

With the above discussion, it is now clear that the major challenge in using wastewater for irrigating the agricultural fields is heavy metal contamination. Particularly, contamination of heavy metals of soil and plant system is one of the major environmental concerns in the world due to its impact on human health (Nkwunonwo et al. 2020). In order to reduce this problem, various remediation options are in practice like pneumatic fracturing, vitrification, chemical reduction/oxidation, and electrokinetics, but they are not found to be very cost-effective and thus not environmentally sustainable techniques. Over to these techniques, there are other cost-effective and less disturbing techniques, which can be easily applied in metal-contaminated system to reduce heavy metal availability. For reducing the availability, many amendments have been used such as addition of lime (Rinklebe et al. 2015), phosphate (Shaheen et al. 2017), and organic and inorganic fertilizers. These amendments lead to the changes in physicochemical properties of the soil

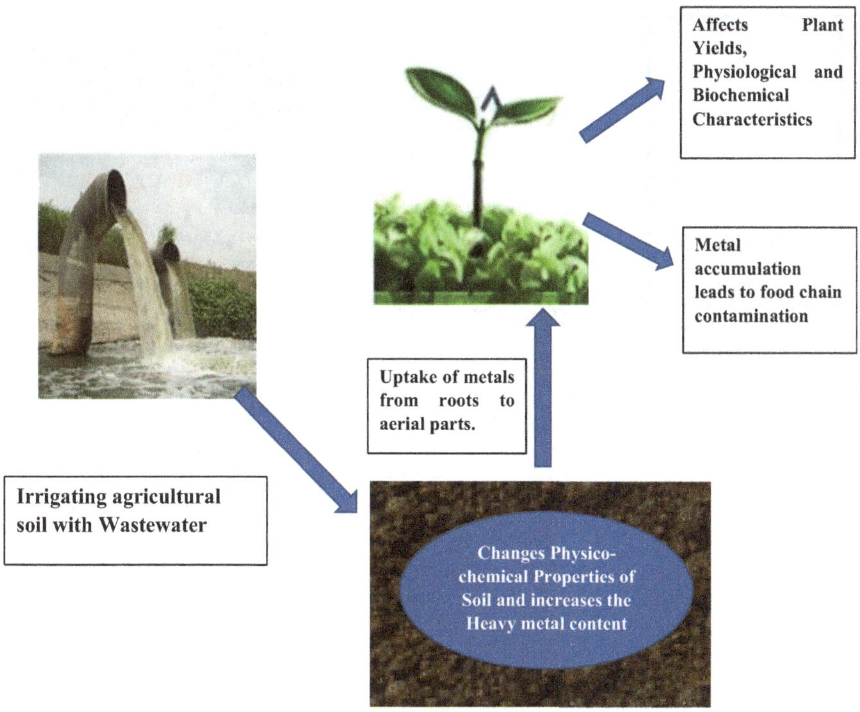

Fig. 8.1 Schematic diagram of reuse of wastewater in agriculture

such as pH, organic matter, etc., which affect the solubility and thus the availability of heavy metals to the plants (Palansooriyaa et al. 2020).

Along with these amendments, now applications of the agricultural residues are also studied to observe their role in reducing the availability of heavy metals. The agricultural residues are the organic materials produced as by-products of the harvesting and processing of agricultural crops such as wheat straw, rice husk, wood fiber, and sugarcane bagasse (Tran et al. 2015). They are preferred because of their low cost, bouncy source, fast recovery period, chemical stability, inexhaustible process, and renewable energy and are environmentally friendly (Alalwan et al. 2020). Such substances have high porosity and higher surface area that make them effective adsorbents for removing toxic heavy metals from the soil and other system. The straw is generated from sugarcane, rice, wheat, cotton, and corn. The generated by-products have been applied as an adsorbent for reducing availability of heavy metals. Agricultural residues mainly consist of compounds such as cellulose, lignin, and hemicelluloses. The composition of such agricultural residues is varied and depends on the forms of the residues and their local environmental situation. The polar functional groups present over these substances such as aldehyde, ketones, alcohol, ether, phenolic, and carboxylic groups are capable of adsorbing heavy metals from the metal-contaminated soils and aqueous system (Fig. 8.2).

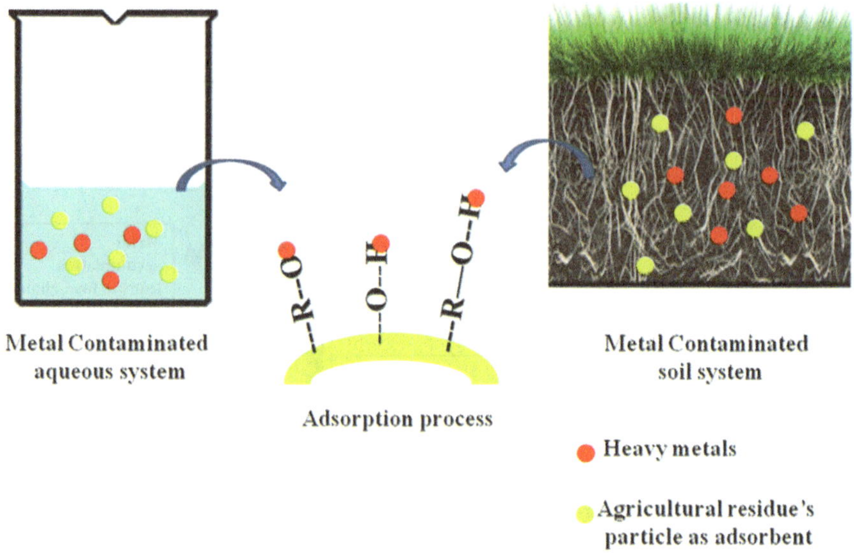

Fig. 8.2 Schematic representation to show the adsorption of metals by the functional group present over the surface of agricultural residue particle

Many of the agricultural residues include rice husk, peels, wheat waste, and bagasses and are commonly used to remove heavy metals from soil. Using agricultural residues in metal immobilization and other uses is strongly encouraged in order to reduce the rate of air pollution and for sustainable agriculture.

Rice husk, sawdust, coconut husk, and other agricultural residues are used for extracting the metal ion from metal-contaminated system. They have several reasons to be good adsorption materials for reducing availability of heavy metals such as low cost, easy to avail, can be reused, easy to separate, and high adsorption capacities with high removal efficiency (Aydin and Aksoy 2009). In recent years, the use of agricultural residues to remove toxic heavy metals has increased to a great extent.

8.5.1 Use of Rice Husk

Rice husk is produced in large quantities as a by-product in rice milling industry. It consists of 35% cellulose, 25% hemicelluloses, 20% lignin, 17% ash (including silica), and 3% crude protein.

These compounds are associated with many chelating ligands and functional groups that help in the fixation of metal ions. The ash produced by the burning of rice husk can also be used as a good adsorptive material and has been reported for the removal of metal ions (Srivastava et al. 2006). Tawwab et al. (2017) have applied rice straw, sugarcane bagasse, and maize stalks in metal (Pb, Cd, Cu, Zn)-

contaminated aquarium and found that among all agricultural residues, rice straw is found to be best for reducing the metals in the polluted aquarium water and consequently reducing their accumulation in the fish body. Lu et al. (2017) have applied bamboo and rice straw biochars in metal-contaminated paddy soil; they are capable of reducing the availability of Cd, Cu, Pb, and Zn by increasing the soil pH.

Shu et al. (2016) applied different forms (dry straw, composted straw, straw biochar, and straw ash) of rice straw to immobilize Hg in soils by forming methyl-Hg compound by binding with organic matter present in amended rice straw. Zhang et al. (2018) have also observed that by incorporating rice straw, there was 28%–136% enhancement in methyl-Hg levels in contaminated soil by reducing their availability to the plants. With the enhancement in the level of methyl-Hg in soils, the microbial activity and dissolved organic matter get increased that consequently induce the formation of Hg-S-DOM complexes to reduce the phyto-availability of Hg (Shu et al. 2016; Zhu et al. 2015). Through batch adsorption technique, Akhtar et al. (2010) have removed Pb(II), Cd(II), Zn(II), and Cu(II) divalent metal ions from aqueous solutions with the help of rice husk (RHA) after chemical and thermal activation with 0.1 M HNO_3 and 1 M K_2CO_3 at 473 K.

Srivastava et al. (2008) have used rice husk ash (RHA) to remove Cd and Zn and reported the competitive adsorptive removal of Cd(II) and Zn(II) ions from binary systems using rice husk ash (RHA). A rice husk after some modification has been used for adsorption of nine heavy metals by Krishnani et al. (2008). Rice husk was treated with 1.5% alkaline solution and autoclaved at 121 °C for 30 min to remove the lignin (low molecular weight) compounds. The characterization through scanning electron microscope and Fourier transform infrared spectroscopy showed that, particularly, the calcium and magnesium available at the surface of biomatrix are responsible for adsorption of heavy metals through ion exchange mechanism. The adsorption capacity of different heavy metals showed increasing trend having the order Ni(II) < Zn(II) < Cd(II) < Mn(II) < Co(II) < Cu(II) Hg(II) < Pb(II) < Cr(III). This treated rice husk has higher adsorption capacity than other kinds of sorbents (Krishnani et al. 2008). Hegazi (2013) have treated rice husk (20 g) with 13 M sulfuric acid (100 ml) and heated at 175–180 °C for 20 min with stirring at regular interval. The mixture obtained was black in color. It was filtered and allowed to cool and finally filtered under vacuum with the help of the Buchner funnel. This treated sorbent can be used to eliminate up to 20–60 mg/l Fe(II), Pb(II), Ni(II), Cd(II), and Cu(II) from wastewater of electroplating industries. Elham et al. (2010) reported that adsorption of Pb and Zn by rice husk is also based upon the amount of adsorbent, contact time, and pH value of wastewater, which were the factors influencing Zn (II) and Pb(II) ion sorption.

8.5.2 Use of Sawdust

Along with the rice husk, for removing heavy metals from metal-contaminated system, sawdust is also used as an attractive low-cost adsorbent. Witek-Krowiak

(2013) has used untreated beech sawdust to adsorb Cu(II) and Cr(III) from water system. It adsorb about 30.22 $mg.g^{-1}$ and 41.86 $mg.g^{-1}$ of Cu(II) and Cr(III), respectively With some modification, Samarghandi et al. (2011) have used holly sawdust to remove Ni(II) from aqueous solutions. At pH 7, the maximum adsorption capacity was found to be 22.47 $mg.g^{-1}$. Sawdust obtained from *Acacia leucocephala* has been applied to reduce the level of Cu(II), Cd(II), and Pb(II) at different pH, i.e., 6.0, 5.0, and 4.0. The adsorption capacity was found to be 147.1 $mg.g^{-1}$, 167.7 $mg.g^{-1}$, and 185.2 $mg.g^{-1}$, at respective pH (Munagapati et al. 2010). Kapur and Mondal (2013) used the sawdust from *Mangifera indica* to remove Cr(VI) with the efficiency of 99.99% at pH 2.0. Larous et al. (2005) has confirmed that untreated sawdust can also be used as adsorbent to remove the heavy metals from wastewater. By converting into activated carbon, *Hevea brasiliensis* sawdust is used to reduce the level of Cr from wastewater (Karthikeyan et al. 2005). Nordine et al. (2016) have used sawdust obtained from pine, beech, and fir tree to reduce the Pb content of metal-contaminated effluents. By analyzing through Freundlich and Langmuir isotherms, the adsorption capacity of maple sawdust was reported to be 1.79 $mg.g^{-1}$ for Cu and 3.19 $mg.g^{-1}$ for Pb (Yu et al. 2001).

Several factors are responsible for adsorption capacity of sawdust and other adsorbent such as time of contact, pH, concentration of adsorbent, initial metal concentrations, adsorbate concentration, temperature, adsorbent particle size, etc. (Park et al. 2010). To understand the principle and mechanism behind the adsorption, different models are done such as Langmuir, Freundlich, Dubinin–Radushkevich isotherm models, etc. (Shukla et al. 2002). They give ideas about sorption capacity, sorption intensity, and energy adsorption. Pretreatment of sawdust and other adsorbent can enhance their adsorption efficiency by increasing the adsorption site amount and exchange properties that favor the metal removal.

Sawdusts of poplar, willow, fir, oak, and black locust wood have been applied for the removal of Cu, Zn, Cd, and Ni from aqueous solutions; along with this, effects of contact time, pH, and particle size have also been investigated. Cu was maximally removed followed by Ni > Zn > Cd. The adsorption of metals increased with decreasing the particle size of sawdust (Sciban and Kasnja 2004).

8.5.3 Use of Other Agricultural Residues and Characterization Process

Other agricultural residues include sugarcane bagasse that is found to be one of the most lucrative sorbents, especially available in tropical regions (Sarker et al. 2017). It is produced as a by-product of the sugar and bioethanol mills in large quantities and can significantly remove a wide range of heavy metal from contaminated system (do Carmo Ramos et al. 2016). Fibrous residue of sugarcane bagasse contains several functional groups that bind with the metals (Abdelhafez and Li 2016). Just like sawdust and rice husk, the sorption efficiency of sugarcane bagasse is also dependent upon pH, temperature, contact time, and the size of sorbent's particle. It has been observed that 96% of Cd can be removed by sugarcane baggage with 25 min of contact time from aqueous solution having pH 7.0 (Moubarik and Grimi 2015). The pretreated sugarcane bagasse can adsorb 92.6, 149.0, and 333.0 mg g^{-1} of Cu, Cd, and Pb, respectively (Karnitz et al. 2010). Esfandiar et al. (2014) have reported that the adsorption efficiency can be increased by increasing adsorbent dosage. Coconut waste can also be used as adsorption material. It can adsorb about 263 and 285 mg g^{-1} of Pb and Cd, respectively. Black oak bark and wheat bran can also adsorb 400 mg g^{-1} and 310 mgg^{-1} of Hg and Cr, respectively (Alalwan et al. 2020).

To determine the physical and chemical properties and to get the information about active sites involved in metal adsorption, the characterization process is performed. It is done by several conventional methods, such as X-ray powder diffraction (XRD), X-ray photoelectron spectroscopy (XPS), X-ray fluorescence (XRF), scanning electron microscope (SEM), energy dispersive X-ray analysis (EDAX), Fourier transform infrared spectroscopy (FTIR), and nuclear magnetic resonance (NMR) (Park et al. 2010). SEM is an efficient technique that can be used for morphological analysis of adsorbent's surface before and after the adsorption of metals (Almendros et al. 2015). Along with this, EDAX technique gives an idea about elemental analysis and characterization of chemical compounds at the surface of the adsorbent (Habib-ur-Rehman et al. 2006). For further study, FTIR spectroscopy can be performed to know about the sorbent functional groups responsible for the adsorption of metal ions. Surface chemistry of materials can be assessed by XPS, i.e., a quantitative technique. It gives the detailed information about the composition of element, empirical formula, and material's electronic state (Michalak et al. 2013). With the help of different characterization techniques, the whole information about the heavy metals adsorption phenomenon can be obtained. So, by using a combination of techniques, the adsorption mechanisms can be explored significantly.

8.6 Conclusions

Wastewater use in agriculture enhances the soil properties such as its aggregation, moisture, and nutrient that consequently enhance the growth and yield of the plants. Wastewater irrigation has multifaceted benefits in sustainable agriculture: (i) reduces the cost of synthetic fertilizers, (ii) improves farmer's income, (iii) continuous water supply in water-deficient regions, (iv) curtails the dependency on the weather for water supply, and (v) upgrades livelihood for poor households, although several risks are also conjugate with the use such as salinization, alkalization, and heavy metal contamination. The irrational and non-judicious use of wastewater recycling in agriculture could provoke change in properties of soils and plants; lastly it shows toxicological response on human health. Several researches do prove that heavy metal enters in the food chain and biomagnifies in every trophic level. Hence, wastewater irrigation should be monitored regularly with irrigation water standards and applied cautiously, in accordance with optimal irrigation-fertilization policies, to decrease leaching and combine with different methods, to decrease salinity and other pollutants. Application of several strategies should be promoted to reduce the metal level in wastewater in order to enhance its beneficial aspects. Among all, the use of agricultural residues is found to be best for reducing the metal availability. With the pretreatment, the adsorption efficiency of different adsorbent can be increased, so future research in this field should be focused in order to find out the best way of reducing the main culprit of wastewater, i.e., heavy metal.

Acknowledgements The authors are thankful to Head and Director of Institute of Environment and Sustainable Development and to the Head, Centre of Advanced Study in Botany, Institute of Science, Banaras Hindu University, Varanasi, for providing necessary facilities. The research is also supported by Science and Engineering Research Board (P07/678) Project. The Banaras Hindu University is also thankfully acknowledged for providing financial assistant to Dr. A. Singh as PI under IoE seed grant scheme (Dev. IoE Scheme No. 6031).

References

AATSE-(Australian Academy of Technological Sciences and Engineering) (2004) Water recycling in Australia Study Director and Authors: Dr. John Radcliffe Victoria

Abbas ST, Sarfraz M, Mehdi SM et al (2007) Trace elements accumulation in soil and rice plants irrigated with the contaminated water. Soil Tillage Res 94:503–509

Abdelhafez AA, Li J (2016) Removal of Pb (II) from aqueous solution by using biochars derived from sugar cane bagasse and orange peel. J Taiwan Inst Chem Eng 61:367–375

Adrover M, Farrús E, Moyà G et al (2012) Chemical properties and biological activity in soils of Mallorca following twenty years of treated wastewater irrigation. J Environ Manag 95:S188–S192

Ahmad MSA, Ashraf M (2012) Essential roles and hazardous effects of nickel in plants. Rev Environ Contam Toxicol. Springer, New York, pp 125–167

Ahmad S, Majhi PK, Kothari R et al (2020a) Industrial wastewater footprinting: a need for water security in Indian context. In: Environment concerns and sustainable development. Springer, Singapore, pp 197–212

Ahmad S, Pandey A, Pathak VV et al (2020b) Phycoremediation: algae as eco-friendly tools for the removal of heavy metals from wastewaters. In: Bioremediation of industrial waste for environmental safety. Springer, Singapore, pp 53–76

Akhtar M, Iqbal S, Kausar A, Bhanger MI, Shaheen MA (2010) An economically viable method for the removal of selected divalent metal ions from aqueous solutions using activated rice husk. Colloids Surf B Biointerfaces 75:149

Akhtar S, Ahmad S, Huifang W, Shahbaz A, Ghafoor A, Imran S, Zafar A (2018) An analysis of wastewater irrigation practices and its impacts on the livelihood generation and food chain contamination in Faisalabad District, Pakistan. ISABB J Health Environ Sci 5(4):33–42

Alalwan HA, Kadhom MA, Alminshid AH (2020) Removal of heavy metal from waste water using agricultural by products. J Water Supply Res Technol AQUA 69:99–112

Alghobar MA, Suresha S (2015) Evaluation of nutrients and trace metals and their enrichment factors in soil and sugarcane crop irrigated with wastewater. J Geosci Environ Protect 3(08):46

Almendros AI, Martín-Lara MA, Ronda A, Pérez A, Blázquez G, Calero M (2015) Physicochemical characterization of pine cone shell and its use as biosorbent and fuel. Bioresour Technol 196:406–412

Angin I, Yaganoglu AV, Turan M (2005) Effects of long-term wastewater irrigation on soil properties. J Sustain Agric 26(3):31–42

AQUASTAT FAO (2012) AQUASTAT Food and Agriculture organization. www.fao.org/aquastat/en

Arif MS, Riaz M, Shahzad SM et al (2016) Contrasting effects of untreated textile wastewater onto the soil available nitrogen-phosphorus and enzymatic activities in aridisol. Environ Monit Assess 188(2):102

Aryal A, Sathasivan A, Zhan W (2011) Effect of suspended solids in secondary wastewater effluent on DOC removal by enhanced coagulation. Desalin Water Treat 32(1–3):27–32

Aydin YA, Aksoy ND (2009) Adsorption of chromium on chitosan: Optimization, kinetics and thermodynamics. Chem Eng J 151:188

Aydin ME, Aydin S, Beduk F et al (2015) Effects of long-term irrigation with untreated municipal wastewater on soil properties and crop quality. Environ Sci Pollut Res 22(23):19203–19212

Belaqziz M, El-Abbassi A, Agrafioti E et al (2016) Agronomic application of olive mill wastewater: effects on maize production and soil properties. J Environ Manag 171:158–165

Bhardwaj RM (2005) Status of wastewater generation and treatment in India. Intersecretariat Working Group on Environment Statistics (IWG–Env) Joint Work Session on Water Statistics Vienna, 20–22

Bolisetty S, Peydayesh M, Mezzenga R (2019) Sustainable technologies for water purification from heavy metals: review and analysis. Chem Soc Rev 48(2):463–487

Boretti A, Rosa L (2019) Reassessing the projections of the world water development report. npj Clean Water 2(1):1–6

Bouwer H (2000) Integrated water management: emerging issues and challenges. Agric Water Manag 45(3):217–228

Brunetti G, Senesi N, Plaza C (2007) Effects of amendment with treated and untreated olive oil mill wastewaters on soil properties, soil humic substances and wheat yield. Geoderma 138 (1–2):144–152

Castillo-Monroy J, Godínez LA, Robles I, Estrada-Vargas A (2020) Study of a coupled adsorption/electro-oxidation process as a tertiary treatment for tequila industry wastewater. Environ Sci Pollut Res:1–8

Chaerun SK, Pangesti NP, Toyota K et al (2011) Changes in microbial functional diversity and activity in paddy soils irrigated with industrial wastewaters in Bandung, West Java Province, Indonesia. Water Air Soil Pollut 217(1–4):491–502

Chopra AK, Pathak C (2015) Accumulation of heavy metals in the vegetables grown in wastewater irrigated areas of Dehradun, India with reference to human health risk. Environ Monit Assess 187(7):445

Chow YN, Lee LK, Zakaria NA et al (2017) New emerging hydroponic system. Symp Innov Creat (iMIT-SIC) 2:353–358

Chow YN, Lee LK, Zakaria NA et al (2018) Phytotoxic effects of trivalent chromium-enriched water irrigation in Vigna unguiculata seedling. J Clean Prod 202:101–108

Christou A, Karaolia P, Hapeshi E et al (2017) Long-term wastewater irrigation of vegetables in real agricultural systems: concentration of pharmaceuticals in soil, uptake and bioaccumulation in tomato fruits and human health risk assessment. Water Res 109:24–34

CPCB (1999) CPCB Bulletin, Central Pollution Control Board, Ministry of Environment and Forest, Govt. of India, Delhi, India. Available at: http://cpcb.nic.in/

CPCB (2007) CPCB Bulletin, Central Pollution Control Board, Ministry of Environment and Forest, Govt. of India, Delhi, India. Available at: http://cpcb.nic.in/

CPCB (2011) CPCB Bulletin, Central Pollution Control Board, Ministry of Environment and Forest, Govt. of India, Delhi, India. Available at: http://cpcb.nic.in/

CPCB (2016) CPCB Bulletin, Central Pollution Control Board, Ministry of Environment and Forest, Govt. of India, Delhi, India. Available at: http://cpcb.nic.in/

Dash AK (2012) Impact of domestic waste water on seed germination and physiological parameters of rice and wheat. IJRRAS 12(2):280–286

Durán–Álvarez JC, Jiménez–Cisneros B (2014) Beneficial and negative impacts on soil by the reuse of treated/untreated municipal wastewater for agricultural irrigation–a review of the current knowledge and future perspectives. In: Environmental risk assessment of soil contamination, pp 137–197

Daud Z, Awang H, Latif AA et al (2015) Suspended solid, color, COD and oil and grease removal from biodiesel wastewater by coagulation and flocculation processes. Procedia Soc Behav Sci 195:2407–2411

De Fraiture C, Wichelns D (2010) Satisfying future water demands for agriculture. Agric Water Manag 97(4):502–511

Dhananjaya BN, Srinivasamurthy CA, Bhaskar S et al (2009) Utilization of coffee pulp effluent in agriculture and its impact on soil and water environment. Environ Ecol 27(4):1612–1624

do Carmo Ramos SN, Xavier ALP, Teodoro FSE, Gil LF, Gurgel LVA (2016) Removal of cobalt (II), copper (II), and nickel (II) ions from aqueous solutions using phthalate-functionalized sugarcane bagasse: mono-and multicomponent adsorption in batch mode. Ind Crop Prod 79:116–130

Drechsel P, Evans AE (2010) Wastewater use in irrigated agriculture. Irrig Drain Syst 24(1–2):1–3

Dučić T, Maksimović V, Radotić K (2008) Oxalate oxidase and non-enzymatic compounds of the antioxidative system in young Serbian spruce plants exposed to cadmium stress. Arch Biol Sci (Belgrade) 60(1):67–76

Durán-Álvarez JC, Jiménez-Cisneros BLANCA (2015) Reuse of municipal wastewater for irrigated agriculture: a review. In: Wastewater management for irrigation: principles and practices, vol 8, pp 147–226

Elgallal M, Fletcher L, Evans B (2016) Assessment of potential risks associated with chemicals in wastewater used for irrigation in arid and semiarid zones. Rev Agric Water Manag 177:419–431

Elham A, Hossein T, Mahnoosh H (2010) Removal of Zn(II) and Pb (II) ions using rice husk in food industrial wastewater. J Appl Sci Environ Manag 14(4):159

Ensink JH, Van der Hoek W, Amerasinghe FP (2006) *Giardia duodenalis* infection and wastewater irrigation in Pakistan. Trans R Soc Trop Med Hyg 100(6):538–542

Esfandiar N, Nasernejad B, Ebadi T (2014) Removal of Mn (II) from groundwater by sugarcane bagasse and activated carbon (a comparative study): application of response surface methodology (RSM). J Ind Eng Chem 20:3726–3736

FAO (1985) Food and Agriculture organization of United Nation Report

Fendri I, Khannous L, Timoumi A et al (2013) Optimization of coagulation-flocculation process for printing ink industrial wastewater treatment using response surface methodology. Afr J Biotechnol 12(30):4819

Garcıa-Gil JC, Plaza C, Soler-Rovira P et al (2000) Long-term effects of municipal solid waste compost application on soil enzyme activities and microbial biomass. Soil Biol Biochem 32 (13):1907–1913

Garg VK, Kaushik P, Dilbaghi N (2006) Vermiconversion of wastewater sludge from textile mill mixed with anaerobically digested biogas plant slurry employing Eisenia foetida. Ecotoxicol Environ Safe 65(3):412–419

Gassama UM, Puteh AB, Abd-Halim MR et al (2015) Influence of municipal wastewater on rice seed germination, seedling performance, nutrient uptake, and chlorophyll content. J Crop Sci Biotechnol 18(1):9–19

Gianfreda L, Rao MA, Piotrowska A et al (2005) Soil enzyme activities as affected by anthropogenic alterations: intensive agricultural practices and organic pollution. Sci Total Environ 341 (1–3):265–279

Guédron S, Duwig C, Prado BL et al (2014) (Methyl) Mercury, Arsenic, and Lead Contamination of the World's Largest Wastewater Irrigation System: the Mezquital Valley (Hidalgo State—Mexico). Water Air Soil Pollut 225(8):2045

Guo Z, Tan H, Zhu Z et al (2005) Effect of intermediates on ascorbic acid and oxalate biosynthesis of rice and in relation to its stress resistance. Plant Physiol Biochem 43(10–11):955–962

Guo TR, Zhang GP, Zhang YH (2007) Physiological changes in barley plants under combined toxicity of aluminum, copper and cadmium. Colloids Surf B Biointerfaces 57(2):182–188

Gupta VK, Gupta B, Rastogi A et al (2011) Pesticides removal from waste water by activated carbon prepared from waste rubber tire. Water Res 45(13):4047–4055

Habib-ur-Rehman, Shakirullah M, Ahmad I, Shah S, Hameedullah (2006) Sorption studies of nickel ions onto sawdust of Dalbergia sissoo. J Chin Chem Soc 53(5):1045–1052

Hanjra MA, Blackwell J, Carr G et al (2012) Wastewater irrigation and environmental health: Implications for water governance and public policy. Int J Hyg Environ Health 215(3):255–269

Henze M, Comeau Y (2008) Wastewater characterization. Biological wastewater treatment: principles modelling and design, pp 33–52

Herpin U, Gloaguen TV, da Fonseca AF et al (2007) Chemical effects on the soil–plant system in a secondary treated wastewater irrigated coffee plantation—a pilot field study in Brazil. Agric Water Manag 89(1–2):105–115

Hegazi HA (2013) Removal of heavy metals from wastewater using agricultural and industrial wastes as adsorbents. HBRC J 9(3):276–282

Hsu YT, Kao CH (2003) Role of abscisic acid in cadmium tolerance of rice (*Oryza sativa* L.) seedlings. Plant Cell Environ 26(6):867–874

Ibekwe AM, Gonzalez-Rubio A, Suarez DL (2018) Impact of treated wastewater for irrigation on soil microbial communities Sci. Total Environ 622:1603–1610

IWMI (2002) Saline water use in agriculture. International Water Management Institute

Jimenez B, Asano T (2008) Water reclamation and reuse around the world. In: Water Reuse: an international survey of current practice, issues and needs, vol 14. IWA, London, pp 3–26

Jin X, Yang X, Islam E et al (2008) Effects of cadmium on ultrastructure and antioxidative defense system in hyperaccumulator and non-hyperaccumulator ecotypes of *Sedum alfredii* Hance. J Hazard Mater 156(1–3):387–397

Jin L, Sun Q, Xu Q et al (2015a) Adsorptive removal of anionic dyes from aqueous solutions using microgel based on nanocellulose and poly vinyl amine. BioresTechnol 197:348–355

Jin R, Liu T, Liu G et al (2015b) Simultaneous heterotrophic nitrification and aerobic denitrification by the marine origin bacterium Pseudomonas sp. ADN-42. Appl Biochem Biotechnol 175 (4):2000–2011

Kaliyamoorthy J, Kannan TS, Nagarajan M (2013) Effect of raw and biologically treated municipal waste water on growth of paddy (*Oryza sativa* L.). Int J Environ Bioenergy 7(1):1–9

Kapur M, Mondal MK (2013) Mass transfer and related phenomena for Cr (VI) adsorption from aqueous solutions onto *Mangifera indica* sawdust. Chem Eng J 218:138–146

Karnitz O, Gurgel LVA, Gil LF (2010) Removal of Ca(II) and Mg(II) from aqueous single metal solutions by mercerized cellulose and mercerized sugarcane bagasse grafted with EDTA dianhydride (EDTAD). Carbohydr Polym 79:184–191

Karthikeyan T, Rajgopal S, Miranda LR (2005) Chromium(VI) adsorption from aqueous solution by *Hevea brasiliensis* sawdust activated carbon. J Hazard Mater 124(1–3):192–199

Kayikcioglu HH (2012) Short-term effects of irrigation with treated domestic wastewater on microbiological activity of a Vertic xerofluvent soil under Mediterranean conditions. J Environ Manag 102:108–114

Keraita BN, Drechsel P (2004) Agricultural use of untreated urban wastewater in Ghana. In: Wastewater use in irrigated agriculture. CABI Publishing, Wallingford, pp 101–112

Kiziloglu FM, Turan M, Sahin U et al (2008) Effects of untreated and treated wastewater irrigation on some chemical properties of cauliflower (*Brassica oleracea* L. var. botrytis) and red cabbage (*Brassica oleracea* L. var. rubra) grown on calcareous soil in Turkey. Agric Water Manag 95 (6):716–724

Klay S, Charef A, Ayed L et al (2010) Effect of irrigation with treated wastewater on geochemical properties (saltiness, C, N and heavy metals) of isohumic soils (Zaouit Sousse perimeter, Oriental Tunisia). Desalination 253(1–3):180–187

Krishnani KK, Meng X, Christodoulatos C, Boddu VM (2008) Biosorption mechanism of nine different heavy metals onto biomatrix from rice husk. J Hazard Mater 153:1222

Kulandaivelu V, Bhat R (2012) Changes in the physico-chemical and biological quality attributes of soil following amendment with untreated coffee processing wastewater. Eur J Soil Biol 50:39–43

Larous S, Meniai AH, Bencheikh Lehocine M (2005) Experimental study of the removal of copper from aqueous solutions by adsorption using sawdust. Desalination 185(1–3):483–490

Lu K, Yang X, Gielen G, Bolan N, Ok YS, Niazi NK, Xu S, Yuan G, Zhang X (2017) Effect of bamboo and rice straw biochars on the mobility and redistribution of heavy metals (Cd, Cu, Pb and Zn) in contaminated soil. J Environ Manag 186:285–292

Lucho-Constantino CA, Álvarez-Suárez M, Beltrán-Hernández RI et al (2005) A multivariate analysis of the accumulation and fractionation of major and trace elements in agricultural soils in Hidalgo State, Mexico irrigated with raw wastewater. Environ Int 31(3):313–323

Ma Shou C, Zhang HB, Ma ST, Wang R, Wang GX, Shao Y, Li CX (2015) Effects of mine wastewater irrigation on activities of soil enzymes and physiological properties, heavy metal uptake and grain yield in winter wheat. Ecotoxicol Environ Saf 113:483–490

Magwaza ST, Magwaza LS, Odindo AO et al (2020) Hydroponic technology as decentralised system for domestic wastewater treatment and vegetable production in urban agriculture: a review. Sci Total Environ 698:134154

Mapanda F, Mangwayana EN, Nyamangara J et al (2005) The effect of long-term irrigation using wastewater on heavy metal contents of soils under vegetables in Harare, Zimbabwe. Agric Ecosyst Environ 107(2–3):151–165

Mateo-Sagasta J, Medlicott K, Qadir M et al (2013) Proceedings of the UN World Water Development Report 2017 UN-Water

Medhi UJ, Talukdar AK, Deka S (2011) Impact of paper mill effluent on growth and development of certain agricultural crops. J Environ Biol 32(2):185–188

Meli S, Porto M, Belligno A, Bufo SA, Mazzatura A, Scopa A (2002) Influence of irrigation with lagooned urban wastewater on chemical and microbiological soil parameters in a citrus orchard under Mediterranean conditions. Sci Total Environ 285:69–77

Mexico CAN (Comision Nacional del Agua) (2004) Water Statistics. National Water Commission, Mexico City

Michalak I, Chojnacka K, Witek-Krowiak A (2013) State of the art for the biosorption process - a review. Appl Biochem Biotechnol 170(6):1389–1416

Morugán-Coronado A, Arcenegui V, García-Orenes F et al (2013) Application of soil quality indices to assess the status of agricultural soils irrigated with treated wastewaters. Solid Earth 4 (1)

Mohammed N, Grishkewich N, Tam KC (2018) Cellulose nanomaterials: promising sustainable nanomaterials for application in water/wastewater treatment processes. Environ Sci Nano 5(3):623–658

Mosse KPM, Patti AF, Christen EW et al (2011) Winery wastewater quality and treatment options in Australia. Aust J Grape Wine Res 17(2):111–122

Moubarik A, Grimi N (2015) Valorization of olive stone and sugar cane bagasse by-products as biosorbents for the removal of cadmium from aqueous solution. Food Res Int 73:169–175

Munagapati VS, Yarramuthi V, Nadavala SK, Alla SR, Abburi K (2010) Biosorption of Cu(II), Cd (II) and Pb(II) by *Acacia leucocephala* bark powder: Kinetics, equilibrium and thermodynamics. Chem Eng J 157(2–3):357–365

Nannipieri P, Ascher J, Ceccherini MT et al (2003) Microbial diversity and soil functions. Eur J Soil Sci 54:655–670

National Statistics Institute, Spain (2012) Survey on water supply and sewage. http://www.ine.es/welcoing.htm. Accessed 21.12.12

Nayak DR, Babu YJ, Adhya TK (2007) Long-term application of compost influences microbial biomass and enzyme activities in a tropical AericEndoaquept planted to rice under flooded condition. Soil Biol Biochem 39(8):1897–1906

Nkwunonwo UC, Odika PO, Onyia NI (2020) A review of the health implications of heavy metals in food chain in Nigeria, Hindawi. Sci World J 2020, Article ID 6594109, 11 pages

Nordine N, El Bahri Z, Sehil H, Fertout RI, Rais Z, Bengharez Z (2016) Lead removal kinetics from synthetic effluents using Algerian pine, beech and fir sawdust's: optimization and adsorption mechanism. Appl Water Sci 46(4):349–358

Nyachhyon BL (2006) Service enhancement and development of sanitary sewerage system in urban and semi-urban setting in Nepal. Policy Paper 23:86–94

Palansooriyaa KN, Shaheen SM, Chene SS, Tsange CWD, HashimotoY HD, Bolanh NS, Rinklebe J, Ok YS (2020) Soil amendments for immobilization of potentially toxic elements in contaminated soils: A critical review. Environ Int 134:105046

Paliwal K, Karunaichamy KSTK, Ananthavalli M (1998) Effect of sewage water irrigation on growth performance, biomass and nutrient accumulation in *Hardwickia binata* under nursery conditions. Bioresour Technol 66(2):105–111

Park D, Yun YS, Park JM (2010) The past, present, and future trends of biosorption. Biotechnol Bioprocess Eng 15(1):86–102

Pomares F, Roca J, Tarazona F et al (1984) Aerobically digested sewage sludge as N and P fertilizer. In: L'Hermite P, Ott H (eds) Processing and use of sewage sludge: proceedings of the third international symposium held at Brighton, September 27–30, 1983. D. Reidel Pub. Co., Dordrecht, p c1984

Qadir M, Ghafoor A, Murtaza G et al (2000) Cadmium concentration in vegetables grown on urban soils irrigated with untreated municipal sewage. Environ Dev Sustain 2(1):13–21

Ramirez-Fuentes E, Lucho-Constantino C, Escamilla-Silva E et al (2002) Characteristics, and carbon and nitrogen dynamics in soil irrigated with wastewater for different lengths of time. Bioresour Technol 85(2):179–187

Rana L, Dhankhar R, Chhikara S (2010) Soil characteristics affected by long term application of sewage wastewater. Int J Environ Res 4(3):513–518

Rinklebe J, Shaheen SM, Frohne T (2015) Amendment of biochar reduces the release of toxic elements under dynamic redox conditions in a contaminated floodplain soil. Chemosphere 142:41–47

Rutkowski T, Raschid-Sally L, Buechler S (2007) Wastewater irrigation in the developing world—two case studies from the Kathmandu Valley in Nepal. Agric Water Manag 88(1–3):83–91

Samarghandi M, Azizian S, Siboni M, Jafari S, Rahimi S (2011) Removal of divalent nickel from aqueous solutions by adsorption onto modified holly sawdust: equilibrium and kinetics. J Environ Health Sci Eng 8(2):167–174

Saha JK, Panwar N, Srivastava A et al (2010) Chemical, biochemical, and biological impact of untreated domestic sewage water use on Vertisol and its consequences on wheat (*Triticum aestivum*) productivity. Environ Monit Assess 161(1–4):403–412

Saravanamoorthy MD, Kumari BR (2007) Effect of textile waste water on morphophysiology and yield on two varieties of peanut (*Arachis hypogaea* L.). J Agric Technol 3(2):335–343

Sarkar A, Pramanik K, Mitra S et al (2018) Enhancement of growth and salt tolerance of rice seedlings by ACC deaminase-producing *Burkholderia* sp. MTCC 12259. J Plant Physiol 231:434–442

Sarker TC, Azam SMGG, Bonanomi G (2017) Recent advances in sugarcane industry solid by-products valorization. Waste Biomass Valoriz 8:241–266

Sciban M, Kasnja M (2004) Wood sawdust and wood originate materials as adsorbents for heavy metal ion. Holz Roh Werkst 62:69–73

Selvamurugan M, Doraisamy P, Maheswari M et al (2010) High rate anaerobic treatment of coffee processing wastewater using upflow anaerobic hybrid reactor. Iran J Environ Health Sci Eng 7 (2):129–136

Sharma HD, De A (2007) Municipal solid waste landfill settlement: Postclosure perspectives. J Geotech Geoenviron 133(6):619–629

Sharma PK, Ayub S, Tripathi CN (2013) Agro and horticultural wastes as low cost adsorbents for removal of heavy metals from wastewater: a review. Int Ref J Eng Sci 2(8):18–27

Shaheen SM, Shams MS, Khalifa MR, Mohamed A, Rinklebe J (2017) Various soil amendments and environmental wastes affect the immobilization and phytoavailability of potentially toxic elements in a sewage effluent irrigated sandy soil. Ecotoxicol Environ Saf 142:375–387

Shu R, Dang F, Zhong H (2016) Effect of incorporating differently-treated rice straw on the phytoavailability of methyl mercury in soil. Chemosphere 145:457–463

Shukla A, Zhang YHH, Dubey P, Margrave JL, Shukla SS (2002) The role of sawdust in the removal of unwanted materials from water. J Hazard Mater 95(1–2):137–152

Siebe C (1998) Nutrient inputs to soils and their uptake by alfalfa through long-term irrigation with untreated sewage effluent in Mexico. Soil Use Manag 14(2):119–122

Singh Y, Singh B, Timsina J (2005) Crop residue management for nutrient cycling and improving soil productivity in rice-based cropping systems in the tropics. Adv Agron 85:269–407

Singh A, Sharma RK, Agrawal M et al (2010) Health risk assessment of heavy metals via dietary intake of foodstuffs from the wastewater irrigated site of a dry tropical area of India. Food Chem Toxicol 48(2):611–619

Singh PK, Deshbhratar PB, Ramteke DS (2012) Effects of sewage wastewater irrigation on soil properties, crop yield and environment. Agric Water Manag 103:100–104

Srivastava VC, Mall ID, Mishra IM (2006) Characterization of mesoporous rice husk ash (RHA) and adsorption kinetics of metal ions from aqueous solution onto RHA. J Hazard Mater B 134:257

Srivastava VC, Mall ID, Mishra IM (2008) Removal of cadmium (II) and zinc (II) metal ions from binary aqueous solution by rice husk ash. Colloids Surf A Physicochem Eng Aspects 312:172

Szymanski N, Patterson RA (2003) Effective Microorganisms (EM) and wastewater systems. Future directions for On-site Systems: Best management practice, pp 347–355

Tandon R, Dipali NT, Vyas M (2020) Plant extracts for purification of industrial waste water-a review. Plant Arch 20(2):3854–3873

Tawwab AM, El-Sayed GO, Shady SHH (2017) Capability of some agricultural wastes for removing some heavy metals from polluted water stocked in combination with Nile tilapia, *Oreochromis niloticus* (L.). Int Aquat Res 9:153–160

The International Water Association, Wastewater Report (2018). www.iwa-network.org

Tran VS, Ngo HH, Guo W, Zhang J, Liang S, Ton-That C, Zhang X (2015) Typical low-cost biosorbents for adsorptive removal of specific organic pollutants from water. Bioresour Technol 182:353–363

UN (2019) United Nation, Department of Economics and Social Affairs Report

Unicef, fao ,SaciWATERs (2013) United Nations International Children's Emergency Fund-Food And Agricultural Organization Report 2013. http://www.indiaenvironmentportal.org

UN WWAP (2003) World Water Assessment Programme. www.unesco.org/new/en/natural-sciences/environment/water/wwap/wwdr/wwdr1-2003/

UN-Water Annual Report (2016) UN - World Water Report

US EPA (2012) Guidelines for water reuse. United National Environmental Protection Agency, National Risk Management Research Laboratory Office of Research and Development Cincinnati, Ohio

Van Lai T (2000a) Perspectives of peri-urban vegetable production in Hanoi. Action Planning Workshop of the CGIAR Strategic Initiative for Urban and Peri-Urban Agriculture (SIUPA), Hanoi, Vietnam, 6–9 June. Convened by International Potato Center (CIP), Lima

Van Lai T (2000b) Perspectives of peri-urban agriculture in Hanoi. In: Proceedings of an Action Plan Development Workshop, pp 6–9

Van-Rooijen DJ, Turral H, Biggs TW (2008) Urban and industrial water use in the Krishna Basin. Irrig Drain. https://doi.org/10.1002/ird.439

Vaverková MD, Zloch J, Adamcová D et al (2019) Landfill leachate effects on germination and seedling growth of hemp cultivars (Cannabis Sativa L.). Waste Biomass Valoriz 10(2):369–376

Vergine P, Salerno C, Libutti A et al (2017) Closing the water cycle in the agro-industrial sector by reusing treated wastewater for irrigation. J Clean Prod 164:587–596

Witek-Krowiak A (2013) Application of beech sawdust for removal of heavy metals from water: biosorption and desorption studies. Eur J Wood Prod 71(2):227–236

WHO (2006) Guidelines for the safe use of wastewater, excreta and greywater. In: Wastewater use in agriculture, vol 2. World Health Organization

World Bank (2012) World-Bank-Development Indicator (WDI). www.worldbank.org

World Bank (2017) World-Bank-Annual-Report. http://pubdocs.worldbank.org/

World Bank (2019) World-Bank-Annual-Report. http://pubdocs.worldbank.org/

World Water Quality Facts And Statistics (2013). https://pacinst.org/

WWAP (2017) World Water Assessment Programme (UNESCO WWAP). http://www.unesco.org/

Xian Y, Wang M, Chen W (2015) Quantitative assessment on soil enzyme activities of heavy metal contaminated soils with various soil properties. Chemosphere 139:604–608

Yang S, Zheng Q, Yuan M, Shi Z et al (2019) Long-term elevated CO2 shifts composition of soil microbial communities in a Californian annual grassland, reducing growth and N utilization potentials. Sci Total Environ 652:1474–1481

Yousaf IMRANA, Ali SM, Yasmin A (2010) Germination and early growth response of Glycine max varieties in textile and paper industry effluents. Pak J Bot 42(6):3857–3863

Yu B, Zhang Y, Shukla a SSS, Dorris KL (2001) The removal of heavy metals from aqueous solutions by sawdust adsorption--removal of lead and comparison of its adsorption with copper. J Hazard Mater 84(1):83–94

Zhang Y, Shen Y (2017) Wastewater irrigation: past, present, and future. Wiley Interdiscip Rev Water 6(3):1234

Zhang Y, Liu Y-R, Lei P, Wang YJ, Zhong H (2018) Biochar and nitrate reduce risk of methyl mercury in soils under straw amendment. Sci Total Environ 619:384–390

Zhu H, Zhong H, Evans D, Hintelmann H (2015) Effects of rice residue in corporation on the speciation, potential bioavailability and risk of mercury in a contaminated paddy soil. J Hazard Mater 293:64–71

Chapter 9
Riparian Ecotones: An Important Derivative for Managing River Pollution

Shikha Sharma, Madhoolika Agrawal, and Arijit Roy

9.1 Introduction

The rivers around the world have been attempted with technological know how while addressing the issues of water pollution. Recently, Rowinski et al. (2018) suggested for nature-based solutions (NBS), a solution which mimics nature while restoring the degrading rivers. The NBS emphasises over the role and importance of riparian vegetation in improving the water quality. The importance of establishing the riparian ecotones for maintaining the water quality has been emphasised since the 1980s (Roberts and Krishnaswami 1982; Sirenko 1981).

The word "riparian" has been defined variedly in the literature and has been associated with variable suffixes. However, riparian vegetation in most simplistic manner designates to the vegetation flanking the riverbank. These vegetation are the diversified ecosystem encompassing the distinct biological diversity, capable of bringing physiochemical and biological changes in the environment (Tabacchi et al. 2000) and are also proving to be efficient solutions for scrubbing harmful pollutants entering the river (Rowinski et al. 2018).

Several researchers via both field and laboratory experiments have established that vegetation along the riverbank has shown equal or better performance than grey infrastructure employed for improving the river water quality (Rowinski et al. 2018; Liquete et al. 2016; Gonzalez et al. 2015). On the broad scale, the vegetation in the riparian ecotone reduces the impact of pollutants on the river environment by acting as a barrier trapping and breaking down the pollutants or sometimes even processing

S. Sharma · A. Roy (✉)
Indian Institute of Remote Sensing, Indian Space Organisation, Dehradun, India
e-mail: arijitroy@iirs.gov.in

M. Agrawal
Centre of Advanced Study in Botany, Institute of Science, Banaras Hindu University, Varanasi, Uttar Pradesh, India

A. Singh et al. (eds.), *Water Pollution and Management Practices*,
https://doi.org/10.1007/978-981-15-8358-2_9

them into other less toxic compounds within the ecotone before the runoff reaches the water body (Rowinski et al. 2018). These systems have been explored in terms of the removal efficiency of pollutants such as nitrogen (N), phosphorus (P) and sediment. Most of the researches pertaining to the role of riparian vegetation in reducing pollutants from river water are confined to the countries of Europe and North America (Vidon et al. 2019) with minimal work being reported from India.

The present scenario of rivers in India demands a research outlook to the role of riparian vegetation as NBS for Indian rivers, as these systems lie at the crossroads of the biosphere, hydrosphere, lithosphere, atmosphere and anthrosphere and are an important part of riverine systems, serving multiple socioecological functions (Dufour et al. 2019; Naiman et al. 2005).

9.2 Defining the Word "Riparian"

The word riparian has a lingual derivation from the Latin word "riparius," meaning from "bank" (https://www.merriam-webster.com/). The origin of this word dates back to the mid-nineteenth century. The initial usage of the term was notably legal rather than scientific. The legal usage of the term dates back to the 1800s confined to political boundaries of the USA for describing the property of citizens adjacent to stream or river (Klett 2002; Baker 2002). However, on the scientific background, it is difficult to trace as to when did the scientists first adopted the term "riparian". Baker (2002) states that this term started appearing in scientific literature since the 1970s. Later during the 1980s, several scientists, agencies and the public were seen managing the riparian areas; thus it was during this decade which an effort was made to define this term; consequently, this decade was seen with conceptualising what "riparian" actually is, and the year 1985 was termed as "the year of the riparian" (Anderson 1987).

In literature the word "riparian" has not been defined as a single entity but in association with various suffixes such as riparian area, riparian zone, riparian reserve, riparian system, riparian ecosystem, riparian corridor and riparian ecotone. Verry et al. (2004) stated that the meanings of all the suffixes are more or less similar. Among the suffixes used, the most common ones are riparian zone and riparian buffer zone. Both of these suffixes have been intermittently used by the researchers, Wegner (1999) puts forth the difference between these two words, explaining that "buffer" is used with riparian as suffix during measurement of the important function of this transition zone, i.e. to shield the rivers/streams from the anthropogenic influences. The differentiation of Wegner (1999) points towards the importance of these areas/zones. Other than these suffixes, the word "riparian zone" has also been synonymously used as stream corridor, river corridor, ribbons of life, riparian woodlands, riparian forests, riparian buffer zones, riparian strips, riparian zones, alluvial floodplains, etc. (Zaimes et al. 2010). This addition of varied suffixes and synonymous usage of the word "riparian" perplexes the researcher as to what to add to riparian while defining it in river science. To make it a little simpler, Ilhardt et al.

(2000) and Verry et al. (2004) suggested to add the term "ecotone" with riparian, because it is not an ecosystem rather a part of the two ecosystems. Thus, it can be stated that in the legacy of the use of varied suffixes the suffix "ecotone" goes much in compliance.

Along with the variability related to the use of suffixes and synonyms against the word riparian, a large amount of variability can also be seen in defining the word "riparian". There is no universally accepted definition of the term riparian by the scientific or regulatory community (Zaimes et al. 2007). The term "riparian" has been defined like a tailor-made approach by the researchers of varied disciplines so far, with each of them emphasising according to their disciplines.

9.2.1 *Criteria of Defining "Riparian" in Literature so Far*

The definition of the word "riparian" has seen a progression from legal aspect to scientific; since the 1980s, several researchers have added different aspects of varied scientific disciplines subsequently seeing a progression.

1. Legal – The term has been used by the US government, in policy and regulations related to water law. The term was used basically to protect the area near the water bodies from anthropogenic activities. Further these zones were considered both hydrologically and ecologically important (FEMAT 1993; Parrott et al. 1997; NRCS 2002).
2. Hydrologic – The authors have defined riparian zone in connection with the river flows indicating towards an area close to the river (Bren 1993; Malanson 1993; Osterkamp 2008; Lovett and Price 1999).
3. Ecology (vegetation) – The word "riparian" has been used over the years by several ecologists to define the vegetated strip near the river (Narumalani et al. 1997; Gregory et al. 1991; Martin et al. 1999).
4. Edaphic – The scientists used an approach defining riparian areas on the basis of soil water availability (Naiman et al. 1993; Naiman and Decamps 1997; Ledesma et al. 2018).

Over the period of time, several authors have attempted to comprehensively discuss the progression and variability in defining the riparian areas. To be mentioned are review papers by Verry et al. (2004), Naiman et al. (2005), Clerici et al. (2011) and Dufour et al. (2019). Baker (2002) asserted that no one definition satisfies more than two to three disciplines, as every researcher emphasises the subject matter while defining the riparian. According to Verry et al. (2004) "the concept of riparian must tie definition, delineation and resource data aggregation together into a logical sequence", and defining "riparian" must include geomorphology, ecology and species association (Verry et al. 2004; Dufour et al. 2019).

The vast array of riparian definitions ranges from simplest to complex, further to structural and functional type. The simplest form of defining the word riparian is considering it as a transition zone, while complexity arises when considering it as the ecotone, wherein the biotic and abiotic interactions are taking place. The simplest definition of riparian comes from Gregory (1991) "as the transition zone, an interface between terrestrial aquatic ecosystems". However, much before Gregory et al. (1991), Lowrance et al. (1985) included organisms while defining riparian and defined riparian as an ecosystem having the complex assemblage of organisms and their environment existing adjacent to and near flowing water. Adding to the definition of riparian was the water extension, which led Naiman and Decamps (1997) to define riparian as a stream channel between the low and high water marks and that portion of the terrestrial landscape from the high water mark towards the upland where vegetation may be influenced by elevated water tables or flooding and by the ability of the soils to hold water. Adding to it is the recent definition by a soil scientist Ledesma et al. (2018) who defined riparian as "the area between the edge of the stream and characteristic transition between organic and mineral soils". The most complicated definition comes by Ilhardt et al. (2000) as 3D ecotones wherein interaction between terrestrial and aquatic ecosystems exists; this area extends from groundwater to over the canopy top, beyond the floodplains, near to the slopes, draining water also having lateral connections with terrestrial ecosystem along the water course, and the width of the zone may vary accordingly.

A review by Dufour et al. (2019) comprehensively categorises riparian definition into functional and structural type. The structural definition defines the riparian from topographical perspectives, whereas functional approach defines considering the interactions between two systems, viz. aquatic and terrestrial, in terms of hydrological, morphological, chemical and biological processes. The authors suggest that most of the definitions of riparian use a functional approach. Defining the riparian areas from structural perspective puts forth the geomorphological perspective of riparian zone which deals with various fluvial features and landforms formed by river deposits (Dufour et al. 2019). Dufour et al. (2019) emphasised that structural approach is more used in delineating the riparian extent.

The diversified definition and suffixes of the word riparian still perplex researchers, so it is important to come on one common note, thus accepting ecotone as suffix to the word riparian and definition by Ilhardt et al. (2000), which not only considers the biotic and abiotic interactions but also gives the spatial dimension to the riparian ecotone, thus defining it completely. So further in the chapter, the authors will be referring to riparian areas as riparian ecotone.

9.3 Components and Characteristics of Riparian Ecotone

Riparian ecotone, being the transition zone between the terrestrial and aquatic ecosystems, comprises of the two important components, viz. land and water. Further, talking in terms of ecology, this transition zones contain both abiotic (soil and water) and biotic (organisms from microscopic to macroscopic fauna and flora) components. These are thus the basic components of riparian areas; however, each river has its unique riparian ecotone with their distinct geomorphology and ecology.

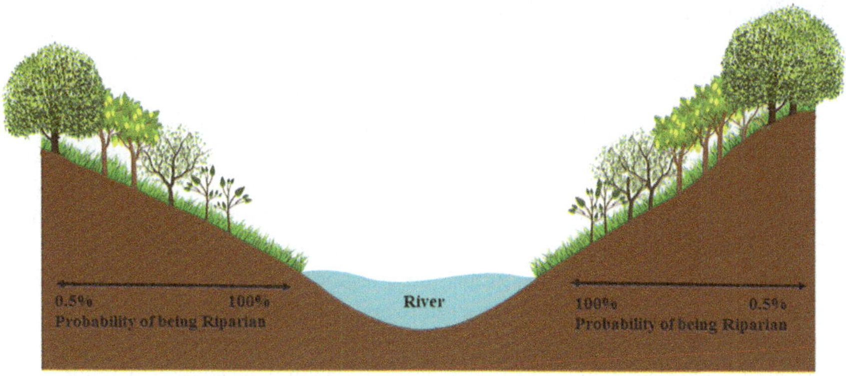

Fig. 9.1 Probability of area to be a riparian ecotone. (Source Ilhardt et al. 2000)

However, in the literature, the riparian ecotone has mostly been discussed in terms of vegetation type, and it is the vegetation which is of importance and mainly researched and discussed. However, Zaimes et al. (2010) ascertain that the riparian ecotone possesses the following common characteristics:

1. The land along the river or fluvial system comprising of distinct vegetation type influencing and being influenced by the various physicochemical and biological interactions between the vegetation, land and water.
2. This ecotone sometime is encompassed within the floodplain or sometime may extend beyond it. However, Ilhardt (2000) stated that the frequency of designating an area to be riparian decreases with distance from the fluvial system (Fig. 9.1).
3. These are the areas through which surface and subsurface hydrology connects water bodies with their adjacent uplands.
4. These areas are distinguished from upland areas via distinct physicochemical, biological and ecological processes and biota.
5. These areas have a distinct mosaic of vegetated patches having different physiognomy, structure and composition.
6. These areas are characterised by varied land use features changing in accordance to the human population.
7. These areas witness local variability in the physical conditions such as flow velocity (flood and drought), water level, etc.

The important thing to note is that areas cannot be categorised or undergo zonation or have similar characteristics at global level because their characteristic features can change in accordance with the region.

9.4 Functions of Riparian Ecotone

Though having characteristics and distinct features, the riparian ecotone delivers an important role in both the terrestrial and aquatic ecosystems, which remains homogenous across the globe. These ecotones play an important role in physical, biological and ecological functions (Klapproth and Johnson 2009), adding to the importance of interactions taking place in this zone between soil, water and biotic communities. The functions of riparian ecotone are dependent on the land use features comprising the riparian ecotone. All over the globe, these regions are experiencing the transformation from natural vegetation to agriculture and subsequently to human habitations (Sharma et al. 2016). However, the role, importance or function of riparian ecotone is discussed and researched mostly in terms of natural vegetation. Certain basic functions of riparian areas are as stated below.

9.4.1 Bank Stabilisation

The vegetation cover over soil has already been known to hold the soil, thus playing the prominent role in the conservation of the top soil. A similar kind of role has been reported long back during the 1980s for establishing the importance of vegetation near rivers towards riverbank stabilisation (Omernik et al. 1981; Smith 1992; Arthington et al. 1997; Townsend and Douglas 2000). Riparian vegetation has shown to play an important role in protecting the river shorelines from damage caused to soil during heavy flows such as runoff from upland or downpours (EPA 2010). Riparian vegetation mediates their input by armouring stream banks against erosion, storing runoff, trapping sediment and transforming nutrients (Omernik et al. 1981; Smith 1992; Osborne and Kovacic 1993; Arthington et al. 1997; Townsend and Douglas 2000) (Fig. 9.2).

The mechanism by which riparian area does is slowing the flow of water, thus ensuring that sediments settle out before they reach the water course (Montgomery et al., 1996). In a naturally vegetated riparian ecotone, floodwater overshoots from the bank and spreads out the broad floodplain, thus reducing the energy of the water (https://www.nrcs.usda.gov/). As floodwater flows through a vegetated area, the plants resist the flow and dissipate the energy, increasing the time available for water to infiltrate into the soil and be stored for use by plants (https://www.nrcs.usda.gov/). The roots of the plants increase the riverbank cohesiveness adding a tensile strength which enables the riverbank soil to resists shear stresses (Castelle and Johnson 2000). Thus, de-voiding the fluvial system from its natural vegetation can have adverse effects on the stability of its bank (Fig. 9.2).

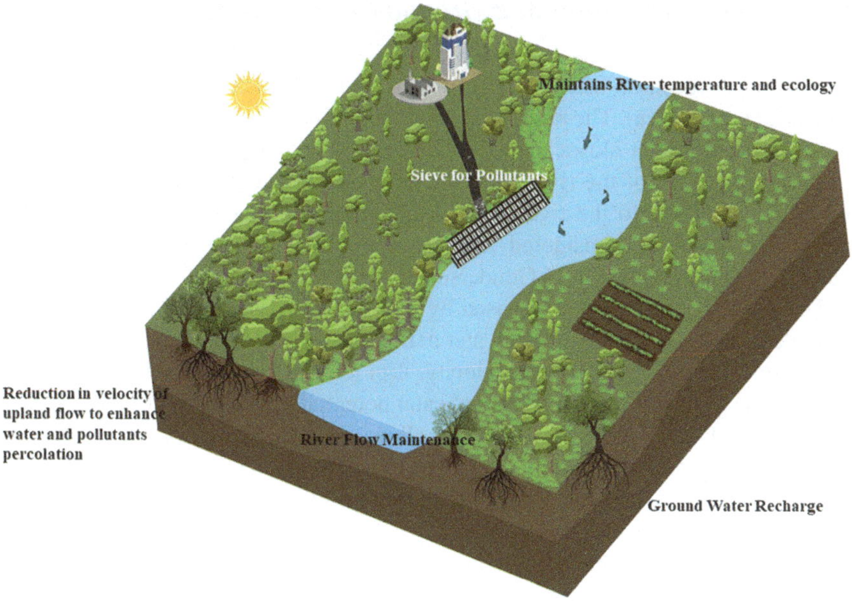

Fig. 9.2 Functions of riparian ecotone

9.4.2 *Filtration*

The riparian ecotone acts as filter for the river; Pinay et al. (2018) have termed it as a "skin for the river". The water draining downwards via riparian ecotone undergoes natural filtration process for the absorption and adsorption of sediments, pollutants, etc. The process of filtration takes place by the vegetation (Luke et al. 2007) and microbial fauna present in the riparian ecotone (Silvan et al. 2002; Palviainen et al. 2004). The vegetation present in the riparian ecotone decreases the export of suspended solids by reducing the velocity of overland flow and thus enhancing water infiltration into soil leading to particle sedimentation (Gundersen et al. 2010). The sedimentation rate depends on flow velocity, particle size and particle density (Gundersen et al. 2010) (Fig. 9.2).

The pollutants such as compounds of nitrogen, phosphorus, potassium, calcium, magnesium, etc. are adsorbed by the plant root system and used as source of nutrients for growth (Mayer et al. 2007). Recently riparian ecotones have also been reported as the carbon pool (Forster et al. 2007). The soil C concentration is largely dependent on soil moisture, and riparian ecotone having high moisture content serves to hold greater carbon per unit area as compared to upland forests (Forster et al. 2007). Thus riparian ecotone comprising of vegetation acts as nutrient sink and sediment entrapment owing to the uptake and sorption processes, thus playing an important role in the protection of river water quality (Fig. 9.2).

9.4.3 Flow Regulation and Groundwater Recharge

The riparian vegetation affects the river flow and is termed as safety valve of the watershed (gov.mb.ca). The reason for designating riparian ecotone as safety valve is the vegetation, which slows the water flow, thus reducing the intensity of flood further downstream of the river channel. This not only controls flood but also the upland water flow into the river, thus maintaining the river flow. Hydrologically, riparian ecotones are considered as discharge areas through which water flows before reaching surface water (Gundersen et al. 2010) (Fig. 9.2).

These areas are often characterised by a high groundwater table and superficial subsurface flow created by groundwater exfiltration (Gundersen et al. 2010). As the vegetation slows down the water velocity, it in turn increases the retention time of water in this zone enhancing the groundwater potential of that region (Swanson et al. 2017). The reduction in the speed of water allows increase in absorption of water into the soil, thus replenishing groundwater reserves (Swanson et al. 2017). The percolation down into the ground is also dependent on soil types of the region. The silt texture of the soil acts as a sponge to aid in groundwater recharge and underground storage. Not only this, the water stored in the ground seeps into the river through lateral connectivity during lean season, thus maintaining river flow throughout the season (https://www.nrcs.usda.gov/). Riparian vegetation in the riparian ecotone is thus important to maintain the river flow, prevent flood and recharge the groundwater (Fig. 9.2).

9.4.4 Maintaining River Channel Environment

The riparian vegetation has proved to play a crucial role in maintaining the river environment, viz. river water temperature and organic-inorganic inputs into the river. The types of vegetation affect the intensity of light reaching the river water; the dense canopy lowers the river water, whereas the open canopy raises the river water (Garner et al. 2017) (Fig. 9.2).

Maintaining the river temperature is essential to prevent the faster rate of chemical speciation subsequently leading to rise in pollutant levels in the river water. The temperature also influences the dissolved oxygen content of the river. The optimised temperature of river water is also essential for the biological activity of the river diversity. The autochthonous input from the vegetation near the river plays an essential role in the river biotic life. Organic matter from riparian ecotones, such as leaves, twigs, logs and stems that fall from the buffer into the water, are the main source of food for aquatic macroinvertebrates (Mayer et al. 2006). The debris or wood from vegetation traps additional leaf litter and wood. Macroinvertebrates use the wood as habitat, living inside the wood, under residual bark and on surfaces that protrude out of the water (Mayer et al. 2006) (Fig. 9.2).

9.5 Riparian Vegetation as Pollutants Scrubber

9.5.1 Ecology of Vegetation

Plant ecology is a wholesome branch of ecological sciences studying how plants interact with their abiotic and biotic environment. These interactions vary in terms of type of species dominating a particular region. The riparian plants acquiring a different physical habitat have altogether different ecological attributes. In order to understand how riparian plants maintain the water quality of the river, it is very important to understand the riparian plant ecology. The likelihood that a particular riparian vegetation type will occur in a given watershed depends on the ecological tolerances of the component plant species and their adaptations for specific hydrogeomorphic conditions, along with water availability, anaerobic soils, surface characteristics and flood disturbance regimes (Steiger et al. 2005). The functionality of riparian vegetation in controlling the water pollutant is dependent over the structural physiological traits of vegetation such as size, form, growth rate, longevity and litter quality (Dosskey et al. 2010). Not only the type but age of vegetation can also play an important role in pollutant removal as the efficiency of the plants declines with age.

The riparian plants vary from region to region so is the ecology; thus, it becomes essential to study the kind of riparian vegetation harbouring the bank of the rivers (Fig. 9.3). Internationally there have been numerous studies both in field and in laboratory to analyse the riparian vegetation for their composition and potency to remove pollutants. The vegetation types analysed are grass riparian ecotone and forested riparian ecotone. Parkyn (2004) reported that most of the researchers have studied the efficiencies of vegetated filter strips of rank paddock grasses. However, in Indian context, the riparian vegetation is studied mostly in terms of determining floristic composition of the species, and very few emphasised their role in removing pollutants. The riparian vegetation reported is diversified across the rivers from plain to mountains (Fig. 9.3). The species reported from plains are *Prosopis* sp., *Acacia* sp., *Azadirachta* sp. and *Jatropha* sp., along the river Yamuna in the floodplain (Jha et al. 2010), whereas *Rhododendron* sp., *Quercus* sp., *Woodfordia* sp., *Oxalis* sp., *Tridax* sp., etc. have been reported from Bhilangna River (tributary of river Bhagirathi) of the Himalayan region.

9.5.2 Mechanism of Pollutant Removal

The plants have their own mechanism and processes for removal of the pollutants, and both above and below ground biomass structure determine the effects of vegetation on river water quality (Dosskey et al. 2010). The riparian vegetation is a community with high productivity resulting in higher nutrient uptake from soil and groundwater via its own growth. Long back in 1985, it has been established that the

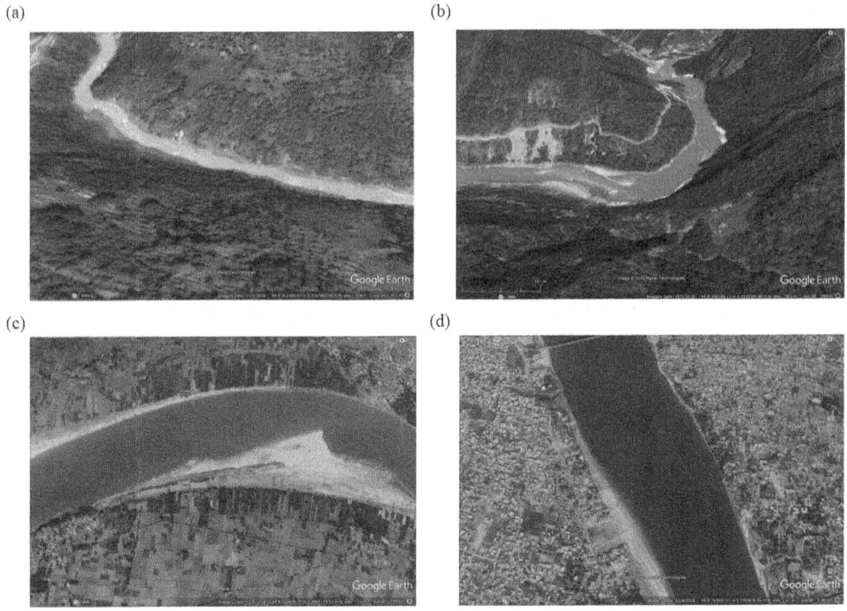

Fig. 9.3 Ecology of riparian vegetation in different regions of Indian river basins. (**a**) Dense vegetation of river Bhagirathi in the hills. (**b**) Dense vegetation of river Ganga in foothills. (**c**) Agricultural practice in the riparian ecotone of river Ganga in plains. (**d**) Human habitation harbouring the riparian ecotone of river Ganga in plains

vegetation near the river or in a watershed spreads and divides the upland flow, thus reducing the velocity (Clinnick 1985) and consequently enhancing the infiltration (Norris 1993). Osborne and Kovacic (1993) explained that major mechanisms involved in the removal of pollutants in the riparian ecotones are physical retention, plant uptake, dilution and chemical transformation.

The riparian vegetation both directly and indirectly improves the water quality (Tabacchi et al. 2000). The shoots and plant litter (above ground biomass) have direct interactions with precipitation and surface runoff in riparian zone (Dosskey et al. 2010), whereas the root systems have indirect interactions with water seeping inside the soil as groundwater (Dosskey et al. 2010). The plants also supplement food in the form of organic matter to autotrophs, which play an important role in transforming the substance dissolved in upland flow, thus contributing in enhancing water quality of the region (Tabacchi et al. 2000).

The stems and litter in the vegetated strip provide resistance to flow of river water via increasing hydraulic roughness (Parkyn 2004; Dosskey et al. 2010). This hydraulic roughness lowers the flow velocity and sediment transport capacity of surface runoff, thus increasing the contact time of water (Parkyn 2004) leading to enhancement of deposition and infiltration of particulates (Gharabaghi et al. 2002). The pollutants which are soluble infiltrate into the ground, and those insoluble deposit above ground (Gharabaghi et al. 2002). Along with this, the root activity, viz. growth and decay, followed by the activity of plant microfauna also increases the soil permeability by creating pores via which water carrying pollutants percolate down (Cardoso et al. 2013). Thus, the more the hydraulic roughness of the riparian ecotone, more will be the infiltration, and less will be the tendency of the surface runoff to input sediments and pollutants into the river.

Among the varied processes of pollutant removal, the absorption by the roots is predominantly researched. The plant roots have been associated with supplying the majority of nutrients to the shoots mostly phosphorus (P), potassium (K), calcium (Ca), magnesium (Mg), sulphur (S) and some other non-nutrients such as Cd, Cr, Hg, Ni, Pb, As, Se, B, Cs and Sr (White and Brown 2010). Many researchers have also reported plants as the important contributors in solubilising pesticides, volatile organic compounds (benzene, trichloroethylene and toluene) and selenium and organo-mercury compounds (Lin et al. 2008; Juraske et al. 2008). But in the water quality studies, the chemical entities of much concern are N and P, K and metals. Not only plant roots but shoots too play an important role in nutrient management. The plant leaves and foliage intercept and evaporate the considerable amount of rainfall and prevent it from reaching the soil (Tabacchi et al. 2000), thus working at canopy level. Further several researchers have established that vegetation increases evapotranspiration from watersheds (Trimble et al. 1987) and riparian zones (Cleverly et al. 2006; Kellogg et al. 2008).

Other than the role of live plant parts, litter too plays an important role in pollutant scrubbing. The plant litter being rich in organic matter retains the pollutants dissolved in percolating water via processes such as ionic attraction, hydrogen-ligand bonding, etc. (Dosskey et al. 2010). There have been report that agricultural pesticides and endocrine disruptors have high binding affinity to soil organic matter and undergo immobilisation (Yamamoto et al. 2003) which can be further degraded by soil microbial fauna. Further the indirect form of pollutant removal by plant is also done by activating the microbes, which switch to alternative electron acceptors such as nitrate, sulphate and oxidised iron and manganese compounds during limited supply of oxygen, thus supporting further decomposition of these compounds (Hill 2000). The riparian soil is often reported with limited oxygen conditions because the decomposition of plant detritus consumes the limited supply of oxygen in wet and saturated soil (Zaimes et al. 2007). The second indirect role includes the activity of organisms such as mycorrhizal fungi or root symbiotic sites. The symbiotic association in plants is associated with increase in the nutrient uptake efficiency by enhancing the volume and surface interaction between the plant and physical environment (Tabacchi et al. 2000).

9.5.3 Types of Riparian Vegetation and Their Pollutant Removal Capacity

There are two most prominent types of vegetation studied in the riparian ecotone, viz. grasses and forest, with their type varying from region to region. The two predominant vegetation types have different roles in the pollutant removal. The grass riparian ecotones are capable of filtering sediments and pollutants associated with sediments such as phosphorus and nitrogen (Parkyn 2004). Further there are reports that grass riparian ecotones are less effective in the removal of soluble nutrients such as nitrate, ammonia and dissolved phosphorus and nitrate removal is greater in forested buffers compared to grasses (Parkyn 2004).

The role of riparian buffer strips in relation to nitrogen has been well studied (Peterjohn and Correll 1984; Daniels and Gilliam 1996). There is a large variability in the removal efficiency of N ranging from 20 to 100% (Valkama et al. 2019). The removal of N and P via vegetation in the riparian ecotone varies widely, and it is difficult to predict (Hoffmann et al. 2009) as the hydrological regimes controlling the biogeochemical processes in the riparian zone change along the river continuum (Hoffmann et al. 2009). None of the riparian ecotone vegetation types, viz. grass or forested, are much effective in reducing concentration of dissolved phosphorus (Parkyn 2004). In most of the research monitoring, the effectiveness of buffer zones for their pollutant removal capacity has focused on vegetated filter strips (VFS) (Parkyn 2004).

Riparian forest reduces the nitrate input into the river via the uptake by plant roots (Silvan et al. 2002; Palviainen et al. 2004), retention to soil and utilisation by the microbes (Silvan et al. 2002; Palviainen et al. 2004), also via gaseous N fluxes from soil to the atmosphere (Silvan et al. 2002). The different fractions of nitrogen, viz. ammonium, nitrate and dissolved organic nitrogen, undergo transformation in the soil for their effective utilisation and assimilation. The riparian forest decreases the concentration of ammonium and dissolved organic nitrogen in surface runoff more than nitrate (Kokkonen et al. 2006).

During the 1980s and 1990s, the mechanism of phosphorus removal by riparian buffers has been studied by several researchers. The mechanisms explained are deposition of phosphorus attached with sediment, settling of particulate phosphorus and adsorption of dissolved phosphorus by clay particles in particular clay containing high levels of aluminium and iron (Reddy et al. 1999). Further there are reports of phosphorus reduction via uptake from understory vegetation and microorganisms (Silvan et al. 2004).

9.6 Other Attributes of Riparian Ecotone Affecting the Pollutant Removal Efficiency

The knowledge of the hydrogeomorphic regimes has been ascertained as the key limitation in the management and restoration of degraded riparian ecotones (Shaw and Cooper 2008). The study of riparian system as a whole requires an understanding of the geomorphic structures and processes that form the bed sediments and maintain flow pathways through them. Further Engelhardt et al. (2012) suggested that it is important to ascertain the relationship between watershed lithology, geomorphology and riparian vegetation for the development, management and restoration of riparian vegetation. Thus, the hydrogeomorphology of the watershed also plays an important role in the pollutant removal activities of the riparian ecotones.

9.6.1 Geomorphology of the Riparian Ecotone

There are various geomorphological attributes associated with riparian ecotone. Morphologically the riparian ecotones differ from other upland features as they receive drainage from large contributing areas, characterised by shallow groundwater table and relatively low slope gradients. The literature suggests that geomorphological attributes affect the pollutant removal functionality of riparian ecotone in two ways: (1) direct and (2) indirect. Certain features, viz. sediment size, sediment load, slope, soil structure, subsurface drainage pattern, etc., of the riparian ecotone affect the pollutant removal, posing direct affect. Indirect effects include the structure, type and distribution of riparian vegetation, which in turn are determined by the geomorphological traits of the region.

The geomorphology of the watershed regulates the movement of both water and sediment, through watershed into the channel and via channel network (Engelhardt et al. 2012). The capacity of riparian ecotone to remove pollutants increases if the flow across the watershed is not canalised rather distributed across wide areas (Polyakov et al. 2005). Further, the slope of the zone is also an important determinant of pollutant trapping efficiency of the ecotone (Jin and Römkens 2001). Another topographic feature of importance in determining the buffering capacity of riparian ecotone is convergence factor, which is the ratio of active area of the riparian buffer to the total area (Polykov et al. 2005). Active area of the riparian buffer according to Polykov et al. 2005 "is the portion of buffer through which runoff flow from the upslope occurs". A low convergence value is indicative of dissected relief, gully and steep valley floors, and higher values (value close to 1) indicate towards plain conditions (Polykov et al. 2005).

The riparian soil-holding capacity, soil type, soil water content, soil particle size and soil aggregate density affect the pollutant removal capacity of the buffer (Munoz-Carpena et al. 1999). The properties of riparian ecotone to trap sediment decrease with reduction in sediment size (Lee et al. 2003). If the riparian ecotone is

Fig. 9.4 Development of river island of the river Ganga (geomorphological feature) and variations in ecology of riparian vegetation

formed of coarse-grained alluvial sediments, it allows complex microbial assemblage and metazoan biota (Huggenberger et al. 1998). It is the three processes, viz. erosion, transportation and deposition, which determine the structure of fluvial forms and hydrogeological properties, thus governing the connectivity between surrounding landscape to the drainage network (Huggenberger et al. 1998). The morphological attributes, viz. length, gradient and shape of the runoff area and that of upstream of riparian ecotone, affect the pollutant removal capacity (Norris 1993). For example, in a convex slope, the upland flow is faster than the concave slope (Norris 1993).

Recently, Pinay et al. (2018) stated that along the river continuum, we could determine three distinct geomorphic characteristics, viz. in the headwater, in the middle reaches and in the lower reaches. These physical or geologic features control the distribution of grain sizes as well as successional dynamics and structure of vegetation communities along the river continuum (Gregory et al. 1991). The headwaters, which are steep, have constrained valleys with high erosional rate and less formation of geological features resulting in minimal riparian vegetation development; in the middle reaches, the slope is comparatively less steep and transports minimum sediment leading to formation of river island and sediment deposition offering sites for generation of riparian vegetation (Pinay et al. 2018). Depositional reaches in the lowland valleys with very fine unstable sediment deposits are important for the development of riparian vegetation (Pinay et al. 2018) (Fig. 9.4).

Several basin properties influence vegetation types because they pose a direct influence on flood regime and water availability. In a study conducted by Engelhardt et al. (2012) in 18 upland watersheds of central Nevada, USA, the following morphometric parameters were studied, viz. watershed area, watershed length, total stream length, drainage density, Shreve magnitude, relief, ruggedness, relief ration, relative stream power, watershed shape and hypsometric integral. Engelhardt et al. (2012) ascertained that composition of riparian ecotone is a function of those geomorphological characteristics of watershed which influence flood regimes, sediment transport and water availability also inclusive of hypsometric integral, relative stream power and topographic relief. It was observed that riparian vegetation distribution and composition were determined by both geologic and morphometric variables. Bedrock lithology is also an important feature determining the riparian vegetation composition.

9.6.2 Hydrology of the Riparian Ecotone

Among the several abiotic (e.g. water chemistry, light and wind) and biotic (e.g. competition, invasive species; see in the succeeding paragraphs) factors that influence riparian vegetation processes, fluvial hydrodynamics (i.e. flow and flood regime and related processes) plays a significant role in all plant life stages, i.e. dispersal, colonisation, recruitment, growth, succession and mortality (Solari et al. 2016). Successful riparian plants often adopt a combination of adaptive strategies during different life stages in order to ensure their survival (e.g. high dispersal rates, adaptations to resist stress and vegetative reproduction; Camporeale et al. 2013).

Norris, 1993, stated that "effectiveness of riparian buffer ecotone depends upon the rate of flow of surface water, hydraulic conductivity and holding capacity of the buffer zone soil". The water or upland flow which passes through the watershed via riparian ecotone exhibits variable random forcing over the vegetation among which the important one is flow variability (Vesipa et al. 2017). The river stage and velocity affect the riparian environment. The variability in the flow regime of the river affects the riparian vegetation from its establishment to its death. The river flow of ample magnitude is needed to remove the vegetation already harbouring the riverbank (Holanda et al. 2005) and further depositing the fresh sediments for the recruitment of new seeds of plants.

The rivers also are known to be associated with the phenomenon of hydrochory. For hydrochory to take place, it is essential that the river shows fluctuations in its flows (Greet et al. 2011). The rise and fall of the river are essential for collection and deposition of seeds (Vesipa et al. 2017). Certain hydraulic conditions such as Froude, Reynolds and Shield number associated with river hydrographs determine the number of seeds scoured or deposited. Other than the recruitment of seeds, the flow of the river also plays an important role in determining whether the seed will germinate and grow into an adult (Fraaije et al. 2015). Both the sudden decline and sudden rise in river water affect the growth of the seedling.

These river flows are also associated with effects on the adult vegetation in their life stage (Garssen et al. 2015). The river flow affects the age and spatial distribution of populations (Mosner et al. 2011). River flow fluctuations affect the stratigraphy of riverbanks (Merritt et al. 2010). The variability in river flow is also responsible for transport of nutrients and salts which are required for the sustenance of vegetation (Asaeda and Rashid 2014). Thus, river flow is the important driver of riparian vegetation.

9.7 Conclusions

The riparian vegetation is turning as one of the natural means of maintaining the river water quality with recently their importance being more emphasised over technical solutions of combating river pollution. The word "riparian" and the suffixes used with it have been associated with the inherent variability. The literature suggests the inherent variability in definitions of the word "riparian" and the suffixes used. The definition of the term and its usage have evolved from legal to scientific, with each branch of science defining it in its own way. The author suggests that the definition given by Ilhardt et al. (2000) is more complete and should be universally used henceforth; further, the most appropriate suffix to explain the functionality of riparian areas is "ecotone," given by Ilhardt et al. (2000).

These areas are to be looked as of utmost importance because of their important role in maintaining the fluvial landscape. The researchers have emphasised over their role in stabilising the riverbank, in filtering the pollutants, in regulating the river flow and groundwater recharge and also for maintaining the river ecosystem. These functions of riparian ecotone are dependent on the type of vegetation harbouring the riverbank. Most of the researches have been done to see the effectiveness of grass and forest riparian ecotone with sediments, nitrate and phosphorus as pollutants. The literature suggests a few studies from India being cited in the context of laying the importance of riparian ecotone, thus requiring directives to research the Indian river from riparian ecotone perspective and including it in river management plans.

Acknowledgements The authors are thankful to the Council of Scientific and Industrial Research, New Delhi, for financial support in the form of research associateship. We also acknowledge Mr. Rahul Kumar, design lead, Defcon group Dehradun, for designing Figs. 9.1 and 9.2. The director, Indian Institute of Remote Sensing, ISRO Dehradun, is also acknowledged for his support and encouragement.

References

Anderson EW (1987) Riparian area definition: a viewpoint. Rangel Arch 9(2):70

Arthington AH, Marshall JC, Rayment GE, Hunter HM, Bunn SE (1997) Potential impact of sugarcane production on riparian and freshwater environments. In: Intensive sugarcane production: meeting the challenge beyond 2000. Proceedings of the Sugar 2000 symposium, Brisbane, Australia, 20–23 August 1996. Cab International, Brisbane, pp 403–421

Asaeda T, Rashid MH (2014) Modelling of nutrient dynamics and vegetation succession in midstream sediment bars of a regulated river. Int J River Basin Manag 12(2):123–133

Baker TT (2002) What is a riparian area? Cooperative Extension Service Animal Resources Department. New Mexico State University, Las Cruces

Bren LJ (1993) Riparian zone, stream, and floodplain issues: a review. J Hydrol 150(2–4):277–299

Camporeale C, Perucca E, Ridolfi L, Gurnell AM (2013) Modeling the interactions between river morphodynamics and riparian vegetation. Rev Geophys 51(3):379–414

Cardoso EJBN, Vasconcellos RLF, Bini D, Miyauchi MYH, Santos CAD, Alves PRL, Paula AMD, Nakatani AS, Pereira JDM, Nogueira MA (2013) Soil health: looking for suitable indicators.

What should be considered to assess the effects of use and management on soil health? Sci Agric 70(4):274–289

Castelle A, Johnson A (2000) Riparian vegetation effectiveness: technical bulletin no. 799. National Council for Air and Stream Improvement, Inc., Research Triangle Park, NC, 36 pp

Cleverly JR, Dahm CN, Thibault JR, McDonnell DE, Allred Coonrod JE (2006) Riparian ecohydrology: regulation of water flux from the ground to the atmosphere in the Middle Rio Grande, New Mexico. Hydrol Process Int J 20(15):3207–3225

Clerici N, Weissteiner CJ, Paracchini LM, Strobl P (2011) Riparian zones: where green and blue networks meet: pan-European zonation modelling based on remote sensing and GIS. Joint Research Centre (JRC), Ispra

Clinnick PF (1985) Buffer strip management in forest operations: a review. Aust For 48(1):34–45

Daniels RB, Gilliam JW (1996) Sediment and chemical load reduction by grass and riparian filters. Soil Sci Soc Am J 60(1):246–251

Dosskey MG, Vidon P, Gurwick NP, Allan CJ, Duval TP, Lowrance R (2010) The role of riparian vegetation in protecting and improving chemical water quality in streams. J Am Water Res Assess 46(2):261–277

Dufour S, Rodríguez-González PM, Laslier M (2019) Tracing the scientific trajectory of riparian vegetation studies: main topics, approaches and needs in a globally changing world. Sci Total Environ 653:1168–1185

Engelhardt BM, Weisberg PJ, Chambers JC (2012) Influences of watershed geomorphology on extent and composition of riparian vegetation. J Veg Sci 23(1):127–139

EPA (2010) Guidance for Federal Land Management in the Chesapeake Bay Watershed Chapter 5. Riparian Area Management, Nonpoint Source Pollution Office of Wetlands, Oceans, and Watersheds U.S. Environmental Protection Agency 841-R-10-002 May 12

Forest Ecosystem Management Assessment Team (FEMAT) (US) (1993) Forest ecosystem management: an ecological, economic, and social assessment: report of the forest ecosystem management assessment team. The service

Forster P, Ramaswamy V, Artaxo P, Berntsen T, Betts R, Fahey DW, Haywood J, Lean J, Lowe DC, Myhre G, Nganga J (2007) Changes in atmospheric constituents and in radiative forcing. In: Climate Change 2007 (ed) The physical science basis, Cambridge University Press, Cambridge

Fraaije RG, terBraak CJ, Verduyn B, Breeman LB, Verhoeven JT, Soons MB (2015) Early plant recruitment stages set the template for the development of vegetation patterns along a hydrological gradient. Funct Ecol 29(7):971–980

Garner G, Malcolm IA, Sadler JP, Hannah DM (2017) The role of riparian vegetation density, channel orientation and water velocity in determining river temperature dynamics. J Hydrol 553:471–485

Garssen AG, Baattrup-Pedersen A, Voesenek LA, Verhoeven JT, Soons MB (2015) Riparian plant community responses to increased flooding: a meta-analysis. Glob Change Biol 21 (8):2881–2890

Gharabaghi B, Rudra R, Whiteley HR, Dickinson WT (2002) Development of a management tool for vegetative filter strips. J Water Manag Model 556:289–302

González E, Sher AA, Tabacchi E, Masip A, Poulin M (2015) Restoration of riparian vegetation: a global review of implementation and evaluation approaches in the international, peer-reviewed literature. J Environ Manag 158:85–94

Greet JOE, Angus Webb J, Cousens RD (2011) The importance of seasonal flow timing for riparian vegetation dynamics: a systematic review using causal criteria analysis. Freshw Biol 56 (7):1231–1247

Gregory SV, Swanson FJ, McKee WA, Cummins KW (1991) An ecosystem perspective of riparian zones. Bioscience 41(8):540–551

Gundersen P, Laurén A, Finér L, Ring E, Koivusalo H, Sætersdal M, Weslien JO, Sigurdsson BD, Högbom L, Laine J, Hansen K (2010) Environmental services provided from riparian forests in the Nordic countries. Ambio 39(8):555–566

Hill AR (2000) Stream chemistry and riparian zones. In: Streams ground waters. Academic Press, New York, pp 83–110
Hoffmann CC, Kjaergaard C, Uusi-Kämppä J, Hansen HCB, Kronvang B (2009) Phosphorus retention in riparian buffers: review of their efficiency. J Environ Qual 38(5):1942–1955
Holanda FSR, Santos LGDC, Santos CMD, Casado APB, Pedrotti A, Ribeiro GT (2005) Riparian vegetation affected by bank erosion in the Lower São Francisco River, Northeastern Brazil. Rev Árvore 29(2):327–336
Huggenberger P, Hoehn E, Beschta R, Woessner W (1998) Abiotic aspects of channels and floodplains in riparian ecology. Freshw Biol 40(3):407–425
Ilhardt BL, Verry ES, Palik BJ (2000) Defining riparian areas. Forestry and the riparian zone. Orono, Maine 26:7–14
Jha P, Mohapatra KP, Dubey SK (2010) Impact of land use on physico-chemical and hydrological properties of ustifluvent soils in riparian zone of river Yamuna, India. Agrofor Syst 80 (3):437–445
Jin CX, Römkens MJ (2001) Experimental studies of factors in determining sediment trapping in vegetative filter strips. Trans ASAE 44(2):277
Juraske R, Antón A, Castells F (2008) Estimating half-lives of pesticides in/on vegetation for use in multimedia fate and exposure models. Chemosphere 70(10):1748–1755
Kellogg DQ, Gold AJ, Groffman PM, Stolt MH, Addy K (2008) Riparian ground-water flow patterns using flownet analysis: evapotranspiration-induced upwelling and implications for N removal. J Am Water Resour Assoc 44(4):1024–1034
Klapproth JC, Johnson JE (2009) Understanding the science behind riparian forest buffers: effects on water quality. Virginia Cooperative Extension, Blacksburg, pp 1–22
Kokkonen T, Koivusalo H, Laurén A, Penttinen S, Starr M, Kellomäki S, Finér L (2006) Implications of processing spatial data from a forested catchment for a hillslope hydrological model. Ecol Model 199(4):393–408
Ledesma JL, Futter MN, Blackburn M, Lidman F, Grabs T, Sponseller RA, Laudon H, Bishop KH, Köhler SJ (2018) Towards an improved conceptualization of riparian zones in boreal forest headwaters. Ecosystems 21(2):297–315
Lee KH, Isenhart TM, Schultz RC (2003) Sediment and nutrient removal in an established multi-species riparian buffer. J Soil Water Conserv 58(1):1–8
Lin CH, Lerch RN, Garrett HE, George MF (2008) Bioremediation of atrazine-contaminated soil by forage grasses: transformation, uptake, and detoxification. J Environ Qual 37(1):196–206
Liquete C, Udias A, Conte G, Grizzetti B, Masi F (2016) Integrated valuation of a nature-based solution for water pollution control. Highlighting hidden benefits. Ecosyst Serv 22:392–401
Lovett S, Price P (1999) Riparian land management technical guidelines. Land and Water Resources Research and Development Corp. (LWRRDC), Canberra
Lowrance RR, Leonard RA, Asmussen LE, Todd RL (1985) Nutrient budgets for agricultural watersheds in the southeastern coastal plain. Ecology 66(1):287–296
Luke SH, Luckai NJ, Burke JM, Prepas EE (2007) Riparian areas in the Canadian boreal forest and linkages with water quality in streams. Environ Rev 15:79–97
Malanson GP (1993) Riparian landscapes. Cambridge University Press, Cambridge
Martin TL, Kaushik NK, Trevors JT, Whiteley HR (1999) Denitrification in temperate climate riparian zones. Water Air Soil Pollut 111(1–4):171–186
Mayer PM, Reynolds SK, McCutchen MD, Canfield TJ (2007) Meta-analysis of nitrogen removal in riparian buffers. J Environ Qual 36(4):1172–1180
Mayer PM, Reynolds SK, McCutchen MD, Canfield TJ (2006) Riparian buffer width, vegetative cover, and nitrogen removal effectiveness: a review of current science and regulations. EPA/600/R-05/118. U.S. Environmental Protection Agency, Cincinnati
Merritt DM, Scott ML, LeRoy PN, Auble GT, Lytle DA (2010) Theory, methods and tools for determining environmental flows for riparian vegetation: riparian vegetation-flow response guilds. Freshw Biol 55(1):206–225

Montgomery DR, Abbe TB, Buffington JM, Peterson NP, Schmidt KM, Stock JD (1996) Distribution of bedrock and alluvial channels in forested mountain drainage basins. Nature 381 (6583):587–589

Mosner E, Schneider S, Lehmann B, Leyer I (2011) Hydrological prerequisites for optimum habitats of riparian Salix communities–identifying suitable reforestation sites. Appl Veg Sci 14(3):367–377

Munoz-Carpena R, Parsons JE, Gilliam JW (1999) Modeling hydrology and sediment transport in vegetative filter strips. J Hydrol 214(1–4):111–129

Naiman RJ, Décamps H, McClain ME (2005) Riparia: ecology, conservation, and management of streamside communities, Aquatic ecology series. Elsevier, Academic Press, Amsterdam

Naiman RJ, Decamps H, Pollock M (1993) The role of riparian corridors in maintaining regional biodiversity. Ecol Appl 3(2):209–212

Naiman RJ, Decamps H (1997) The ecology of interfaces: riparian zones. Ann Rev Ecol Syst 28 (1):621–658

Narumalani S, Zhou Y, Jensen JR (1997) Application of remote sensing and geographic information systems to the delineation and analysis of riparian buffer zones. Aquat Bot 58 (3–4):393–409

National Research Council (2002) Riparian areas: functions and strategies for management. National Academies Press, Washington, DC

Norris VO (1993) The use of buffer zones to protect water quality: a review. Water Resotr Manag 7:257–272

Omernik J, Abernathy AR, Male LM (1981) Stream nutrient levels and proximity of agricultural and forest land to streams: some relationships. J Soil Water Conserv 36(4):227–231

Klett CT (2002) New Mexico water rights. 2 nd update. New Mexico Water Resources Research Institute. New Mexico State, Las Cruces

Osborne LL, Kovacic DA (1993) Riparian vegetated buffer strips in water-quality restoration and stream management. Freshw Biol 29(2):243–258

Osterkamp WR (2008) Annotated definitions of selected geomorphic terms and related terms of hydrology, sedimentology, soil science and ecology. Geological Survey (US), Reston

Palviainen M, Finér L, Kurka AM, Mannerkoski H, Piirainen S, Starr M (2004) Decomposition and nutrient release from logging residues after clear-cutting of mixed boreal forest. Plant Soil 263 (1):53–67

Parkyn S (2004) Review of riparian buffer zone effectiveness. Ministry of Agriculture and Forestry, Wellington

Parrott HA, Marions DA, Perkinson RD (1997) A four-level hierarchy for organizing stream resources information. In: Proceedings, Headwater Hydrology Symposium, pp 41–44

Peterjohn WT, Correll DL (1984) Nutrient dynamics in an agricultural watershed: observations on the role of a riparian forest. Ecology 65(5):1466–1475

Pinay G, Bernal S, Abbott BW, Lupon A, Marti E, Sabater F, Krause S (2018) Riparian corridors: a new conceptual framework for assessing nitrogen buffering across biomes. Front Environ Sci 6:47

Polyakov V, Fares A, Ryder MH (2005) Precision riparian buffers for the control of nonpoint source pollutant loading into surface water: a review. Environ Rev 13(3):129–144

Reddy KR, Kadlec RH, Flaig E, Gale PM (1999) Phosphorus retention in streams and wetlands: a review. Crit Rev Environ Sci Technol 29(1):83–146

Roberts SA, Krishnaswami SK (1982) Protecting the source. Water/Eng Manage 129(3):28–31

Rowiński PM, Västilä K, Aberle J, Järvelä J, Kalinowska MB (2018) How vegetation can aid in coping with river management challenges: a brief review. Ecohydrol Hydrobiol 18(4):345–354

Sharma S, Roy A, Agrawal M (2016) Spatial variations in water quality of river Ganga with respect to land uses in Varanasi. Environ Sci Pollut Res 23(21):21872–21882

Shaw JR, Cooper DJ (2008) Linkages among watersheds, stream reaches, and riparian vegetation in dryland ephemeral stream networks. J Hydrol 350(1–2):68–82

Silvan N, Regina K, Kitunen V, Vasander H, Laine J (2002) Gaseous nitrogen loss from a restored peatland buffer zone. Soil Biol Biochem 34(5):721–728
Silvan N, Vasander H, Laine J (2004) Vegetation is the main factor in nutrient retention in a constructed wetland buffer zone. Plant Soil 258:179–187
Sirenko LA (1981) Possibilities of optimizing processes forming the quality of natural water. Hydrotech Constr 15(6):319–323
Smith CM (1992) Riparian afforestation effects on water yields and water quality in pasture catchments. J Environ Qual 21(2):237–245
Solari L, Van Oorschot M, Belletti B, Hendriks D, Rinaldi M, Vargas-Luna A (2016) Advances on modelling riparian vegetation—hydromorphology interactions. River Res Appl 32(2):164–178
Steiger J, Tabacchi E, Dufour S, Corenblit D, Peiry JL (2005) Hydrogeomorphic processes affecting riparian habitat within alluvial channel–floodplain river systems: a review for the temperate zone. River Res Appl 21(7):719–737
Swanson S, Kozlowski D, Hall R, Heggem D, Lin J (2017) Riparian proper functioning condition assessment to improve watershed management for water quality. J Soil Water Conserv 72 (2):168–182
Tabacchi E, Lambs L, Guilloy H, Planty-Tabacchi AM, Muller E, Decamps H (2000) Impacts of riparian vegetation on hydrological processes. Hydrol Process 14(16–17):2959–2976
Townsend SA, Douglas MM (2000) The effect of three fire regimes on stream water quality, water yield and export coefficients in a tropical savanna (northern Australia). J Hydrol 229 (3–4):118–137
Trimble SW, Weirich FH, Hoag BL (1987) Reforestation and the reduction of water yield on the Southern Piedmont since circa 1940. Water Resour Res 23(3):425–437
Valkama E, Usva K, Saarinen M, Uusi-Kämppä J (2019) A meta-analysis on nitrogen retention by buffer zones. J Environ Qual 48(2):270–279
Verry ES, Dolloff CA, Manning ME (2004) Riparian ecotone: a functional definition and delineation for resource assessment. Water Air Soil Pollut Focus 4(1):67–94
Vesipa R, Camporeale C, Ridolfi L (2017) Effect of river flow fluctuations on riparian vegetation dynamics: processes and models. Adv Water Resour 110:29–50
Vidon PG, Welsh MK, Hassanzadeh YT (2019) Twenty years of riparian zone research (1997–2017): where to next? J Environ Qual 248(2):248–260
Wegner S (1999) A review of the scientific literature on riparian buffers width, extent and vegetation. Institute of Ecology, University of Georgia, Athens
White PJ, Brown PH (2010) Plant nutrition for sustainable development and global health. Ann Bot 105(7):1073–1080
Yamamoto H, Liljestrand HM, Shimizu Y, Morita M (2003) Effects of physical– chemical characteristics on the sorption of selected endocrine disruptors by dissolved organic matter surrogates. Environ Sci Technol 37(12):2646–2657
Zaimes G, Nichols M, Green D, Crimmins M (2007) Understanding Arizona's riparian areas. College of Agriculture and Life Sciences, University of Arizona, Tucson
Zaimes GN, Iakovoglou V, Emmanouloudis D, Gounaridis D (2010) Riparian areas of Greece: their definition and characteristics. J Eng Sci Technol Rev 3(1):176–183. https://www.merriam-webster.com/. Assessed date 2 Jan 2020. https://www.nrcs.usda.gov/. Assessed date assessed date 2 Jan 2020

Chapter 10
Ecosystem Responses to Pollution in the Ganga River: Key Issues to Address River Management

Deepa Jaiswal, Usha Pandey, and Jitendra Pandey

10.1 Introduction

During the recent few decades, the anthropogenic activities have dramatically altered the water quality of the Ganga River and its tributaries. Point and nonpoint source-driven inputs contain a variety of organic and inorganic substances including nutrients, dissolved and particulate organic carbon, and metal toxicants. Nutrient enrichment enhances productivity (autochthonous carbon), and this together with atmospheric, terrigenous, and anthropogenic carbon (allochthonous carbon) input causes C eutrophication (Pandey et al. 2014a). The coupled effects of eutrophy and metal pollution are expected to cause long-term consequences, higher in the order of magnitude than could be predicted from short-term, small-scale studies. Despite this, most of the studies conducted on the river Ganges focus generally on eutrophication and some on metal pollution. Accordingly, most of the water treatment technologies consider removal of biological oxygen demand (BOD), i.e., removal of carbon load only. However, recent studies show that eutrophication and metal pollution in the Ganga River most often occur simultaneously (Jaiswal and Pandey 2019a). For instance, sewage-associated inputs contain large amount of carbon and nutrients leading to eutrophication, whereas industrial effluents add quantifiable amount of metals and other pollutants causing toxic effects.

The average population density in the Ganges basin is 712 persons/km^2 as compared to 382 persons/km^2 for India. The river along its 2525 km course flows through 29 megacities, 23 small cities, and 48 townships and receives massive fluxes

D. Jaiswal · J. Pandey (✉)
Ganga River Ecology Research Laboratory, Environmental Science Division, Centre of Advanced Study in Botany, Institute of Science, Banaras Hindu University, Varanasi, India

U. Pandey
Department of Botany, Faculty of Science and Technology, Mahatma Gandhi Kashividyapith University, Varanasi, India

A. Singh et al. (eds.), *Water Pollution and Management Practices*,
https://doi.org/10.1007/978-981-15-8358-2_10

of nutrients and other pollutants. The total wastewater generation in the basin is ~8250 million liters per day (MLD), out of which, 2538 MLD is discharged directly into the river, 4491 MLD into tributaries, and 1220 MLD on land or low-lying areas (CPCB 2013). The Assi drain at Varanasi discharges over 66 MLD of sewage leading to ~156 mg L^{-1} chemical oxygen demand (COD) and 4060 kg day^{-1} biological oxygen demand (BOD) load to the river in this region. Additionally, the drain adds over 535 tons of dissolved inorganic nitrogen (DIN) and 133 tons of dissolved reactive phosphorus (DRP) annually (Yadav and Pandey 2017a). In addition to point sources, the nonpoint sources, such as agricultural runoff from 73.44% agricultural land of the basin containing residues of ~10 million tons of fertilizers and 9000 tons of pesticides, add a sizable amount of nutrients and pesticide to the river. A sub-watershed-scale study by Yadav and Pandey (2017a) shows that the river in Varanasi region alone receives ~289.69 tons of DIN, 47.1 tons of DRP, and 1421.55 tons of dissolved organic carbon (DOC) through surface runoff and ~15.23 tons of DIN and 1.19 tons of DRP through atmospheric deposition (AD) annually. A watershed-scale study by Pandey et al. (2016a) shows that the Ganges basin receives ~3.32 Tg reactive nitrogen (NO_3^- + NH_4^+) and ~173.20 Gg phosphorus (PO_4^{3-}) annually, and the inputs were relatively higher in the middle segment which is considered as the most polluted stretch of the river (CPCB 2013). All these sources add large amount of oxygen-demanding substances and pollutants into the river throughout the year creating tremendous pressure on water quality and ecosystem services such as drinking water supply, recreation, and fisheries (Pandey et al. 2017; Jaiswal and Pandey 2019b; Siddiqui et al. 2019a).

Studies conducted on the Ganga River generally select parameters bifurcating eutrophy and metal pollution and even without considering ecosystem-level consequences. For instance, to address trophic state, biological oxygen demand is generally considered where the sampling is restricted to the upper water column only. The issues such as stratification of dissolved oxygen in the water column, sediment oxygen demand (SOD), dissolved oxygen deficit (DOD), nature of oxygen-demanding substances (ODS), benthic hypoxia, and ecosystem feedbacks have been altogether ignored for this major river system of India. Similarly, for metal pollution, the analysis of pelagic water and freshly deposited sediments are generally considered (Siddiqui and Pandey 2019a). Some of the studies conducted in earth sciences context have taken into account the deep sediment analysis (Verma and Pandey 2019). Studies considering the factors of in situ metal release and its bioavailability are altogether lacking. In particular, no data so far, except few studies conducted in our laboratory, are available on changing state of ecosystem functions coupling feedbacks and ecological assimilation capacity of the Ganga River. The present review is an effort to make a critical analysis on the need for understanding ecosystem responses coupling eutrophy and metal pollution in the Ganga River. This has relevance because our multi-temporal, multi-scale studies suggest the need for identifying the determinants of ecosystem responses to metal pollution and eutrophy for integrated management of the Ganga River (Jaiswal and Pandey 2019a, c).

10.2 Causes of Pollution and Eutrophy in Ganga River

10.2.1 Metal Pollution

The heavy metal pollution is gaining global attention because of their abundance, toxicity, nonbiodegradable nature, and persistence in the environment (Fu et al. 2014). The unprecedented increase in urbanization, industrialization, and economic development enhances the addition of heavy metals to water bodies worldwide. Both the natural and anthropogenic sources, including rock weathering, soil erosion, domestic and industrial effluents, agricultural leachate, coal burning, etc., contribute to water pollution (Singh and Pandey 2014; Alvarez-Vazquez et al. 2017). Both point and nonpoint sources contribute to metal pollution in the Ganga River (Yadav and Pandey 2017a).

In the Ganges basin, ~2500 MLD industrial waste is generated (Trivedi 2010), and the middle segment (from Kannauj to Varanasi) adds the largest amount of industrial effluents to the river. The industrial sectors, including tannery, battery, and electroplating and heavy duty in Kannauj and Kanpur city; engineering in Allahabad; and locomotives and carpet industries in Varanasi city, are considered to be main sources of metal pollution. Tanneries alone constitute ~58% of grossly polluting industries in the middle stretch of the Ganga River. Besides this, a large amount of metals are added through tributaries in the main river. For example, over 359 industries add their effluents to the Yamuna tributary which is finally released to the Ganga River (CPCB 2016). Recent studies conducted in our laboratory show long-distance atmospheric transport as well as deposition of metals in the Ganga River (Pandey and Pandey 2009; Pandey et al. 2010). A recent study (Pandey et al. 2010) shows that the Ganga River receives 0.78–18.65 g ha^{-1} $year^{-1}$Cd, 0.48–6.28 g ha^{-1} $year^{-1}$ Cr, and 152.7–447.50 g ha^{-1} $year^{-1}$ Zn in Varanasi region only through atmospheric deposition. A basin-scale study by Siddiqui and Pandey (2019a) shows that only ~44.44% samples of Cr and 33.33% of Cd are below their respective permissible limits of 50 and 3.0 μg L^{-1} (Bureau of Indian Standards, BIS 2012). Similarly, ~54.63% samples of Cu, ~66.66% of Pb, ~44.44% of Mn, ~66.66% of Fe, and ~33.33% of Ni samples along the basin have exceeded their permissible limits.

The overall concentrations of metals in the river are influenced by multiple factors such as the magnitude of external loading, source partitioning, chemical composition, adsorption, and episodic events driven by fluctuating river flow, sediment delivery, and urban-industrial discharge (Yadav and Pandey 2017a; Jaiswal and Pandey 2018). The sediment composition, redox potential, pH, and cation exchange capacity of the system are the major determinants that regulate dissolution, precipitation, absorption, and complexation of metals in the sediment and, consequently, their bioavailability, associated toxicity to aquatic organisms, and impact on the food web (Jaiswal and Pandey 2018, 2019d; Siddiqui and Pandey 2019a; Verma and Pandey 2019).

10.2.2 Nutrient Loading

Nitrogen (N) and phosphorus (P) are considered as most critical nutrients because their demand to supply ratios generally lie at unity or below. Because of this reason, they become limiting nutrients for phytoplankton growth. Excessive input of these nutrients, although increases food production and stimulates plant growth, causes detrimental effects on surface waters including rivers. Rivers receive nutrients from the airshed, in the form of atmospheric deposition (AD), and watershed in the form of leaching and runoff (Pandey et al. 2014a; Bellmore et al. 2018). A global-scale study has shown that during the period of 1860–2005, the reactive N (Nr) deposition has increased from ~15 Tg to 187 Tg which could be doubled by 2050 (Galloway et al. 2008). Further, AD adds over 3.7 Tg of P annually at global scale (Tipping et al. 2014). Studies show AD-N input range between 1.54 and 40 kg ha^{-1} $year^{-1}$ in inland water bodies of the world (Edokpa et al. 2015; Bellmore et al. 2018) and between 12 and 38 kg ha^{-1} $year^{-1}$ in many parts of India (Pandey 2011; Pandey and Pandey 2013; Siddiqui et al. 2019b). It varies between 9.72 and 42.85 kg ha^{-1} $year^{-1}$ in the Ganga River, an amount high enough to change the Ganga River ecology (Singh and Pandey 2019). Basin-scale extrapolation of a recent study shows that through AD, ~2.77 Tg DIN and ~0.13 Tg DRP are deposited in the basin and ~ 5.31 Gg DIN and 0.37 Gg DRP are added directly on the Ganga River surface annually (Singh and Pandey 2019). Additionally, the river receives a large amount of nutrients from tributaries and sub-tributaries (Mishra et al. 2016; Yadav and Pandey 2017a).

In Ganges basin, over 2723 MLD municipal sewage is generated, out of which only ~1200 MLD is flushed to the river after treatment. These urban effluents add ~13.28 Gg DIN and 5.29 Gg of DRP annually into the river. Surface runoff also adds huge amount of nutrient to surface waters. A sub-watershed-scale estimation in Varanasi region shows that agricultural land adds ~403, 186, and 24 Gg $year^{-1}$; woodland adds ~67, 30, and 19 Gg year¯1; and built-up area adds ~11, 55, and 34 Gg $year^{-1}$of NO_3^-, NH_4^+, and DRP, respectively, through surface runoff. The basin-scale extrapolation of these data shows that the river receives ~193 to 1181 Gg DIN and ~59 to 218 Gg DRP annually through surface runoff. The river transports ~31.58 Gg of DIN and 3.97 Gg of DRP annually to the Bay of Bengal (Singh and Pandey 2019). These overall quantitative estimations show that, along with the point sources, the nonpoint sources also add a large amount of nutrients to the Ganga River across its length which needs proper attention and management.

10.2.3 Carbon Enrichment

Along with point sources, the atmospheric deposition and surface runoff are important regulators of autochthonous and allochthonous C pool in the Ganga River (Pandey et al. 2014a, 2015a; Siddiqui et al. 2019b; Siddiqui and Pandey 2019b).

In Varanasi region, the autochthonous carbon varies between 256.6 and 567.8 kg C ha^{-1}. Across the river length, the autochthonous input contributes ~102.58 Gg of organic C annually (Singh and Pandey 2019). Comparative studies of Goldstein and Galbally (2007) and Hallquist et al. (2009) show that, at global scale, the total organic carbon (TOC) deposition increased from 305–645 to 950 Tg over a period of only 2 years. The Ganga River in Varanasi region receives ~110.8–558.6 tons organic carbon through atmospheric deposition, ~1421.55 tons DOC through surface runoff, and 364.0–1456.3 tons organic carbon through Assi drain (Yadav and Pandey 2017a). Basin-scale extrapolation showed that the basin receives ~1.81 Tg TOC each year through atmospheric deposition of which ~4.26 Gg is added directly on to the river surface (Singh and Pandey 2019). Surface runoff exports 16.16 to 26.90 Tg TOC and 9.19 to 16.46 Tg DOC, whereas the point sources add over 110 Gg of TOC to the river annually (Singh and Pandey 2019).

Studies show that a large amount of carbon in rivers and streams comes from terrigenous sources (Pandey et al. 2014a; Siddiqui et al. 2019b). Because nonpoint source C of terrestrial origin is flushed mainly through monsoon season runoff, this C is largely transported to the sea. The Ganga River leads to a burial of ~1.1×10^{12} mol sediment-driven organic carbon each year in the Bengal Fan which is ~10% of the global organic carbon burial flux in the continental margins (France-Lanord and Derry 1997). At global scale, rivers export ~611 Tg carbon per year (Cole et al. 2007) from which the contribution from South Asian rivers alone is ~7% (42.9 Tg $year^{-1}$) (Patra et al. 2013). The Ganga-Brahmaputra with ~7 Tg C $year^{-1}$ contributes the largest share in global DOC export to the oceans (361 Tg $year^{-1}$) (Patra et al. 2013).

10.3 Ecosystem Responses to Pollution and Eutrophy

10.3.1 Shifts in Microbial Enzyme Activity

The riverbed sediments, an important component of riverine ecosystem, are a biologically active and comparatively stable zone and support benthic communities which play an important role in ecosystem functions including biogeochemical cycling, secondary production, carbon metabolism, and sedimentation of carbon, nutrients, and heavy metals (Covitch et al. 2004). Recent studies have established that sediment microbial extracellular enzymes (EEs) can be used as an indicator of carbon and nutrient limitation/acquisition and to uncover the influence of regional-scale anthropogenic stressors (Sinsabaugh et al. 2009; Yadav and Pandey 2017b; Jaiswal and Pandey 2018, 2019e). The substrate-specific nature of EEs makes them important tools to investigate the functional profile of microbial communities as influenced by human-induced alterations (Sinsabaugh and Linkins 1990). The EE activities show quantifiable and instantaneous response (for instance, toward substrates and toxicants) even to small alterations in the ecosystems (Paerl et al. 2003). As the microbes play vital role in mediating the biogeochemical cycles and regulating the ecosystem structure and functioning, these have been proved to be the most suitable parameters for quantification of shift in ecosystem responses toward alteration in organic matter (substrate) and heavy metals (inhibitors) (Sinsabaugh et al. 2008; Jaiswal and Pandey 2018). Long-term sustainability and stability of an ecosystem depends, in a major way, on functionally active microbial communities, and such measurements provide the actual picture of health condition of an ecosystem. Further, because they constitute the key node connecting detritus trophic chain, any change in their metabolism affects the whole ecosystem processes including organic matter decomposition, nutrient cycling, and food web. Measurement of enzyme activities requires a small quantity of sample, generally proved to be simple, accurate, cost-effective, and rapid. Studies have shown that EE activities can be used as an index of microbiological functional diversity and integration of these with other physical and chemical measurements can provide important information on which ecosystem management strategies can be keyed (Jaiswal and Pandey 2019a).

As the microbial functional diversity involves various metabolic processes, a representative set of enzymes that control the key metabolic pathways/processes can be used to assess the microbial response to changing carbon, nutrients, and heavy metal concentrations. The extracellular enzyme β-D-glucosidase can be used as a measure of C acquisition (Sinsabaugh et al. 2009) and alkaline phosphatase (AP) as a proxy of P starvation (Duhamel et al. 2010), while protease can be used as an indicator of N mineralization (Rejsek et al. 2008). Further, because the fluorescein diacetate hydrolytic bioassay (FDAase) involves three major groups of enzymes (lipases, esterases, and proteases), which contribute to organic matter decomposition, it can be used as an indicator of total microbial activity (Schnürer and Rosswall 1982). The FDAase can directly be correlated with biomass and ATP content (Fontvieille et al. 1992). The microbial metabolic quotient (ratio of basal respiration to substrate-induced respiration) is used as an index to measure adversities in environmental conditions for soil microbes (Wardle 1993).

The microbial structure and functioning in aquatic ecosystems are generally influenced by episodic events driven by stormwater and intermittent flushing of urban-industrial effluents. Anthropogenic input of nutrients increases the autochthonous C which, along with the allochthonous C, induces EE activities. Studies have shown that EE activities can be used as a better substitute of sediment and water quality variables because of their direct linkages with carbon and nutrients as well as with the concentration of toxicants in water and sediments (Hill et al. 2006; Jaiswal and Pandey 2018). In addition to carbon and nutrients, rivers receive huge amount of metals from natural and anthropogenic sources, a large fraction of which is deposited in the bed sediments. Heavy metals are potentially toxic to microbial community and influence the overall ecological structure and functioning (Jaiswal and Pandey 2018, 2019e). Organic carbon in sediment forms complex with metals and often buffers their bioavailability and, consequently, the toxicity (Jaiswal and Pandey 2019e). Thus, the investigations on relationships between EE activities, substrate (carbon), and total and bioavailable metal concentration (toxicants) can provide important cues through which the state of river health and ecotoxicological implications can be appropriately assessed. In a recent study, we found positive correlations of EE activities (FDAase, β-D-glucosidase, protease) with C, N, and P and negative correlations with heavy metals (Jaiswal and Pandey 2018, 2019e). Earlier studies show significant positive correlations of heavy metals with microbial metabolic quotient (qCO_2) indicating further that heavy metals are major stressors affecting microbial activity (Wardle 1993).

These results suggest that EE activities in the riverbed sediment can be a sensitive indicator of C, N, and P status in the river. Further, the metals accumulated in the sediment inhibit EE activities, although C-rich sediment can reduce the toxic effects probably by reducing the bioavailability (Jaiswal and Pandey 2018). A recent study conducted in the Ganga River has shown that when total heavy metal concentration (ΣTHM; a sum of six most common heavy metals (Cd, Cr, Cu, Ni, Pb, and Zn) of human-impacted riverine systems) exceeds 347.44 $\mu g\ g^{-1}$, it becomes detrimental to EE activities. However, if a site has very high concentration of TOC, this limit may reach 472.53 $\mu g\ g^{-1}$ indicating a modulatory effect of TOC in metal toxicity (Jaiswal and Pandey 2019a). The study further shows that the stimulatory effect of substrate on EE activities declines when ΣTHM exceeds 284.73 $\mu g\ g^{-1}$ (Jaiswal and Pandey 2019e). Thus, the EE activities can be a good indicator of river response toward metal pollution and C eutrophy in large rivers, and the accuracy has displayed to be higher than other biological variables (Pandey and Yadav 2017; Jaiswal and Pandey 2018, 2019e).

10.3.2 Shift in Aquatic Diversity

One of the major impacts of human-induced environmental change is the loss of biodiversity and altered ecosystem processes. A recent global-scale study by Hooper et al. (2012) reveals that biodiversity loss can be considered as one of the major drivers of ecosystem change. Accordingly systematic database on interaction among biodiversity loss, environmental change, and associated impacts on ecosystem functioning is necessary. The Ganga River, which is under strong influence of anthropogenic perturbations, has been identified as a system with a massive shift in aquatic diversity and abundance. The Ganges dolphin has been an indicator species of this river, but due to increasing pollution load, this species is now at the boundary of extinction (Bashir et al. 2010). Dolphins are now found in some restricted locations of the upper stretch, and even there they are under threat due to increasing pollution and decreasing deepwater pools (Bashir et al. 2010). A sharp decline in abundance of major carp and spawn has been reported due to massive exploitation, habitat fragmentation, water abstraction, and increasing pollution load (Jhingran and Ghosh 1978). Besides this, large-scale reduction in clupeid fisheries has been reported due to increasing pollution in the river. This has increased the pressure on other commercially important fishes, such as *Aorichthys seenghala* (Sykes) and *Aorichthys aor* (Ham.), causing a threat to these species as well (Seth and Katiha 2003). A study conducted by Singh and Sharma (1998) reported 14 abundant, 7 vulnerable, 15 low-risk, 1 data-deficient, and 2 endangered fish species in

Alaknanda of upper Ganges. The exotic species are comparatively stronger, cause habitat modification, and generate competition for food, light, and other resources. Singh et al. (2010) have reported that abundance of exotic fishes such as *Cyprinus carpio* and *Oreochromis niloticus* in the Ganga River has negatively influenced the economically important indigenous fishes such as *Catla*, *Labeo rohita*, and *Cirrhinus mrigala*. Khanna et al. (2007) have reported that prolonged exposure of fishes to deteriorated environmental conditions causes severe irreversible damage to scale circuli and lepidonts resulting in loosening of scales.

The excessive inputs of nutrients promote phytoplankton growth. Although they are the basis of food chain, they can be harmful to human and other vertebrates as they release toxic substances (Ariyadej et al. 2004). At Kanpur (Jajmau), *Phormidium*, *Aphanocapsa*, and *Oscillatoria* have been reported in abundance (Singh et al. 2014). These species generally occur at places with high concentration of nutrients. A study conducted in Varanasi region has reported negative relationship between benthic algal biomass and DOC and a positive correlation with Secchi depth suggesting the role of water transparency (Pandey 2013). Furthermore, large biomass of *Phormidium uncinatum* was reported in high DOC region. The study suggests that the increasing DOC concentration is changing the light climate and fate of benthic producers in the Ganga River (Pandey 2013). A point source-linked study at Bijnor (UP) has reported the prevalence of pollution-tolerant species *Synedra*, *Cocconeis*, *Spirulina*, *Botryococcus*, and *Chlamydomonas*. Sensitive species such as *Cladophora* and *Euglena* disappeared from these sites (Negi and Rajput 2013). In another study of Kanpur region, abundance of Chlorococcales members such as *Microcystis aeruginosa*, *M. flos-aquae*, *Chroococcus minutus*, *C. varius*, *C. minor*, *Aphanocapsa grevillei*, *A. montana*, *Aphanothece microscopica*, and *Coelosphaerium kuetzingianum* has been reported (Singh et al. 2014). The prevalence of these species indicates conditions less conducive for sensitive species.

A shift in microbial diversity, in response to human perturbations, has also been reported in the Ganga River (Singh et al. 2013; Reddy et al. 2019). For example, multiple drug resistance has been observed in coliform, fecal coliform, fecal streptococci, and *E. coli* between Fatehgarh and Kannauj (Malik et al. 1995). A large number of multiple antimicrobial-resistant, Shiga toxin and enterotoxin-producing *E. coli* and virulence traits of *Enterococci* have been reported in the Ganga River near Kanpur city between Bithoor to Shuklaganj region (Lata et al. 2009). Due to increasing human intervention, a large number of harmful bacterial species including *Bacillus*, *Escherichia coli*, *Staphylococcus aureus*, *Pseudomonas aeruginosa*, and *Salmonella typhi* have been reported in the most polluted stretch of the river between Kanpur and Varanasi (Singh et al. 2013).

10.3.3 Shifts in Diatom Diversity

Being fast growing, phytoplankton responds quickly to nutrient, temperature, and light regimes. Because light and temperature do not generally observe as limiting in tropics, nutrients are considered as the principal determinants. For this reason, phytoplankton is considered as an indicator of nutrient pollution and eutrophy. Nutrients influence phytoplankton growth in terms of absolute concentration and in specific elemental ratios. The specific stoichiometric ratio (Redfield ratio) of N, P, and Si (16:1:16) is important for the balanced growth of phytoplankton specifically for diatoms (Turner et al. 2003). Recent studies have reported a shift in this ratio due to disproportionate input of nutrients in lakes and rivers (Pandey et al. 2014b, 2016b). This shift in stoichiometry would alter the ecosystem structure and functioning in due course of time with long-term irrecoverable consequences (Pandey et al. 2014b). The shifts in nutrient stoichiometry affect a diverse group of primary producers, specifically diatoms which are R-strategic (high reproductive potential) phytoplankton and show rapid change in their growth with changes in aquatic ecosystems. Besides nutrient stoichiometry, the changes in ionic strength, pH, light penetration, and temperature also affect the diversity and abundance of diatoms (Potapova and Charles 2003).

Benthic diatoms are an important group of protists that show niche specialization against nutrient pollution and light regime shifts. Because this group grows by attaching to certain substrate, they can be used as a selected group to assess responses especially in lotic systems. In response to atmospheric deposition of N in oligotrophic alpine lakes and P enrichment in temperate lakes, species such as *Asterionella* and *Fragilaria* have been reported to constitute the dominant group (Saros et al. 2005). The Ganga River receives, along with point and nonpoint sources, huge amount of N and P through atmospheric deposition (Pandey et al. 2013, 2014a, 2016a) suggesting that these inputs will also have strong effect on the distribution of diatoms. In an earlier study conducted along 268 km stretch from Allahabad to Varanasi downstream, we found a significant shift in diatom abundance and diversity in concordance with the changing state of carbon and nutrient pollution (Pandey et al. 2017). A significant decline in species diversity was observed with decreasing N/P stoichiometry, and, except for a few species which are adapted to high nutrient concentrations, the sites with nutrient-rich condition showed less overall diversity of diatoms. Species adapted to nutrient pollution, especially to P pollution, are accommodated in heterogeneous habitats through dominance transference (Fig. 10.1 and Table 10.1), which helps in the restoration of ecosystem functioning under adverse conditions (Pandey et al. 2017). Using an appropriate statistical tool, P-loving species can easily be grouped in one cluster (Fig. 10.2).

The N- and Si-limiting condition could shift the diatom assemblages from P-sensitive to P-tolerant species. The freshwater diatom species have high content of Si compared to the marine species (Conley et al. 1989). Thus, the relative availability of dissolved silica (DSi) regulates the proportion of siliceous diatoms in freshwater bodies (Conley 1997). Centric diatom bloom farmer, *Cyclotella,* a highly silicified diatom, is of growing concern in eutrophied rivers (Tavernini et al. 2011) and is reported to enhance the export of biogenic silica and C to the river sediment (Pandey et al. 2015a, b). The shift in environmental variables such as light and nutrients changes diatom species composition (Bere and Tundisi 2011). A study conducted in Varanasi region has shown increasing abundance of *Cocconeis,*

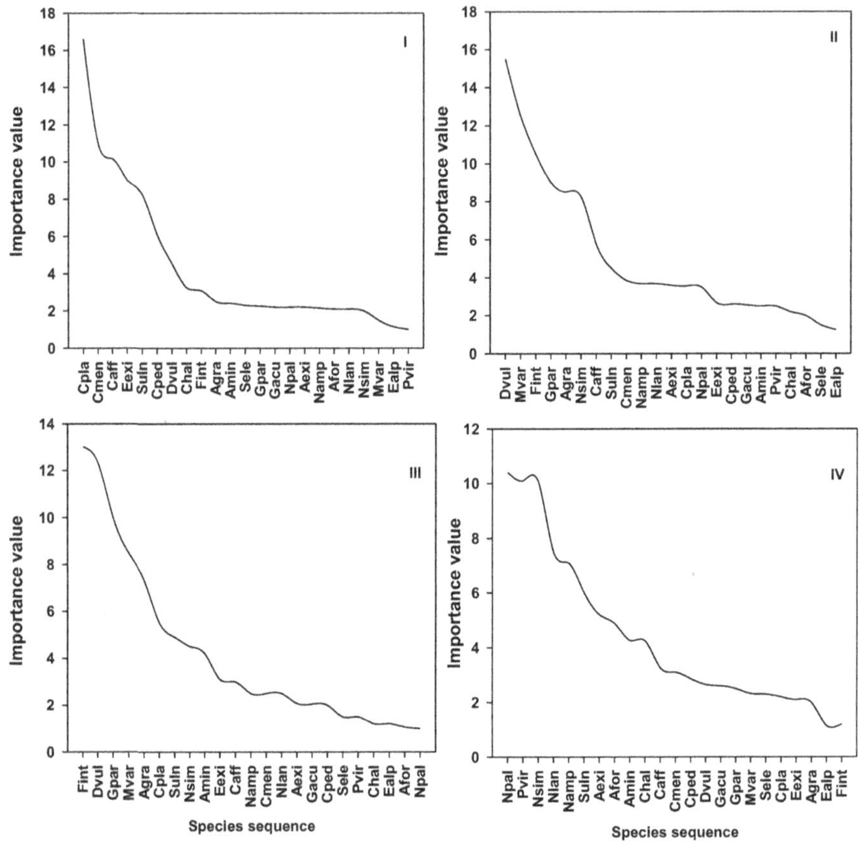

Fig. 10.1 Dominance-diversity curve of benthic diatoms at four study sites (I, Yamuna; II, Assi; III, Varuna; and IV, Gomti confluences with the Ganga River). (*Reprinted from Pandey et al. 2017, with permission from Current Science)

Cyclotella, *Hyalodiscus*, and *Aulacoseira* and declining abundance of *Diatoma* (Pandey et al. 2015b). Species such as *Achnanthidium*, *Amphipleura*, *Asterionella*, *Craticula*, and *Cyclotella* were found to be more abundant at highly polluted downstream sites; and nutrients and DOC were found to be the major regulator of abundance variabilities. At high N/P and Si/P stoichiometric ratios, species such as *Hyalodiscus*, *Navicula*, *Nitzschia*, etc. have been found in abundance (Pandey et al. 2015b).

Further, the abundance of *Aulacoseira* has been shown to correlate with high P loading (Walsh and Wepener 2009), while that of *Synedra* and *Cocconeis* has been linked with P-limiting condition (Walsh and Wepener 2009). The species such as *Pinnularia* and *Diatoma* are able to tolerate mesotrophic to eutrophic conditions (Muscio 2002; Walsh and Wepener 2009), and *Gomphonema* can tolerate eutrophic to hypereutrophic condition (Walsh and Wepener 2009). Ponader and Potapova

Table 10.1 Explanation of abbreviations used in Figs. 10.1 and 10.2

Abbreviation	Full form
Aexi	*Achnanthes exigua*
Amin	*Achnanthidium minutissimum*
Afor	*Asterionella formosa*
Agra	*Aulacoseira granulata*
Cped	*Cocconeis pediculus*
Cpla	*Cocconeis placentula* var. *lineata*
Chal	*Craticula halophila*
Cmen	*Cyclotella meneghiniana*
Caff	*Cymbella affinis*
Dvul	*Diatoma vulgare*
Eexi	*Eunotia exigua*
Ealp	*Eunotia alpine*
Fint	*Fragilaria intermedia*
Gpar	*Gomphonema parvulum*
Gacu	*Gyrosigma acuminatum*
Mvar	*Melosira varians* Agardh
Nlan	*Navicula lanceolata*
Nsim	*Navicula simplex*
Namp	*Nitzschia amphibia*
Npal	*Nitzschia palea*
Pvir	*Pinnularia viridis*
Sele	*Surirella elegans*
Suln	*Synedra ulna*

(2007) have reported high growth of *Achnanthidium* in polluted water including those affected by acid mine drainage. Similarly, studies have reported high growth of *Surirella*, *Amphipleura*, and *Craticula* at sites affected with urban discharge and are known to be tolerant to high concentration of nutrients (Muscio 2002). Studies on heavy metal-diatom responses are limited. Species such as *Nitzschia* has been shown to grow under metal contamination if coupled with nutrient-rich condition (Trobajo et al. 2013).

10.4 Dissolved Oxygen Deficit and Ecosystem Feedbacks

One of the most general ways of aquatic ecosystems to respond to increasing anthropogenic perturbations is the reduction in dissolved oxygen (DO) level. Development of hypoxic/anoxic zones in aquatic ecosystems is among the most widespread and deleterious effects of anthropogenic influences on global scale (Diaz and Rosenberg 2008). High input of oxygen-demanding substances (ODS) increases the rate of decomposition to an extent greater than the rate of oxygen supply (Rabalais et al. 2010) leading to dissolved oxygen deficit (DOD) in the system. Besides this,

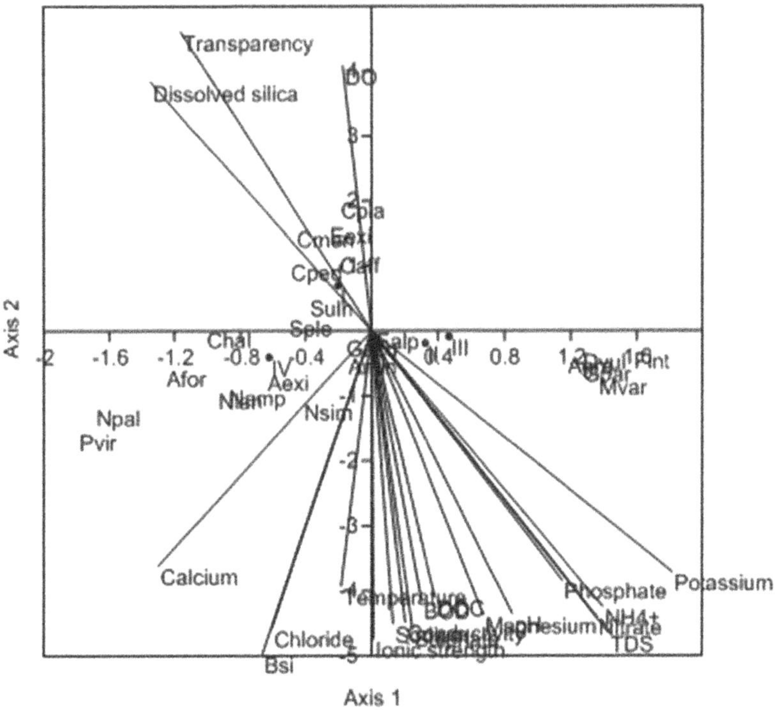

Fig. 10.2 The CCA bi-plot showing diatom species and environmental variables in the ordination space of four different quadrants. (*Reprinted from Pandey et al. 2017, with permission from Current Science)

increasing autochthonous carbon and temperature coupled with reduced flow also contribute to oxygen depletion in riverine ecosystems (Jaiswal and Pandey 2019b; Pandey et al. 2019). Accordingly, the variations in DO and associated shift in DOD can be used to get insight into the human-induced alterations in biological and chemical processes operating in river ecosystems. Studies have shown that the riverbed sediment consumes a large amount of DO from the overlying water leading to increased dissolved oxygen deficit at sediment-water interface (DOD_{SW}) (Jaiswal and Pandey 2019b). Benthic hypoxia/anoxia has been reported in the Gulf of Mexico (Rabalais et al. 2001), in the St. Lawrence River (Gilbert et al. 2005), and in the Arabian Sea and the Bay of Bengal (McCreary et al. 2013). The DOD_{SW} affects microhabitats (Mackenzie et al. 2000) and biogeochemical processes controlling nutrient cycling (Diaz and Rosenberg 2008) and causes habitat fragmentation leading to a shift in the benthic community and trophic cascade (Rabalais et al. 2001). At low DO (<2.0 mg L^{-1}), benthic organisms start showing abnormal behavior, and mass mortality and a shift in community structure can be observed if the level below 0.5 mg L^{-1} persists for a longer time (Diaz and Rosenberg 2008).

Most of the studies conducted so far on the Ganga River consider only biological oxygen demand (BOD) and chemical oxygen demand (COD) to describe oxygen linkages and measure the organic pollution load and health status of the river (Tare et al. 2003; Dwivedi et al. 2018). The sediment oxygen demand (SOD) has been reported to share >50% of total oxygen demand (MacPherson et al. 2007) but is a more neglected cause of increasing DOD in the Ganga River. The SOD has two components: biological sediment oxygen demand (BSOD), which is associated with benthic respiration and microbial decomposition of organic matter, and chemical sediment oxygen demand (CSOD), the oxygen consumed in the oxidation of Fe, Mn, and NH_4^+ (Higashino et al. 2004). A study conducted along a 518-km-long segment of the middle stretch of the Ganga River and downstream of two point sources showed high SOD and DOD_{sw} and development of hypoxia/anoxia at many locations of the river (Jaiswal and Pandey 2019b). This merits attention because wastewater treatment technologies generally address removal of BOD only. The increasing load of other chemicals in the river is also causing a greater risk to DO and needs proper management.

Episodic development of hypoxic/anoxic zones is considered as a signal of critical health of an aquatic ecosystem as the DO less than 2 mg L^{-1} can cause lethal to sublethal effects on benthic organisms and fish (Diaz and Rosenberg 2008). Hypoxia/anoxia-induced ecosystem consequences include loss of biodiversity, changes in organic matter processing and nutrient cycling, decreased resistance to invasion and ecosystem functioning, and ultimately reduced resistance to natural and human perturbations (Solan et al. 2004; Carstensen et al. 2014). Further, the hypoxic/anoxic conditions affect secondary production which ultimately influences the food web (Carstensen et al. 2014). These conditions lead to ~75% reduction in burrow productions which reduces the sediment oxygenation exacerbating the problem further (Middelburg and Levin 2009).

The benthic hypoxia/anoxia generates positive feedbacks such as sediment-P release, sediment-metal release, and denitrification (Conley et al. 2002; Eyre and Ferguson 2009) which result in massive alteration in nutrient and metal cycling (Hu et al. 2001; Villnäs et al. 2012). In a study conducted in the Ganga River, we found a high rate of sediment-P release at sites with high dissolved oxygen deficit (Jaiswal and Pandey 2019b; Pandey et al. 2019). The phosphate bounded to iron oxides in oxygenated sediments is released to the overlying waters when the oxides are reduced under hypoxic/anoxic condition (Middelburg and Levin 2009). The increased flux of P from bed sediment alleviates P limitation (Conley et al. 2002) and propels phytoplankton production (Cardinale 2011) leading to DOD which delays the recovery further (Jaiswal and Pandey 2019b). The oxygen deficit in the benthic region accelerates denitrification (Eyre and Ferguson 2009). Studies have reported severe hypoxia/anoxia at the Bay of Bengal, and it has been urged that even a small change in oxygen level will lead to a drastic change in nitrogen balance (Bristow et al. 2017). We have reported a high rate of denitrification downstream of point sources and tributary confluences at sites where DO_{sw} was below 2.0 mg L^{-1} (Jaiswal and Pandey 2019f; Pandey et al. 2019). The study sites with high denitrification also have high concentrations of total organic carbon (TOC), nitrate, and total nitrogen (TN) which are known to enhance the rate of denitrification (Piña-Ochoa and Alvarez-Cobelas 2006). The organic carbon concentration affects denitrification as it generates electron source for denitrifying enzymes (Dodla et al. 2008). The increase in NH_4^+ efflux ultimately stimulates phytoplankton growth and in turn increases hypoxia/anoxia at sediment water interface making the recovery of ecosystem even more difficult (Middelburg and Levin 2009).

About 85% of metal inputs to the river deposit in the bed sediments (Zhang et al. 2016). However, the benthic hypoxic/anoxic condition leads to enhanced efflux of metals from sediment to overlying waters (Jaiswal and Pandey 2019d). The hypoxia/anoxia affects the redox condition of sediments and in turn provides positive feedback to sediment-metal release (Jaiswal and Pandey 2019d). The metals in the sediment are adsorbed onto organic carbon, iron, and manganese oxides and clay particles (Eggleton and Thomas 2004) in different ways such as occlusion in amorphous materials, absorption at the surface of oxy-hydroxides of Fe and Mn, complexation with organic matter, and incorporation with sulfides (Zhang et al. 2014). A high DOD_{sw} causes reduction of nitrate and oxy-hydroxides of iron and manganese leading to increased release of metals from riverbed sediment (Eggleton and Thomas 2004). We found a high rate of sediment-metal release in the Ganga River at sites associated with hypoxia/anoxia such as the locations close to drain mouth, tributary confluences, and sites downstream of cities (Jaiswal and Pandey 2019d). Significant increase in Fe and Mn releases at DO < 2.0 mg L^{-1} has been reported by Banks et al. (2012) in an incubation experiment, by Fu et al. (2014) for Jialu River, and by Liu et al. (2019) in microcosm experiments. Banks et al. (2012) found increased dissolved fraction of Zn, Pb, Cd, and Cu from contaminated sediments even at very short-time hypoxic conditions. Liu et al. (2019) have reported significant increase in bioavailability of Zn, Pb, Cd, Cu, and Cr under severe hypoxic condition (DO, 0–2.0 mg/l). For the Ganga River, an increase in benthic metal bioavailability has been reported at locations with $DO_{sw} < 2.0$ mg L^{-1} (Jaiswal and Pandey 2019d).

Enhanced release of metals at sediment-water interface has toxicological implications. Different forms of the metals show different degree of mobility, chemical interactions, biological availability, and toxicity (Xu et al. 2017). The mobile and bioavailable fractions of metals cause greater toxicity to aquatic organisms (Eggleton and Thomas 2004). Thus, assessing the bioavailable fractions and the factors affecting their concentration and release can be helpful in understanding transformations, transport, and impact of metals in the aquatic environment (Morelli and Gasparon 2014). Various studies have shown the impact of metal toxicity on benthic ecosystems in terms of increased mortality and biodiversity loss (Stark et al. 2004), effect on colonization and dispersal (Stark et al. 2004), and reduced reproduction rate and population growth (Vicente-Martorell et al. 2009). Increased fraction of bioavailable metals have been shown to cause more negative impacts on epibenthic and pelagic invertebrates and fishes (Vicente-Martorell et al. 2009). Given that the Ganga River is the home of a diversity of economically important fisheries (Rao 2001) and other aquatic organisms, a decrease in DO, development of hypoxic zones, and consequent release of bioavailable metals will lead to death of these organisms, influencing trophic cascade and the nutritional and livelihood security to human consumers.

10.5 Shifts in Carbon Sequestration

Although freshwater ecosystems cover relatively less geographical area (less than 4% of the Earth's surface), they play a critical role in the global carbon cycle because of the high rate of respiration and carbon sequestration (Cole et al. 2007). The inland water bodies transport huge amounts of carbon from land to the ocean and perform a major role in carbon transit. Recent studies have shown that the inland water bodies emit the carbon in the amount close to those absorbed by organisms on Earth's land surface and in oceans (Raymond et al. 2013). Further, in freshwater bodies, more carbon is buried each year than those in vast oceanic floor (Aufdenkampe et al. 2011). Global studies have shown that every year about 2.7 billion metric tons of carbon reaches to the inland water bodies through different sources (Raymond et al. 2013). Half of this carbon is respired and returned back to the atmosphere as CO_2 (Bastviken et al. 2011; Raymond et al. 2013), ~0.4 billion tons of carbon is buried in bed sediments, and ~0.9 billion tons is exported to oceans (Aufdenkampe et al. 2011). Human interference such as land use change is dramatically affecting the carbon cycle in freshwater bodies. Nutrient input to freshwater bodies increases the algal blooms which absorb carbon from the atmosphere and increase the carbon sequestration (Pacheco et al. 2013). However, on decomposition and respiration, this carbon is released to the atmosphere and the cycle goes on (Bastviken et al. 2011; Borges et al. 2015).

A worldwide study by Cole et al. (2007) reveals that ~87% of the lakes are CO_2-supersaturated and the average pCO_2 is about three times higher than the overlying atmosphere. Thus, due to increasing human pressure, the surface waters may become a source rather than a sink of atmospheric CO_2. Recent studies have established that this CO_2 flux from rivers and stream is enough to affect the regional C budget at landscape scale (Raymond et al. 2013; Jaiswal et al. 2018). Amazonian rivers, for instance, have been reported to emit CO_2 more than ten times of the amount of C exported to the ocean (Richey et al. 2002). One of the main reason of high CO_2 efflux from rivers is in situ breakdown of young organic matter (Richey et al. 2002). Recent studies have indicated that the human-impacted Ganga River is receiving an increasingly high amount of carbon from terrestrial sources (Pandey et al. 2014a). In an earlier study, conducted at land-water interface (LWI) of the Ganga River, we found that the LWI is outgassing a huge amount of CO_2 into the atmosphere indicating that due to increasing human perturbations many parts of the Ganga River are now converted into a source of CO_2 (Jaiswal et al. 2018; Jaiswal and Pandey 2019e).

The organic matter degradation and carbon cycling are controlled by microbial extracellular enzyme (EE) activities (Sinsabaugh et al. 2009). The anthropogenic input of carbon (allochthonous C) causes a shift in microbial community structure and functioning, including organic matter degradation and carbon cycle. The EE activity is influenced by human inputs such as carbon, nutrients, and heavy metals (Pandey and Yadav 2017; Jaiswal and Pandey 2018). Carbon and nutrients act as substrate and enhance the EE activities, while heavy metals act as toxicant and thus reduce the EE activities. As most of the human-impacted rivers receive a high amount of toxicants along with carbon and nutrients (Jaiswal and Pandey 2019a), a deviation has also been reported regarding the carbon sequestration in these rivers. Studies have reported a reduced rate of decomposition and C mineralization even at low concentration of metals (Nwachukwu and Pulford 2011). In a study, we found a contrasting result regarding the carbon decomposition and sequestration along the middle stretch of the Ganga River (Verma et al. 2019). In the sites rich in carbon and nutrients and where metal concentrations did not exceed the toxic threshold (ΣTHM $< 360\ \mu g\ g^{-1}$), an increase in CO_2 emission was observed. These sites were also reported with high EE activities as organic C enhances microbial activity. However, at sites with high metal concentration (ΣTHM $> 360\ \mu g\ g^{-1}$), a significant reduction in CO_2 emission was recorded (Jaiswal and Pandey 2019e). This could be linked with the fact that increased metal concentration negatively influences the microbial activity and carbon degrading enzymes (β-D-glucosidase and FDAase) leading to reduction in microbial ability to metabolize carbon sources (Jaiswal and Pandey 2018). A high C/N ratio was also reported at these sites, further indicating higher accumulation of C relative to release (Verma et al. 2019). Overall, our studies with the Ganga River clearly indicate that metal pollution in eutrophic lotic systems enhances C storage relative to flux (Jaiswal and Pandey 2019e; Verma et al. 2019).

These results indicate that the metals cause physiological constraints in carbon breakdown and consequently enhance C accumulation. If the heavy metal concentration continues to increase, as expected in future, the CO_2 emission and C degradation may not be proportionate to the amount of carbon the human-impacted rivers are receiving. This will lead to enhanced C accumulation relative to flux in anthropogenically impacted large rivers.

10.6 Alternative Alert Systems

10.6.1 Extracellular Enzymes

Biomonitors play an important role in identifying shifts in ecosystem structure and functioning, recognizing the causal factors, and understanding the consequences. Unlike terrestrial ecosystems, where most of the shifts are quantitatively detectable (Oliver et al. 2015), scientists often face a number of limitations in identifying specific and universal biomonitors for changes in water quality and trophic status of riverine ecosystems (Lafont 2001). The major challenges for lotic ecosystems are the recurrence of variable and multiple anthropogenic perturbations, climate change, hydrological forcing, and connectivity with other domains such as watershed and airshed, which influence the universality and specificity of a biomonitoring tool (Pearson et al. 2016).

Studies generally use variables such as biological oxygen demand (BOD), chlorophyll **a** biomass (Gholizadeh et al. 2016), phycocyanin (Ahn et al. 2007), microinvertebrates (Turley et al. 2016), and diatom indices (Potapova et al. 2004;

Rimet and Bouchez 2012) as biomonitoring tools for assessment of water quality in different parts of world. The benthic diatoms have been established as a more stable predictor of trophic state and human perturbations as they respond directly to nutrient and carbon inputs (Pan et al. 1996). This merits attention because benthic diatoms grow attached to a certain substrate and thereby are less influenced by lotic forces of river ecosystems (Pandey et al. 2017). Wide distribution of diatoms supports the suitability and universality of diatom indices to be an indicator of eutrophy. However, such indices have been reported to be less suitable in geographical regions other than those where these were actually developed reducing the universal applicability of diatom-based tools across the globe. Other biomonitoring tools that have been generally used in the assessment of human impact on river waters are biological measures of eutrophication such as algal growth and pigments (Potapova et al. 2004). However, the relationships of algal community, chlorophyll **a**, biomass, and nutrient concentrations are often influenced by environmental factors such as climate, upstream basin size, river width, and flow (Pan et al. 1996) which question the applicability of these determinants in lotic ecosystems. The suitability of animal organisms as biomonitor also is hampered by factors such as mobility, feeding behavior, and position in trophic state (Rimet and Bouchez 2012).

The riverbed sediments, an important component of riverine ecosystems, are a biologically active and comparatively stable zone, which plays an important role in ecosystem functions including biogeochemical cycling, carbon metabolism and sedimentation, secondary production, and nutrient and heavy metal removal from the water column (Covitch et al. 2004). Because microbial community constitutes the key component of detritus system and any change in their metabolism affects the whole ecosystem processes including organic matter decomposition and nutrient cycling, their functional and structural attributes provide an actual picture of the health condition of an ecosystem. Studies have shown that enzyme activities can be used as an index of microbiological functional diversity and combining enzyme activities with other physical and chemical measurements can provide important information regarding ecosystem stability (Nannipieri et al. 2002). The sediment-based determinants such as the extracellular enzymes have been proved to be the most suitable parameter for quantification of shift in ecosystem responses toward alteration in organic matter (substrate) and heavy metals (inhibitors) (Sinsabaugh et al. 2008; Jaiswal and Pandey 2018, 2019a).

The extracellular enzymes β-D-glucosidase, alkaline phosphatase, and protease are used as indicator of C acquisition, P starvation, and N mineralization, respectively (Rejsek et al. 2008; Sinsabaugh et al. 2009; Duhamel et al. 2010). Similarly the fluorescein diacetate hydrolytic essay (FDAase) is used as an indicator of overall microbial activities (Schnürer and Rosswall 1982). The latter involves all the three major group of enzymes (lipases, esterases, and proteases) that mediate organic matter decomposition (Fontvieille et al. 1992). The substrate such as carbon and nutrients cause stimulatory effect, while toxicants such as heavy metal inhibit the activities (Sinsabaugh et al. 2008). Our multi-year and multi-scale studies confirm these relationships validating them for the riverbed sediment of the Ganga River

(Yadav and Pandey 2017b; Jaiswal and Pandey 2018, 2019a) indicating that extracellular enzymes can be used as an alternative alert system against increasing human pressure.

10.6.2 Elemental Stoichiometry

Elemental stoichiometry, the mass balance of key elements (C, N, P, and Si) in an ecosystem (Elser et al. 2009), has a central role in the theory of resource ratio competition between alga (Makulla and Sommer 1993), consumer-driven nutrient recycling (Elser et al. 2009), and food chain efficiency (Sterner et al. 1998). The N/P/Si Redfield ratio (16:1:16) is essential for balanced growth of phytoplankton specifically for diatoms (Turner et al. 2003). The changes in relative proportion of these nutrients define which nutrient to limit phytoplankton growth (Elser et al. 2009). A major factor to influence this ratio in the Ganga River is the increasing use of N and P fertilizers to meet the demand of food of overpopulated Ganges basin. About 10 million tons of chemical fertilizers are applied in the Ganges basin, which represent 45% of India's total annual fertilizer consumption. These N and P fertilizers reach to the river through atmospheric deposition, leaching, and runoff in the form of highly mobile NO_3^- and PO_4^{3-} ions. Unlike human-induced increases in the concentration of N and P at global scale, the Si concentration in most cases is either stable or declining. Hydrologic shifts in the watershed may reduce Si concentration by as much as 50% (Correll et al. 2000). A shift in this ratio causes cellular nutrient imbalances and induces a change in phytoplankton composition, biogeochemical cycles, carbon sequestration, biological diversity, and trophic cascades (Elser et al. 2009; Pandey and Yadav 2015). Therefore, the N/P/Si ratio is considered as a sensitive indicator of aquatic health and food web structure. The absolute concentration of nutrients and their stoichiometric ratios can be used together as a comprehensive predictor of eutrophy across broad landscapes as represented by large rivers.

Studies show that the disproportionate nutrient loading and management efforts have changed the canonical N/P stoichiometric ratios in many aquatic ecosystems of the world including India (Pandey and Yadav 2015; Pandey et al. 2016a). The anthropogenic causations that enhance N and P input do not generally lead to a proportionate increase in the concentration of Si in rivers. Indeed, some of the anthropogenic activities such as river damming decrease the amount of Si reaching to the coast. Further, low flow season induced increase in riverine primary productivity and nutrient uptake, and subsequently the sedimentation of diatoms leads to loss of adsorbed silicate from the water column (Conley 1997). Since Si is essential for the growth of diatoms, a deviation in the supply of Si may change the magnitude of diatom-driven C sequestration. Studies have shown that a decrease in N/P ratio causes a shift in the dominance in phytoplankton assemblage toward diazotrophic cyanobacteria (Elser et al. 2000; Pandey et al. 2017). A shift in N/P ratio toward <16:1 changes the phytoplankton community and promotes P-favored taxa.

Similarly, the Si/N ratio below 1 leads to reducing the proportion of diatoms/siliceous algae in the phytoplankton assemblage and consequently causes a shift in the higher trophic levels (Gilpin et al. 2004). Further, the Si-limiting condition leads to enhance non-diatom algal growth (Pandey et al. 2017). The excessively higher concentration of N and P compared to Si is causing a dramatic shift in the phytoplankton composition, changing the pattern of community dominance toward green or blue-green algae (Teubner and Dokulil 2002) including those in the Ganga River as indicated also by high concentration of phycocyanin at nutrient-rich sites (Pandey et al. 2016b). Further, as the system moves toward eutrophy, feedbacks at sediment-water interface may increase P supply and consequently promote the growth of P-favored harmful algal species (Pandey et al. 2017). Thus, the changing pattern of nutrient limitation and the resulting competition for resources in phytoplankton would decrease the proportion of less adapted algal species, increasing the share of non-siliceous diatoms in the community and consequently decreasing the C sequestration and compromising the ecological assimilation capacity of the river.

The atmospheric deposition (AD) of N and P has increased tremendously in various parts of the world (Galloway et al. 2008) and is continuing to rise in the Ganges basin (Siddiqui et al. 2019b). Since N and P are the major component of AD, a potential shift in AD-N/P ratio will alter N/P ratios of surface waters which would shift phytoplankton composition. Many of the temperate European and North American lakes have been reported to be suffering with this problem (Bergstrom and Jansson 2006). As the Ganga River receives large but disproportionate input of nutrients through point and nonpoint sources including atmospheric deposition, at many locations, the river experiences shifts in N/P/Si stoichiometry and the proportion of specific nutrient availability (Pandey et al. 2016b). A watershed-scale study from Devprayag to Ganga Sagar (Pandey et al. 2016a) has reported that at polluted sites, the ratio of N/P remained below 16:1 indicating that P is no more a limiting nutrient in the river and concordantly the abundance of dominant diatom genera has also changed. Diatom species such as *Diatoma vulgaris*, *Fragilaria intermedia*, and *Gomphonema parvulum* were found abundantly at sites characterized by high P, whereas species such as *Cocconeis placentula*, *Cyclotella meneghiniana*, and *Cymbella affinis* were found at sites with low phosphorus concentration (Pandey et al. 2017). The results further revealed a relatively lower proportion of Si ($Si/P < 16:1$) in the Ganga River than the required ratio to meet cellular Si for diatoms. A higher N/Si ratio shows that if this condition is continued, it will lead to Si limitation in the long-term future (Pandey et al. 2016a).

A study of Mississippi River has shown that excessive N and P input as compared to Si is causing severe eutrophication in the river (Turner et al. 2003). Shifts in stoichiometric ratios affect the quantity as well as quality of primary production. With increasing N input, an increase in the cellular N/P ratio of terrestrial and aquatic plants has been reported (Elser et al. 2009). This change in cellular N/P ratio affects various metabolic processes ultimately leading to a cascade of effects ranging from shift in growth of individual organism to alteration in species composition and community functioning (Pandey et al. 2017). A classic example of this type of ecosystem response is the shift in population of *Daphnia* (a freshwater zooplankton).

Daphnia enjoys a P-rich lifestyle and encounters potential P deficiency when cellular P declines (Elser et al. 2000). Thus, the population of *Daphnia*, which can be used as an indicator of P eutrophy, will decline sharply under N-rich condition. These studies have led to conclude that the shifts in nutrient stoichiometry can be used as an alternative response indicator of shifting ecosystem structure and functioning driven by increasing natural and anthropogenic perturbations.

10.6.3 Diatom-TEP Linkages

Diatoms, a highly diverse group of photosynthetic protists and widely used indicators of environmental shifts, are producers of transparent exopolymeric particles (TEP). The diversity and abundance of these primary producers are affected by absolute nutrient concentrations, stoichiometry, ionic strength, pH, light penetration, and temperature (Potapova and Charles 2003). The abundance of diatoms is often negatively influenced by high concentrations of nutrients because only some species can grow in nutrient-rich condition. The species adapted to high nutrient concentrations remain generally accommodative to heterogeneous habitats by dominance transference (Pandey et al. 2017). As already mentioned, the N/P/Si ratio is the most important factor that drives the diatoms diversity and abundance. The decrease in N/P ratio decreases the species diversity although P-loving diatoms proliferate rapidly (Pandey et al. 2017). Increased N/Si ratio leads to Si limitation with potential effects on the quantity (cell number and biomass) as well as quality (composition of biomass) of diatom assemblages (Davidson and Gurney 1999). Alterations in the abundance of specific diatom species, assemblage, and cellular metabolic states affect the microbial trophic transfer and population of meso-zooplankton (Miralto et al. 1999) with overall effects on carbon export and biogeochemical cycling.

Diatoms produce acidic polysaccharides in the form of transparent exopolymeric particles (TEP). The size of TEP ranges from >0.4 μm to <200 μm and is stained with the Alcian blue. Diatoms with C_4 photosynthetic pathway are prolific in carbon capture and storage under excess N supply and may accumulate excessive carbon (Riebesell et al. 2007). To maintain a normal physiological state (C/N ratio, for instance), the excessive carbon is excreted in the form of acidic polysaccharides which are ultimately converted into TEP. Because of their ability to form coagulates and aggregates, the TEP play an important role in the regulation of DOC-POC pump and carbon sequestration. Additionally, high-density particles such as heavy metals get aggregated enhancing the density of TEP and consequent sedimentation of nutrients, metals, and pathogens (Passow et al. 2001). The results of a recent study (Pandey et al. 2017) reveal that to cope up with changing nutrient concentrations and their stoichiometric ratio, the diatoms increase the production of TEP which enhances the sedimentation and removal of turbidity and other harmful components. This is an important mechanism responsible for high self-purification capacity of the Ganga River (Pandey et al. 2017). The study further shows that to compensate the reduction in TEP under excessive human pressure, the diatom tends to accommodate through dominance transference. The low-profile guilds representing species such as *Cocconeis placentula*, *Cymbella affinis*, *Cyclotella meneghiniana*, and *Synedra ulna* were found abundantly at P-poor sites, and high-profile guilds representing species such as *Diatoma vulgaris*, *Gomphonema parvulum*, and *Fragilaria intermedia* were present at P-rich sites (Pandey et al. 2017).

The excessive nutrient loading alters the diatom dominance pattern (Fig. 10.1 and Table 10.1), and a marked skewness in diatom dominance-diversity linkages has been observed in the Ganga River (Pandey et al. 2017). The synchrony between skewness and altered water quality shows the ability of diatoms to cope with nutrient stressors and disturbances. Among the TEP producers, species such as *Cocconeis placentula*, *Cyclotella meneghiniana*, and *Cymbella affinis* have been reported to flourish at nutrient-poor sites, while *Aulacoseira granulata*, *Diatoma vulgaris*, *Melosira varians*, and *Fragilaria intermedia* show extensive growth in nutrient-rich condition (Pandey et al. 2017; Fig. 10.2). These results indicate that the diatom ecological guilds can be used as holistic and alternative indicators of short-term changes or disturbances in the aquatic environment. Additionally, the dependence of TEP on Chl **a** biomass and N/P stoichiometry makes it an indicator of trophic status and nutrient pollution. Because the TEP production is maintained partly by changes in diatom dominance-diversity linkages despite variable ecological conditions and human perturbations, the TEP coupled diatom dominance transference can be used as a key node to cue nutrient pollution and ecological assimilation capacity of anthropogenically impacted large rivers.

10.6.4 Ecological Response Index

Quantitative estimation of ecosystem responses against increasing human perturbations has become a growing research area in aquatic pollution control. Despite urgent need, only few studies so far are available, providing a universal index to quantify holistic changes in the water quality (Satyamurthy 2017). For a universal applicability, an index should have intricate links with ecosystem structure and functioning (Peterson and Stevenson 1992). Sediment-based biomonitoring tools are now being suggested to be more accurate in designing empirical relationships to uncover the ecosystem responses and magnitude of degradation (Turley et al. 2016; Pandey and Yadav 2017).

For lotic ecosystems, where hydrologic forcing drives unpredictability, selecting a suitable response variable is difficult. There is no study so far available, except Jaiswal and Pandey (2019a), linking simultaneously the carbon-heavy metal-ecosystem responses to quantitatively predict the human-driven alterations in large rivers (Table 10.2). The indices developed so far for the assessment of pollution load, toxicity, and trophic status, such as enrichment factor (Buat-Menard and Chesselet 1979), trophic state index (Carlson 1977), potential ecological risk index (Håkanson 1980), pollution load index (Tomilson et al. 1980), pollution index (Nemerow 1991), and geoaccumulation index (Müller 1969), consider the

Table 10.2 Indicators/indices of eutrophy and metal pollution used in water quality assessment

Index/indicator	Water body	References
FDAase	Ganga River	Jaiswal and Pandey (2019a)
Dissolved oxygen deficit	Ganga River	Jaiswal and Pandey (2019b)
CO_2 emission coupled extracellular enzyme activities	Ganga River	Jaiswal and Pandey (2019e)
N/P/Si ratio	Oceans	Redfield (1958)
Diatom indices	Rivers	Potapova et al. (2004) and Rimet and Bouchez (2012)
Transparent exopolymeric particles (TEP)	Ganga River	Pandey et al. (2017)
Daphnia	Freshwater systems	Elser et al. (2000)
Phormidium uncinatum	Ganga River	Pandey (2013)
Microinvertebrates	Rivers and streams	Turley et al. (2016)
Chlorophyll **a** biomass	Ganga River, lake	Tare et al. (2003) and Pandey and Pandey (2013)
Phycocyanin	Rivers	Ahn et al. (2007)
Light penetration	Ganga River	Pandey (2013)
Alkaline phosphatase	Ganga River	Pandey and Yadav (2017)
Water quality index	Inland water bodies	Brown et al. (1972)
Diatom dominance transference	Ganga River	Pandey et al. (2017)
Biological oxygen demand	Ganga River	Dwivedi et al. (2018)
Diatom pollution tolerance index	Freshwater systems	Muscio (2002)
Trophic state index	Lake	Carlson (1977)
Enrichment factor	Tropical North Atlantic Ocean	Buat-Menard and Chesselet (1979)
Potential ecological risk index	Lakes and other limnetic systems	Håkanson (1980)
Pollution load index	Estuaries	Tomilson et al. (1980)
Pollution index	Inland water bodies	Nemerow (1991)
Geoaccumulation index	Rhine River	Müller (1969)
Modified pollution index	Estuarine and marine environment	Brady et al. (2015)
Contamination factor	Lakes and other limnetic systems	Håkanson (1980)
Modified ecological risk index	Brisbane River	Duodu et al. (2016)
Ecological response index[a]	Ganga River	Jaiswal and Pandey (2019a)

[a]First index that simultaneously predicts C eutrophy and metal pollution in large rivers

concentration of individual variable/metal pollutant at a time. These indices generally address the "status" and not the functionality of how the ecosystem is exactly responding to human perturbations (Table 10.2). Further, the versatility of these indices is influenced by hydrological factors, sediment load, and sensitivity of the chosen determinants. The natural ecosystem receives a number of toxicants so the effect should be governed by their cumulative concentration (Håkanson 1980). Also, the organic carbon in a water body may reduce the toxic effect of metals making it difficult to establish a linear dose-response relationship under in situ condition (Håkanson 1980). In human-impacted rivers, eutrophication and metal pollution generally occur simultaneously. Thus, for an accurate measurement of river health, it is important to understand how the ecosystem responds toward human perturbations.

As a solution of this problem, Jaiswal and Pandey (2019a) developed an "ecological response index (ERI)" which is able to quantitatively predict ecosystem response to C eutrophication and metal pollution in large rivers. The index was developed using carbon and its response determinant (fluorescein diacetate hydrolytic assay, FDAase) and a sum of six heavy metal concentrations in an empirical relationship as below (Jaiswal and Pandey 2019a):

$$\text{Ecological Response Index} = \frac{\text{FDAase} \times qM}{\sum_{i=1}^{n} (M_i)}$$

where FDAase = fluorescein diacetate hydrolytic activity; qM = microbial quotient; and (M_i) = concentration of ith metal.

The ERI, as a quantitative predictor of eutrophy and metal pollution, was validated using Carlson's trophic state index (TSI) (Carlson 1977), Håkanson's risk index (RI) (Håkanson 1980), and Duodu's modified ecological risk index (MRI) (Duodu et al. 2016). Strong relationships among these indices indicated that the ERI is highly appropriate for quantitative prediction of metal pollution and trophic state of lotic ecosystems. Based on ERI, Jaiswal and Pandey (2019a) concluded that the ERI between 0 and 24 represents extreme metal pollution (toxic condition), between 25 and 38 indicates combination of eutrophy and metal pollution, 39 to 77 represents hypereutrophic condition and low metal pollution, 78 to 155 indicates eutrophic condition, and 156 to 320 represents the oligotrophic state of the river ecosystem. The ERI can be used as an alternative ecological tool for appropriately addressing concordant changes in ecological functioning with stronger mechanistic linkages between the causal factor and associated responses. Besides this, it is stable, widely applicable, and cost-effective in addressing the impact of multiple human stressors on functional shifts in riverine ecosystems.

10.6.5 Ecosystem Feedbacks

A positive feedback in an ecosystem begins when a driving force is likely to cause a transitional shift (Jaiswal and Pandey 2019f). Positive feedbacks are self-enhancing and are among the most prominent signature of transitional shift where the changes in ecosystem processes are expected to lead to development of alternative stable states (Scheffer et al. 2001). It occurs in a feedback loop leading to enhance the magnitude of even small perturbations. Positive feedbacks are widespread in nature ranging from cellular- to system-level processes and may lead to ecosystem-level consequences if it persists for longer duration (Dai et al. 2013). These nonlinear responses of ecosystems to environmental changes cause destabilization of nutrient cycling and other ecosystem processes leading to delay in the recovery from disturbances. Although various studies have emphasized the importance and relevance of understanding these processes in ecosystem restoration (Scheffer et al. 2001), identification and quantification of positive feedbacks for riverine ecosystems are very difficult.

High input of oxygen-demanding substances leads to development of hypoxic/anoxic zones which trigger positive feedbacks in aquatic environments. The dissolved oxygen (DO) plays the most important regulatory role in aquatic ecosystem structure and functioning. The oxygen deficiency causes dramatic changes in benthic communities and thus leads to noticeable changes on ecosystem functioning, species richness, and abundance, ultimately leading to development of alternative stable states. Some of the important feedbacks associated with benthic hypoxia/anoxia include the denitrification, sediment-P release, and sediment-metal releases. These functional shifts coupled with increased dissolved oxygen deficit (DOD) lead the system toward alternate stable states and development of fragmented patches with varying species composition, nutrient and metal concentrations, and shift in trophic state and food webs (Scheffer et al. 2001; Pandey et al. 2019).

Studies conducted by Jaiswal and Pandey (2019d, 2019f) reveal a high rate of denitrification, sediment-P release, and sediment-metal release at sites downstream of point sources, tributary confluences, and downstream cities along the Ganga River. These locations did show DO_{sw} below 1.5 mg L^{-1} indicating that the benthic hypoxia has generated positive feedbacks leading to the release of sediment-bound nutrients and metals to the overlying water. The release of nutrients and metals from the sediment accelerates eutrophication enhancing the DOD further. Also, the ecological communities and ecosystems respond to human perturbations in different ways. For instance, some processes are enhanced but respond smoothly and gradually to changes in environmental conditions, whereas some variables show instantaneous response and others may remain inert until a threshold condition is reached at which the ecosystem responds abruptly (Scheffer et al. 2009). Studies show that the extracellular enzymes can be used as a suitable predictor of ecosystem responses toward carbon, nutrient, and metal enrichment (Sinsabaugh et al. 2009; Jaiswal and Pandey 2019a). Because, the carbon and nutrients stimulate microbial activities, a concordant increase in the enzyme activity can be used as an indicator of eutrophy. Opposite to this, the heavy metal pollution causes toxic effects leading to a sudden decrease in the enzyme activity even in the presence of high concentration of carbon and nutrients (Jaiswal and Pandey 2018, 2019a). Because the carbon chelates metals leading to a reduction in toxicity, in ecosystems with prevalence of toxicants and stimulants, the toxic impact is reduced to a certain extent. The shifts in these responses can be used as alternative alert systems for understanding the health and level of deterioration of human-impacted riverine ecosystems.

10.7 Conclusions and Suggestions

Quantitative estimation of ecosystem-level responses against increasing human perturbations has become a growing research area in aquatic pollution control. However, despite urgent need, only few studies so far are available providing a universal index able to quantify holistic changes in the water quality. Studies conducted on the Ganga River generally consider parameters bifurcating eutrophy and metal pollution and, in most cases, without considering ecosystem-level consequences. A critical analysis of available studies/data and emerging trends leads to the following conclusions/suggestions to cue management strategies on the Ganga River:

1. Future research needs to focus on the changing state of ecosystem functions coupling human perturbations with ecosystem feedbacks such as denitrification, sediment-P and sediment-metal release, and ecological assimilation capacity of the Ganga River.
2. Because in human-impacted rivers eutrophication and metal pollution generally occur simultaneously, multi-temporal and multi-scale data base is needed for understanding ecosystem responses coupling eutrophy and metal pollution in the river.
3. The wastewater treatment technologies, available so far, generally address removal of biological oxygen demand (BOD) only. Because the increasing load of other oxygen-demanding substances (ODS) is contributing substantially to dissolved oxygen deficit (DOD) causing a greater pressure on overall dissolved oxygen (DO) levels, these need proper control and management. Also, advance metal removal technologies are needed because metal pollution in the river is continuing to rise.
4. Because sediment-based biomonitors are relatively stable and because microbial extracellular enzymes (EE) respond concordantly to carbon and metal enrichment, the EE activities such as β-D-glucosidase and FDAase can be a good indicator of river responses toward metal pollution and C eutrophy.
5. Shifts in nutrient stoichiometry can be used as an alternative response indicator of shifting ecosystem structure and functioning, driven by increasing anthropogenic perturbations. Diatom ecological guilds, rather than individual species, can be used as a holistic indicator of short-term disturbances in the aquatic environment. Initiatives can be taken to enhance micro-niches to increase the diversity of diatoms, the most important water purifiers in the Ganga River. Furthermore, because the production of transparent exopolymeric particles (TEP) is maintained partly by changes in diatom dominance-diversity linkages, the TEP coupled diatom dominance transference can be used as a key node to cue nutrient pollution and ecological assimilation capacity of the river.

Acknowledgments We acknowledge CSIR, New Delhi, India, for financial report as fellowship to DJ.

References

Ahn CY, Joung SH, Yoon SK, Oh HM (2007) Alternative alert system for cyanobacterial bloom, using phycocyanin as a level determinant. J Microbiol 45:98–104

Alvarez-Vazquez MA, Prego R, Caetano M, De Una-Alvarez E, Doval M, Calvo S, Vale C (2017) Contributions of trace elements to the sea by small uncontaminated rivers: effects of a water reservoir and a wastewater treatment plant. Chemosphere 178:173–186

Ariyadej C, Tansakul R, Tansakul P, Angsupanich S (2004) Phytoplankton diversity and its relationships to the physico-chemical environment in the Banglang Reservoir, Yala Province. J Sci Technol 26:595–607

Aufdenkampe A et al (2011) Riverine coupling of biogeochemical cycles between land, oceans, and atmosphere. Front Ecol Environ 9:53–60

Banks J, Ross DJ, Keough MJ (2012) Measuring hypoxia induced metal release from highly contaminated estuarine sediments during a 40 day laboratory incubation experiment. Sci Total Environ 420:229–237

Bashir T, Khan A, Gautam P, Behera SK (2010) Abundance and prey availability assessment of Ganges River Dolphin (*Platanista gangetica gangetica*) in a stretch of upper Ganges River, India. Aquat Mamm 36:19–26

Bastviken D et al (2011) Freshwater methane emissions offset the continental carbon sink. Science 331:50

Bellmore R, Compton J, Brooks J et al (2018) Nitrogen inputs drive nitrogen concentrations in US streams and rivers during summer low flow conditions. Sci Total Environ 639:1349–1359

Bere T, Tundisi JG (2011) Diatom-based water quality assessment in streams influenced by urban pollution: effects of natural and two selected artificial substrates, São Carlos-SP, Brazil. Braz J Aquat Sci Technol 15:54–63

Bergstrom AK, Jansson M (2006) Atmospheric nitrogen deposition has caused nitrogen enrichment and eutrophication of lakes in the northern hemisphere. Glob Chang Biol 12:635–643

Borges A et al (2015) Globally significant greenhouse-gas emissions from African inland waters. Nat Geosci 8:637–642

Brady JP, Ayoko GA, Martens WN, Goonetilleke A (2015) Development of a hybrid pollution index for heavy metals in marine and estuarine sediments. Environ Monit Assess 187:306

Bristow LA, Callbeck CM, Larsen M, Altabet MA, Dekaezemacker J, Forth M, Gauns M, Glud RN, Kuypers MM, Lavik G (2017) N_2 production rates limited by nitrite availability in the Bay of Bengal oxygen minimum zone. Nat Geosci 10:24–29

Brown RM, McClelland NI, Deininger RA, O'Connor MF (1972) A Water Quality Index – crashing the psychological barrier. In: Thomas WA (ed) Indicators of environmental quality. Environmental science research, vol 1. Springer, Boston

Buat-Menard P, Chesselet R (1979) Variable influence of atmospheric flux on the trace metal chemistry of oceanic suspended matter. Earth Planet Sci Lett 42:398–411

Bureau of Indian Standards (BIS) (2012) Specification for drinking water. IS: 10500:2012. Revised (2012) Bureau of Indian Standards, New Delhi

Cardinale BJ (2011) Biodiversity improves water quality through niche partitioning. Nature 472:86–89

Carlson RE (1977) A trophic state index for lakes. Limnol Oceanogr 22:361–369

Carstensen J et al (2014) Hypoxia in the Baltic Sea: biogeochemical cycles, benthic fauna, and management. Ambio 43:26–36

Central Pollution Control Board (CPCB) (2013) Pollution Assessment: River Ganga. Ministry of Environment and Forests, Govt. of India, Parivesh Bhawan, Delhi

Central Pollution Control Board (CPCB) (2016) Conservation of water quality of Ganga River, CPCB. Ministry of Environment and Forest, Government of India

Cole JJ et al (2007) Plumbing the global carbon cycle: integrating inland waters into the terrestrial carbon budget. Ecosystems 10:172–185

Conley DJ (1997) River contribution of biogenic silica to the oceanic silica budget. Limnol Oceanogr 42:774–777
Conley DJ, Kilham SS, Theriot E (1989) Differences in silica content between marine and freshwater diatoms. Limnol Oceanogr 34:205–213
Conley DJ, Humborg C, Rahm L, Savchuk OP, Wulff F (2002) Hypoxia in the Baltic Sea and basin scale changes in phosphorus biogeochemistry. Environ Sci Technol 36:5315–5320
Correll DL, Jordan TE, Weller DE (2000) Dissolved silicate dynamics of the Rhode River watershed and estuary. Estuaries 23:188–196
Covitch AP et al (2004) The role of biodiversity in the functioning of freshwater and marine benthic ecosystems. Bioscience 54:767–775
Dai L, Korolev K, Gore J (2013) Slower recovery in space before collapse of connected populations. Nature 496:355–358
Davidson K, Gurney WSC (1999) An investigation of non-steady state algal growth II. Mathematical modelling of co-nutrient limited algal growth. J Plankton Res 21:839–858
Diaz RJ, Rosenberg R (2008) Spreading dead zones and consequences for marine ecosystems. Science 321:926–929
Dodla SK, Wang JJ, DeLaune RD, Cook RL (2008) Denitrification potential and its relation to organic carbon quality in three coastal wetland soils. Sci Total Environ 407:471–480
Duhamel S, Dyhrman ST, Karl DM (2010) Alkaline phosphatase activity and regulation in the North Pacific subtropical gyre. Limnol Oceanogr 55:1414–1425
Duodu GO, Goonetilleke A, Ayoko GA (2016) Comparison of pollution indices for the assessment of heavy metal in Brisbane River sediment. Environ Pollut 219:1077–1091
Dwivedi S, Mishra S, Tripathi RD (2018) Ganga water pollution: a potential health threat to inhabitants of Ganga basin. Environ Int 117:327–338
Edokpa D, Evans M, Rothwell J (2015) High fluvial export of dissolved organic nitrogen from a peatland catchment with elevated inorganic nitrogen deposition. Sci Total Environ 532:711–722
Eggleton J, Thomas KV (2004) A review of factors affecting the release and bioavailability of contaminants during sediment disturbance events. Environ Int 30:973–980
Elser JJ et al (2000) Nutritional constraints in terrestrial and fresh water food webs. Nature 408:578–580
Elser JJ et al (2009) Shifts in lake N:P stoichiometry and nutrient limitation driven by atmospheric nitrogen deposition. Science 326:835–837
Eyre BD, Ferguson AJP (2009) Denitrification efficiency for defining critical loads of carbon in shallow coastal ecosystems. Hydrobiologia 629:137–146
Fontvieille DA, Outaguerouine A, Thevenot DR (1992) Fluorescein diacetate hydrolysis as a measure of microbial activity in aquatic systems: application to activated sludges. Environ Technol 13:531–540
France-Lanord C, Derry LA (1997) Organic carbon burial forcing of the carbon cycle from Himalayan erosion. Nature 390:65–67
Fu J et al (2014) Heavy metals in surface sediments of the Jialu River, China: their relations to environmental factors. J Hazard Mater 270:102–109
Galloway JN et al (2008) Transformation of the nitrogen cycle: recent trends, questions, and potential solutions. Science 320:889–892
Gholizadeh MH, Melesse AM, Reddi L (2016) A comprehensive review on water quality parameters estimation using remote sensing techniques. Sensors (Basel) 16:1–43
Gilbert D, Sundby B, Gobeil C, Mucci A, Tremblay GH (2005) A seventy two year record of diminishing deep water oxygen in the St. Lawrence estuary: the Northwest Atlantic connection. Limnol Oceanogr 50:1654–1666
Gilpin LC, Davidson K, Roberts E (2004) The influence of changes in nitrogen: silicon ratios on diatom growth dynamics. J Sea Res 25:21–35
Goldstein AH, Galbally IE (2007) Known and unexplored organic constituents in the earth's atmosphere. Environ Sci Technol 41:1514–1521

Håkanson L (1980) An ecological risk index for aquatic pollution control: a sedimentological approach. Water Res 14:975–1001
Hallquist M et al (2009) The formation, properties, and impact of secondary organic aerosol: current and emerging issues. Atmos Chem Phys 9:5155–5236
Higashino M, Gantzer CJ, Stefan HG (2004) Unsteady diffusional mass transfer at the sediment/water interface: theory and significance for SOD measurement. Water Res 38:1–12
Hill BH, Elonen CM, Jicha TM, Cotter AM, Trebitz AS, Danz NP (2006) Sediment microbial enzyme activity as an indicator of nutrient limitation in Great Lakes coastal wetlands. Freshw Biol 51:1670–1683
Hooper DU et al (2012) A global synthesis reveals biodiversity loss as a major driver of ecosystem change. Nature 486:105–108
Hu WF, Lo W, Chua H, Sin SN, Yu PHF (2001) Nutrient release and sediment oxygen demand in a eutrophic land-locked embayment in Hong Kong. Environ Int 26:369–375
Jaiswal D, Pandey J (2018) Impact of heavy metal on activity of some microbial enzymes in the riverbed sediment: ecotoxicological implications in the Ganga River (India). Ecotoxicol Environ Saf 150:104–115
Jaiswal D, Pandey J (2019a) An ecological response index for simultaneous prediction of eutrophication and metal pollution in large rivers. Water Res 161:423–438
Jaiswal D, Pandey J (2019b) Anthropogenically enhanced sediment oxygen demand creates mosaic of oxygen deficient zones in the Ganga River: implications for river health. Ecotoxicol Environ Saf 171:709–720
Jaiswal D, Pandey J (2019c) Investigations on peculiarities of land-water interface and its use as a stable test bed for accurately predicting changes in ecosystem responses to human perturbations: a sub-watershed scale study with the Ganga River. J Environ Manag 238:178–193
Jaiswal D, Pandey J (2019d) Benthic hypoxia in anthropogenically-impacted rivers provides positive feedback enhancing the level of bioavailable metals at sediment-water interface. Environ Pollut. https://doi.org/10.1016/j.envpol.2019.113643
Jaiswal D, Pandey J (2019e) Carbon dioxide emission coupled extracellular enzyme activity at land-water interface predict C-eutrophication and heavy metal contamination in Ganga River, India. Ecol Indic 99:349–364
Jaiswal D, Pandey J (2019f) Hypoxia and associated feedbacks at sediment-water interface as an early warning signal of resilience shift in an anthropogenically impacted river. Environ Res 178:108712
Jaiswal D, Siddiqui E, Verma K, Pandey J (2018) Carbon dioxide emission and its regulation at land-water interface downstream of a point source at Ganga River (India). Water Environ J 32:351–357
Jhingran AG, Ghosh KK (1978) The fisheries of the Ganga river system in the context of Indian aquaculture. Aquaculture 14:141–162
Khanna DR, Sarkar P, Gautam A, Bhutiani V (2007) Fish scales as bio-indicator of water quality of River Ganga. Environ Monit Assess 134:153–160
Lafont M (2001) A conceptual approach to the biomonitoring of freshwater: the ecological ambience system. J Limnol 60:17–24
Lata P, Ram S, Agrawal M, Shanker R (2009) Enterococci in river Ganga surface waters: propensity of species distribution, dissemination of antimicrobial-resistance and virulence-markers among species along landscape. BMC Microbiol 9:140–149
Liu JJ, Diao Z, Xu XR, Xie Q (2019) Effects of dissolved oxygen, salinity, nitrogen and phosphorus on the release of heavy metals from coastal sediments. Sci Total Environ 666:894–901
Mackenzie BR, Hinrichsen HH, Plikshs M, Wieland K, Zezera AS (2000) Quantifying environmental heterogeneity: habitat size necessary for successful development of cod Gadus morhua eggs in the Baltic Sea. Mar Ecol Prog Ser 193:143–156
MacPherson TA, Cahoon LB, Mallin MA (2007) Water column oxygen demand and sediment oxygen flux: patterns of oxygen depletion in tidal creeks. Hydrobiologia 586:235–248

Makulla A, Sommer U (1993) Relationships between resource ratios and phytoplankton species composition during spring in five North German lakes. Limnol Oceanogr 38:846–856
Malik A, Qadri SA, Musarrat J, Ahmad M (1995) Studies on the water quality of River Ganga at Fatehgarh and Kannauj, U.P., India. Environ Toxicol Water Qual 10:91–95
McCreary JP Jr, Yu Z, Hood RR, Vinaychandran PN, Furue R, Ishida A, Richards KJ (2013) Dynamics of the Indian-Ocean oxygen minimum zones. Prog Oceanogr 112–113:15–37
Middelburg JJ, Levin LA (2009) Coastal hypoxia and sediment biogeochemistry. Biogeosciences 6:1273–1293
Miralto A et al (1999) The insidious effect of diatoms on copepod reproduction. Nature 402:173–176
Mishra S, Kumar A, Shukla P (2016) Study of water quality in Hindon River using pollution index and environmetrics, India. Desalin Water Treat 57:19121–19130
Morelli G, Gasparon M (2014) Metal contamination of estuarine intertidal sediments of Moreton Bay, Australia. Mar Pollut Bull 89:435–443
Müller G (1969) Index of Geoaccumulation in sediments of the Rhine River. J Geo 2:108–118
Muscio C (2002) The diatom pollution tolerance index: assigning tolerance values. City of Austin-Water Protection and Development Review Department
Nannipieri P, Kandeler E, Ruggiero P (2002) Enzyme activities and microbiological and biochemical processes in soil. In: Burns RG, Dick RP (eds) Enzymes in the environment: activity, ecology and applications. Marcel Dekker, Inc, New York, pp 1–34
Negi RK, Rajput A (2013) Impact of pulp and paper mill effluents on phytoplanktonic community structure in Ganga River at Bijnor (UP), India. J Entomol Zool Stud 1:70–73
Nemerow NL (1991) Stream, lake, estuary, and ocean pollution. Van Nostrand Reinhold, New York
Nwachukwu OI, Pulford ID (2011) Microbial respiration as an indication of metal toxicity in contaminated organic materials and soil. J Hazard Mater 185:1140–1147
Oliver TH et al (2015) Biodiversity and resilience of ecosystem functions. Trends Ecol Evol 30:673–684
Pacheco F, Roland F, Downing J (2013) Eutrophication reverses whole-lake carbon budgets. Inland Waters 4:41–48
Paerl HW et al (2003) Microbial indicators of aquatic ecosystem change: current applications to eutrophication studies. FEMS Microbiol Ecol 46:233–246
Pan Y, Stevenson RJ, Hill BH, Herlihy AT, Collins GB (1996) Using diatoms as indicators of ecological conditions in lotic systems – a regional assessment. J North Am Benthol Soc 15:481–495
Pandey J (2011) The influence of atmospheric deposition of pollutants on cross–domain causal relationships for three tropical freshwater lakes in India. Lakes Reserv Res Manage 16:113–121
Pandey U (2013) The influence of DOC trends on light climate and periphyton biomass in the Ganga River, Varanasi, India. Bull Environ Contam Toxicol 90:143–147
Pandey J, Pandey U (2009) Accumulation of heavy metals in dietary vegetables and cultivated soil horizon in organic farming system in relation to atmospheric deposition in a seasonally dry tropical region of India. Environ Monit Assess 148:61–74
Pandey U, Pandey J (2013) Impact of DOC trends resulting from changing climatic extremes and atmospheric deposition chemistry on periphyton community of a freshwater tropical lake of India. Biogeochemistry 112:537–553
Pandey J, Yadav A (2015) The changing state of N: P: Si stoichiometry in Ganga River: possible implications for production and fate of phytoplankton biomass. Water 6:91–99
Pandey J, Yadav A (2017) Alternative alert system for Ganga river eutrophication using alkaline phosphatase as a level determinant. Ecol Indic 82:327–343
Pandey J, Shubhashish K, Pandey R (2010) Heavy metal contamination of Ganga River at Varanasi in relation to atmospheric deposition. Tropic Ecol 51:63–71

Pandey J, Singh AV, Singh A, Singh R (2013) Impacts of changing atmospheric deposition chemistry on nitrogen and phosphorus loading to Ganga River (India). Bull Environ Contam Toxicol 91:184–190

Pandey J, Pandey U, Singh AV (2014a) Impact of changing atmospheric deposition chemistry on carbon and nutrient loading to Ganga River: integrating land–atmosphere–water components to uncover cross-domain carbon linkages. Biogeochemistry 14:179–198

Pandey J, Pandey U, Singh AV (2014b) The skewed N: P stoichiometry resulting from changing atmospheric deposition chemistry drives the pattern of ecological nutrient limitation in the Ganges. Curr Sci 107:956–958

Pandey J, Singh AV, Singh R (2015a) Atmospheric deposition coupled terrestrial export of organic carbon in Ganga River (India): linking cross-domain carbon transfer to river DOC. Int Aquat Res 7:273–285

Pandey J, Yadav A, Singh R (2015b) The N: P: Si stoichiometry and relative abundance of diatoms in Ganga River. Br J Appl Sci Technol 9:137–147

Pandey J, Pandey U, Singh AV, Tripathi S, Mishra V (2016a) Atmospheric N and P deposition in the Ganges Basin. Curr Sci 110:974–976

Pandey J, Tripathi S, Pandey U (2016b) Anthropogenically induced shifts in N: P: Si stoichiometry and implications in Ganga River. Air Soil Water Res 9:35–43

Pandey U, Pandey J, Singh AV, Mishra A (2017) Anthropogenic drivers shift diatom dominance–diversity relationships and transparent exopolymeric particles production in River Ganga: implication for natural cleaning of river water. Curr Sci 113:959–964

Pandey J, Jaiswal D, Pandey U (2019) Point source-driven seasonal hypoxia signals habitat fragmentation and ecosystem change in River Ganga. Curr Sci 117:1947–1949

Passow U, Shipe RF, Murray A, Pak DK, Brzezinski MA, Alldredge AL (2001) The origin of transparent exopolymer particles (TEP) and their role in the sedimentation of particulate matter. Cont Shelf Res 21:327–346

Patra P et al (2013) The carbon budget of South Asia. Biogeosciences 10:513–527

Pearson CE, Ormerod SJ, Symondson WO, Vaughan IP (2016) Resolving large-scale pressures on species and ecosystems: propensity modelling identifies agricultural effects on streams. J Appl Ecol 53:408–417

Peterson CG, Stevenson RJ (1992) Resistance and resilience of lotic algal communities: importance of disturbance timing and current. Ecology 73:1445–1461

Piña-Ochoa E, Álvarez-Cobelas M (2006) Denitrification in aquatic environments: a cross-system analysis. Biogeochemistry 81:111–130

Ponader KC, Potapova MG (2007) Diatoms from the genus *Achnanthidium* in flowing waters of the Appalachian Mountains (North America): ecology, distribution and taxonomic notes. Limnologica – Ecol Manage Inland Waters 37:227–241

Potapova M, Charles DF (2003) US Rivers in relation to conductivity and ionic composition. Freshw Biol 48:1311–1328

Potapova MG, Charles DF, Ponader KC, Winter DM (2004) Quantifying species indicator values for trophic diatom indices: a comparison of approaches. Hydrobiology 517:25–41

Rabalais NN, Turner RE, Wiseman WJ (2001) Hypoxia in the Gulf of Mexico. J Environ Qual 30:320–329

Rabalais NN, Díaz RJ, Levin LA, Turner RE, Gilbert D, Zhang J (2010) Dynamics and distribution of natural and human-caused hypoxia. Biogeosciences 7:585–619

Rao RJ (2001) Biological resources of Ganga River, India. Hydrobiologia 458:159–168

Raymond P et al (2013) Global carbon dioxide emissions from inland waters. Nature 503:355–359

Reddy B, Pandey J, Dubey SK (2019) Assessment of environmental gene tags linked with carbohydrate metabolism and chemolithotrophy associated microbial community in River Ganga. Gene 704:31–41

Redfield AC (1958) The biological control of chemical factors in the environment. Am Sci 46:205–221

Rejsek K, Formanek P, Pavelka M (2008) Estimation of protease activity in soils at low temperatures by case in amendment and with substitution of buffer by demineralized water. Amino Acids 35:411–417

Richey JE, Melack JM, Aufdenkampe AK, Ballester VM, Hess LL (2002) Outgassing from Amazonian rivers and wetlands as a large tropical source of atmospheric CO_2. Nature 416:617–620

Riebesell U et al (2007) Enhanced biological carbon consumption in a high CO_2 ocean. Nature 450:545–548

Rimet F, Bouchez A (2012) Biomonitoring river diatoms: implications of taxonomic resolution. Ecol Indic 15:92–99

Saros JE, Michel TJ, Interlandi SJ, Wolfe AP (2005) Resource requirements of *Asterionella formosa* and *Fragilaria crotonensis* in oligotrophic alpine lakes: implications for recent phytoplankton community reorganizations. Can J Fish Aquat Sci 62:1681–1689

Satyamurthy N (2017) An obsession with numbers: quantifying quality. Curr Sci 113:7–8

Scheffer M, Carpenter S, Foley JA, Folke C, Walker B (2001) Catastrophic shifts in ecosystems. Nature 413:591–596

Scheffer M et al (2009) Early-warning signals for critical transitions. Nature 461:53–59

Schnürer J, Rosswall T (1982) Fluorescein diacetate hydrolysis as a measure of total microbial activity in the soil and litter. Appl Environ Microbiol 43:1256–1261

Seth RN, Katiha PK (2003) Riverine fishing methods with special reference to catfishes *Aorichthys seenghala* (Sykes) and *Aorichthys aor* (Ham.). Ind J Fish 50:125–130

Siddiqui E, Pandey J (2019a) Assessment of heavy metal pollution in water and surface sediment and evaluation of ecological risks associated with sediment contamination in the Ganga River: a basin-scale study. Environ Sci Pollut Res. https://doi.org/10.1007/s11356-019-04495-6

Siddiqui E, Pandey J (2019b) Temporal and spatial variations in carbon and nutrient loads, ion chemistry and trophic status of the Ganga River: a watershed-scale study. Limnology 20:255–266

Siddiqui E, Verma K, Pandey U, Pandey J (2019a) Metal contamination in seven tributaries of the Ganga River and assessment of human health risk from fish consumption. Arc Environ Contam Toxicol 77:263–278

Siddiqui E, Pandey J, Pandey U, Mishra V, Singh AV (2019b) Integrating atmospheric deposition-driven nutrients (N and P), microbial and biogeochemical processes in the watershed with carbon and nutrient export to the Ganga River. Biogeochemistry 147(2):149–178. https://doi.org/10.1007/s10533-019-00634-w56789

Singh AV, Pandey J (2014) Heavy metals in the mid-stream of the Ganges River: spatiotemporal trends in a seasonally dry tropical region (India). Water Int 39:504–516

Singh R, Pandey J (2019) Non-point source-driven carbon and nutrient loading to Ganga River (India). Chem Ecol 35:344–360

Singh D, Sharma RC (1998) Biodiversity, ecological status and conservation priority of the fish of the River Alaknanda, a parent stream of the River Ganges (India). Aquatic conservation: marine and freshwater ecosystems. Aquat Conser Mar Freshw Ecosys 8:761–772

Singh AK, Pathak AK, Lakra WS (2010) Invasion of an exotic fish—common carp, Cyprinus carpio L. (Actinopterygii: Cypriniformes: Cyprinidae) in the Ganga River, India and its impacts. Acta Ichthyol Piscat 40:11–19

Singh VK, Chaturvedi A, Shukla S, Gautam A (2013) Bacterial surveillance associated with water from River Ganga. Trends Biosci 6:802–804

Singh A, Tiwari V, Mohan J (2014) Chroococcales in River Ganga at Jajmau Ghat, Kanpur. Trop Plant Res 1:28–30

Sinsabaugh RL, Linkins AE (1990) Enzymic and chemical analysis of particulate organic matter from a boreal river. Freshw Biol 23:301–309

Sinsabaugh RL et al (2008) Stoichiometry of soil enzyme activity at global scale. Ecol Lett 11:1252–1264

Sinsabaugh RL, Hill BH, Shah JJF (2009) Ecoenzymatic stoichiometry of microbial organic nutrient acquisition in soil and sediment. Nature 462:795–798
Solan M, Cardinale BJ, Downing AL, Engelhardt KAM, Ruesink JL, Srivastava DS (2004) Extinction and ecosystem function in the marine benthos. Science 306:1177–1180
Stark J, Riddle M, Smith S (2004) Influence of an Antarctic waste dump on recruitment to nearshore marine soft-sediment assemblages. Mar Ecol Prog Ser 276:53–70
Sterner RW, Classen J, Lampert W, Weisse T (1998) Carbon: phosphorus stoichiometry and food chain production. Ecol Lett 1:146–150
Tare V, Yadav AVS, Bose P (2003) Analysis of photosynthetic activity in the most polluted stretch of river Ganga. Water Res 37:67–77
Tavernini S, Pierobon E, Viaroli P (2011) Physical factors and dissolved reactive silica affect phytoplankton community structure and dynamics in a lowland eutrophic river (Po River, Italy). Hydrobiologia 669:213–225
Teubner KD, Dokulil M (2002) Ecological stoichiometry of TN: TP: SRSi in freshwaters: nutrient ratios and seasonal shifts in phytoplankton assemblages. Arch Hydrobiol 154:625–646
Tipping E et al (2014) Atmospheric deposition of phosphorus to land and freshwater. Environ Sci Proc Impacts 16:1608–1617
Tomilson DL, Wilson J, Harris CR, Jeffrey DW (1980) Problem in assessment of heavy metals in estuaries and the formation of pollution index. Helgol Wiss Meeresunlter 33:566–575
Trivedi RC (2010) Water quality of the Ganga River – an overview. Aquat Ecosyst Health Manag 13:347–351
Trobajo R, Rovira L, Ector L, Wetzel CE, Kelly M, Mann DG (2013) Morphology and identity of some ecologically important small *Nitzschia* species. Diatom Res 28:37–59
Turley MD, Bilotta GS, Chadd RP, Extence CA, Brazier RE, Burnside NG, Pickwell AGG (2016) A sediment-specific family-level biomonitoring tool to identify the impacts of fine sediment in temperate rivers and streams. Ecol Indic 70:151–165
Turner RE, Rabalais NN, Justic D, Dortch Q (2003) Global patterns of dissolved N, P and Si in large rivers. Biogeochemistry 64:297–317
Verma K, Pandey J (2019) Heavy metal accumulation in surface sediments of the Ganga River (India): speciation, fractionation, toxicity, and risk assessment. Environ Monit Assess 191:414. https://doi.org/10.1007/s10661-019-7552-7
Verma K, Pandey J, Siddiqui E (2019) Heavy metal pollution in the Ganga River enhances carbon storage relative to flux. Bull Environ Contam Toxicol. https://doi.org/10.1007/s00128-019-02761-4
Vicente-Martorell JJ, Galindo-Riaño MD, García-Vargas M, Granado-Castro MD (2009) Bioavailability of heavy metals monitoring water, sediments and fish species from a polluted estuary. J Hazard Mater 162:823–836
Villnäs A, Norkko J, Lukkari K, Hewitt J, Norkko A (2012) Consequences of increasing hypoxic disturbance on benthic communities and ecosystem functioning. PLoS One 7:e44920
Walsh G, Wepener V (2009) The influence of land use on water quality and diatom community structures in urban and agriculturally stressed rivers. Water SA 35:579–594
Wardle DA (1993) Response of microbial biomass and metabolic quotient to leaf litter succession in some New Zealand forest and scrubland ecosystem. Funct Ecol 7:346–355
Xu L, Wang T, Wang J, Lu A (2017) Occurrence, speciation and transportation of heavy metals in 9 coastal rivers from watershed of Laizhou Bay, China. Chemosphere 173:61–68
Yadav A, Pandey J (2017a) Contribution of point sources and non-point sources to nutrient and carbon loads and their influence on the trophic status of the Ganga River at Varanasi, India. Environ Monit Assess 189:475–493
Yadav A, Pandey J (2017b) Water quality interaction with alkaline phosphatase in the Ganga River: implications for river health. Bull Environ Contam Toxicol 99:75–82
Zhang C et al (2014) Effects of sediment geochemical properties on heavy metal bioavailability. Environ Int 73:270–281

Zhang Z, Juying L, Mamat Z, Qingfu Y (2016) Sources identification and pollution evaluation of heavy metals in the surface sediments of Bortala River, Northwest China. Ecotoxicol Environ Saf 126:94–101

Chapter 11
An Overview on Heavy Metal Contamination of Water System and Sustainable Approach for Remediation

Indrani Mukherjee, Umesh Kumar Singh, and Rajeev Pratap Singh

11.1 Introduction

Contamination of freshwater systems with toxic heavy metal(loid)s (HMs) is a major global issue owing to its adverse impact on human health, dependent ecology/aquatic habitats, and agriculture. Certain HMs such as zinc (Zn), boron (B), molybdenum (Mo), copper (Cu), iron (Fe), and cobalt (Co) at their lower concentrations act as cofactors in various metabolic and other biological processes, whereas several HMs such as arsenic (As), cadmium (Cd), chromium (Cr), lead (Pb), mercury (Hg), and selenium (Se) are found to be nonessential and deleterious for humans, animals, and plants, even at their trace levels (Edelstein and Ben-Hur 2018; Kim et al. 2019). Freshwater systems can be perturbed by the HMs at varying degree from both natural and human-induced activities. Industrial activities such as thermal power plants, coal/mineral mines, electroplating, metal smelting, textiles, leather, e-waste processing, and chemical industries discharge several HMs into the environment such as soil, atmosphere, surface water courses, and eventually groundwater systems through atmospheric deposition and leaching during subsequent groundwater

I. Mukherjee
Integrated Science Education and Research Centre (ISERC), Institute of Science, Visva-Bharati, Santiniketan, Birbhum, West Bengal, India

U. K. Singh (✉)
Integrated Science Education and Research Centre (ISERC), Institute of Science, Visva-Bharati, Santiniketan, Birbhum, West Bengal, India

Department of Environmental Science, School of Earth, Biological and Environmental Sciences, Central University of South Bihar, Gaya, Bihar, India
e-mail: umeshsingh@cusb.ac.in

R. P. Singh
Department of Environment and Sustainable Development, Institute of Environment and Sustainable Development, Banaras Hindu University, Varanasi, Uttar Pradesh, India

A. Singh et al. (eds.), *Water Pollution and Management Practices*,
https://doi.org/10.1007/978-981-15-8358-2_11

recharge (Mukherjee et al. 2020). Inadequately treated industrial, domestic, and agricultural wastewater contains higher concentrations of HMs, which are usually discharged into the river water in many developing countries (Das Gupta 2008; Liu et al. 2011, 2012; Barsova et al. 2019). Such practices contaminate the river water and the groundwater systems of the river discharge zones. Chemical weathering of HM(s)-containing minerals in soils, igneous, and sedimentary rocks (basalt, granite, shales, clays, and sandstones) and abiotic redox processes can mobilize the HMs into the environment (Caporale and Violante 2016; Edelstein and Ben-Hur 2018). HMs can also be released from the soil layers by biogeochemical processes to the surface water systems through soil runoff and groundwater systems through leaching and percolation (Haferburg and Kothe 2012). Humans, animals, and aquatic habitants are exposed to the water HMs through ingestion, dermal contact, and inhalation pathways which can cause several carcinogenic and noncarcinogenic diseases (Singh and Kumar 2017; Mukherjee et al. 2019; Mukherjee and Singh 2020). Exposure to the HMs through water may cause tremors; headaches; mental fogginess; anxiety and depression; infertility; deteriorating eye health; memory losses; kidney dysfunction; digestive problems; tingling sensations in the hands, feet, and/or around the mouth; poor immune function; recurrent infections; skin diseases; infertility; or even cancers to humans. Therefore, it is necessary to remove those contaminants in order to produce relatively safe water. Several technologies are available for HM removal/ recovery from water which include ion exchange, adsorption, membrane filtration, chemical precipitation, coagulation-flocculation, and electrochemical treatment. Several researchers have explored phytoremediation for removal of the toxic HMs. The objectives of the present chapter are to emphasize (a) the sources, mobility, and fate of the HMs into the water systems and (b) sustainable remediation approaches of the HMs.

11.2 Sources of Heavy Metal(loid)s in Water Systems

Heavy metal contaminations in water systems of several regions are primarily affected by the anthropogenic activities. However, these elements may also get released from the natural sources.

11.2.1 Natural Sources

Heavy metals (HMs) and metalloids are the naturally occurring elements in the earth's crust (Ahmad and Bhattacharya 2019; Barsova et al. 2019). HMs are present in the constituent matrix of several minerals of the soils and several rocks. HM-containing minerals undergo dissolution caused by chemical weathering due to rock-water interaction and continually release HMs into the water systems and other environments at various degrees and concentrations. Mining process releases

HMs by the pounding of ore and gangue materials, which may further distribute into the other systems of environment though chemical weathering like dissolution, oxidation, hydrolysis, acidolysis, and alcalinolysis. Oxidation of sulfide minerals present in the heavy metal(loid) ore deposits increases the acidity of the water systems which further accelerates the mobility of the HMs in potentially bioavailable forms (Cravotta 1993; Jambor et al. 2000; Revan et al. 2014; Shengo et al. 2019), as expressed by following Eqs. (11.1, 11.2, 11.3, and 11.4):

$$\underset{(pyrite)}{FeS_2} + \frac{7}{2}O_2 + H_2O \rightarrow Fe^{+2} + 2H^+ + 2SO_4^{-2} \tag{11.1}$$

$$\underset{(chalcopyrite)}{\mathrm{CuFeS}} + 4O_2 \rightarrow Cu^{+2} + Fe^{+2} + 2SO_4^{-2} \tag{11.2}$$

$$\underset{(sphalerite)}{\mathrm{ZnS}} + 2O_2 \rightarrow Zn^{+2} + SO_4^{-2} \tag{11.3}$$

$$\underset{(galena)}{\mathrm{PbS}} + 2O_2 \rightarrow Pb^{+2} + SO_4^{-2} \tag{11.4}$$

Clay and sediments can adsorb/desorb the HMs significantly depending upon the phyllosilicate properties, clay minerals, available organic matters, binding energies, and microorganisms. Sorption capacity of the clay minerals and sediments is highly controlled by the pH values of water systems and increases with higher pH values. Furthermore, surface coverage, ionic strength, nature of the sorbent, and residence time also control the mobility/immobility of the HMs from the clay minerals and sediments to the circulating water (Caporale and Violante 2016).

Arsenic occurs within the chemical structure of several minerals such as clinoclase [$Cu_3(AsO_4)(OH)_3$], adelite [$CaMgAsO_4(OH)$], hoernesite [$Mg_3(AsO_4)_2{\cdot}8H_2O$], chalcophyllite [$Cu_{18}Al_2(AsO_4)_3(SO_4)_3(OH)_{24}{\cdot}36H_2O$], scorodite [$FeAsO_4{\cdot}2H_2O$]), duftite [$CuPb(AsO_4)(OH)$], ecdemite [$Pb_6As_2O_7Cl_4$], armangite [$Mn_{26}(As_{18}O_{50})(OH)_4CO_3$], paulmooreite [$Pb_2As_2O_5$], finnemanite [$Pb_5(AsO_3)_3Cl$], trippkeite [$CuAs_2O_4$], trigonite [$Pb_3Mn(AsO_3)_2(AsO_2)(OH)$], lautite [CuAsS], and arsenopyrite [FeAsS] (Brown et al. 1999; Craw et al. 2003; DeSisto et al. 2017; Walker et al. 2006). Chemical weathering of the Fe and As sulfides releases As(III) into the water system as illustrated by Eqs. (11.5, 11.6 and 11.7), which can be further oxidized to As(V) as expressed by Eqs. (11.8 and 11.9):

$$\underset{(arsenopyrite)}{\mathrm{FeAsS}} + 7H_2O \rightarrow Fe^{+2} + \underset{(arsenous\ acid)}{H_3AsO_3} + 11H^+ + 11e^- + SO_4^{-2}\ (\text{at pH} = 4 \text{ to } 10) \tag{11.5}$$

$$\underset{(scorodite)}{\mathrm{FeAsO_4} \cdot 2H_2O} + 3H^+ \rightarrow Fe^{+3} + \underset{(arsenic\ acid)}{H_3AsO_4} + 2H_2O \tag{11.6}$$

$$\underset{(scorodite)}{FeAsO_4 \cdot 2H_2O} + H_2O \rightarrow Fe(OH)_3 + \underset{(dihydrogen\ arsorate)}{H_2AsO_4^-} + H^+ \quad (11.7)$$

$$\underset{(arsenous\ acid)}{2H_3AsO_3} + O_2 \rightarrow \underset{(hydrogen\ arsenate)}{2HAsO_4^{2-}} + 4H^+ \quad (11.8)$$

$$\underset{(arsenous\ acid)}{2H_3AsO_3} + O_2 \rightarrow \underset{(dihydrogen\ arsorate)}{2H_2AsO_4^-} + 2H^+ \quad (11.9)$$

Lead can occur in several minerals such as pyromorphite [$Pb_5(PO_4)_3$(Cl, OH, F)], Pb-bearing apatites and jarosites, anglesite ($PbSO_4$), galena (PbS), and cerussite ($PbCO_3$). Lead can be mobilized into the water systems from these minerals. Fe(III) and Mn(III/IV) oxides and hydroxides are abundant in most soils which are efficient sorbents of lead. A number of studies have demonstrated that Mn oxides in particular are very efficient sorbents of Pb. The presence of such minerals in aquifer or soil can adsorb the lead from the water and can control its mobility. Selenium is present in several minerals, including molybdomenite ($PbSeO_3$), olsacherite [$Pb_2(SeO_4)(SO_4)$], chalcomenite ($CuSeO_3 \cdot 2H_2O$), sofiite [$Zn_2SeO_3Cl_2$], umangite [Cu_3Se_2], schmiederite [$Pb_2Cu^{(II)}{}_2(Se^{(IV)}O_3)Se^{(VI)}O_4(OH)_4$]), berzelianite [$Cu_2Se$], and hannebachite [$Ca_2(SeO_3)_2 \cdot H_2O$], and mobile in the aqueous systems in oxidized forms.

In Mekong Delta region, Vietnam, reductive dissolution of sediment minerals and release of surface-bound arsenic and phosphate from the anoxic aquifers were reported as the principal source of dissolved arsenic in groundwater system where lower dissolved arsenic was detected in the groundwater with higher manganese as As is less mobile under Mn-reducing conditions (Buschmann et al. 2008). Occurrences of Ba, Cd, Pb, Se, and Ni at trace levels in the groundwater system of the region have been also reported by geogenic activities (Buschmann et al. 2008). Cobalt (Co) and Cr in surface water and groundwater systems were reported to occur predominantly by the natural processes at Patancheru region of Medak district, Andhra Pradesh, India (Krishna et al. 2009). Groundwater of the semiarid Birbhum district, West Bengal, India, was also found to be contaminated by Cd, Cr, Cu, Fe, Mn, Ni, Pb, Sr, and Zn due to geogenic sources such as with lateritic and laterite soils, sedimentary and clay, and granitic gneiss-dominated terrains (Mukherjee et al. 2020). The concentrations of trace elements in groundwater of the granitic gneiss region were higher through the several deep-seated faults, folds, and lineaments, but their occurrence was higher at the sedimentary formations of the region where ion exchange and sorption/desorption processes play important roles to control the HM levels. Mixing of long and deep-circulating hot spring water (Bakreswar hot springs) with the local groundwater, longer residence time of the circulating water in the aquifer, and mixing of the meteoric fresh water with deep groundwater also attributed certain HMs to the local groundwater systems. Occurrences of Zn, Cu, Mn, Cd, Pb, and As were also reported in the groundwater systems of Kamrup district, Assam, India (Chakrabarty and Sarma 2011). In another study, Khdary and Gassim (2014) reported the occurrences of Ba, Se, Cr, As, Co, Cu, Hg, Pb, and Cd in groundwater of Mecca city, Saudi Arabia, due to geogenic sources.

11.2.2 Anthropogenic Sources

Water systems at global scale are largely contaminated with the HMs due to anthropogenic activities. Livestock manure, atmospheric deposition, usage of wastewater and polluted water for irrigation, use of sewage sludge-based amendments, HM-containing pesticides or herbicides, and phosphate-based fertilizers in the irrigation fields contribute several HMs to the soil and irrigation land significantly which further contaminate the water systems through runoff and leaching (Elgallal et al. 2016; Mukherjee and Singh 2018a). Furthermore, HMs are ubiquitous in the sludge produced from the chemical textile, tannery, distilleries, and electroplating industries (Chowdhary et al. 2018; Kumar and Chandra 2020). Coal-fired industries such as thermal power plants and vehicles (burning of fossil fuels) emit heavy metal-containing particulates into the environment which ultimately accumulate on the topsoil through atmospheric deposition during subsequent rainfall, which eventually can contaminate the subsurface water through leaching and percolation. Groundwater contamination with HMs due to the combustion residues of the thermal power plant and unscientific dumping has been reported by several studies (Sushil and Batra 2006; Gaikwad et al. 2009; Mukherjee et al. 2020). Surface water bodies of the Ranipet industrial area of Tamil Nadu, India, were found to be contaminated with elevated Cd, Cr, Cu, Ni, Pb, and Zn due to the industrial activities such as tannery and ceramic factories, refractory, boiler auxiliary plant, and chromium chemicals (Srinivasa Gowd and Govil 2007). Furthermore, landfill leachate is usually enriched with several HMs, and its inadequate management practices can contaminate the groundwater and surface water bodies. Such incidents were reported in groundwater of several Indian cities such as Ahmedabad, Bangalore, Chennai, Dehradun, Kolkata, and Delhi (Rawat et al. 2008); Nakhon Si Thammarat, Songkha, and Phatthalung provinces of southern Thailand (Decharat 2016); near Oti landfill site of Kumasi, southern Ghana (Boateng et al. 2019); Bulawayo, Zimbabwe (Teta and Hikwa 2017); Vientiane, Laos (Vongdala et al. 2019); and Warsaw, Poland (Koda et al. 2018). Occurrences of Fe, Mn, Pb, Ni, Cr, Zn, Cu, Co, and Cd in surface water of River Nile distributaries of Dakahlia Governorate, Egypt, were reported to originate from the steel, plastic, and battery industries (Mandour and Azab 2011). Direct disposal of the industrial and agricultural wastes into the waterways and massive use of pesticides and fertilizers were reported as primary sources of elevated As and Sr in the groundwater of the Sindh and Punjab provinces of Pakistan (Ali et al. 2019). Paint and cement industries have also been reported as the sources of Zn, Cu, Pb, and Cd in river water (Singh and Kumar 2017; Kumar and Singh 2018). Contamination of the surface and groundwater systems by the HMs due to mineral and coal mining have been widely reported at different parts of the world (Liang et al. 2011; Demková et al. 2017; Gabrielyan et al. 2018; Wei et al. 2018). The contaminated mine tailing water can be leached to more than 10 km from its source and might impact on the soil and water bodies of neighboring area due to leaching process, increased permeability caused by excavation of aquifer materials, and extensional fracturing induced by blasting (Singh et al. 2018).

Several studies reported the occurrences and sources of different HMs in water systems as illustrated in Table 11.1. The sources of As in water systems were mainly geogenic and associated with natural anoxic conditions in the aquifers and subsurface contamination due to floodplain deposits. Furthermore, Cd, Cr, Cu, Fe, Hg, Ni, and Pb are primarily released from agricultural and industrial activities, domestic and tannery effluents, corrosion of coolers, water tanks, pipes and pipe coatings, wastewater, mining, and incineration.

11.3 Toxicity of the HMs due to Their Exposures in Water

Carcinogenicity of arsenic has been reported by several studies (Smith et al. 2000; Litter et al. 2019). Chronic exposure to arsenic through ingestion route may lead to lung, bladder, and kidney cancers. It was estimated that about 20×10^4 to 27×10^4 deaths from cancer were associated with As exposure via drinking water in Bangladesh (Chowdhury et al. 2016). Acute toxicity of As via drinking water exhibits excess risks of gastrointestinal symptoms, severe disturbances of the cardiovascular and central nervous systems, hypertension, bone marrow depression, hemolysis, hepatomegaly, melanosis, polyneuropathy, encephalopathy and peripheral vascular disease with gangrenous changes (known as black foot disease), diabetes, and reproductive and neurological effects (Fowler et al. 2007, 2015). Furthermore, dermal exposure to arsenic in water poses an increased risk of skin lesions such as hyperkeratosis and pigmentation changes and skin cancer (WHO 2001). Long-term exposure to Cd at higher level via ingestion may cause renal tubular dysfunction, kidney malfunction bronchitis, anemia, skeletal damage (*itai-itai* disease, which is a permutation of osteoporosis and osteomalacia), and fractures (Järup 2003; Wu et al. 2016; Mukherjee et al. 2020). Animal and human studies reported higher risks of cardiovascular disease (CVD) due to Cd exposure in water (Alissa and Ferns 2011; Tellez-Plaza et al. 2013; Hudson et al. 2019). The cancer risks associated with Cd in drinking water is relatively insignificant compared to its occupational exposure on humans (WHO 2010; Nordberg et al. 2018). The oral exposure of Cr can affect the gastrointestinal, immune, kidney, and liver systems (Saha et al. 2011). The exposure may also induce hematological effects such as decreasing total red blood cell count and mean corpuscular hemoglobin (MCH) and microcytic anemia to the vulnerable populations (Mohammed et al. 2011; Ray 2016). Dermal exposure of Cr causes skin ulcers, increased skin sensitivity, and dermatitis. The exposure of Pb may cause damages to the central and peripheral nervous systems resulting in anxiety, depression, and increased aggression; gastrointestinal and urinary tracks resulting in bloody urine; dysfunctions in the cardiovascular, reproductive, and endocrine systems and kidneys and joints; or even neurological disorder and brain damage to the vulnerable populations (Mohammed et al. 2011; Wang et al. 2017). Development (stunted growth), behavioral (such as inattentiveness, hyperactivity, and irritability), and emotional problems can also be

Table 11.1 Heavy metal(loid) concentration in water systems and sources at different parts of the globe

S. no.	Location	Water system	Source(s)	Concentrations (μg/L)	Reference (s)
1	Cairo and Giza, Egypt	Harvested rainwater	Air pollution caused by industrial activities	Al: 3.95–4.95 Cr: 0.23–0.96 Mn: 0.16–0.27 Ni: 0.60–0.63 Cu: 0.74–3.52 Zn: 15.1–39.2 As: 0.23–0.71 Cd: 0.05–0.06 Hg: 0.02–0.15 Pb: 0.03–0.04	Saleh et al. (2001)
2	Northern Greece	Surface water/drinking water	Heavy metal activities Soil leaching Industrial effluents Domestic wastewaters Agricultural runoff Animal husbandry	Cu: 2.0–7.0 Cr: 1.0–18.0 Ni: 2.0–12.0 Mn: 45.0–291.0 Fe: 113.0–833.0 Pb: 1.0–16.0 Zn: 20.0–157.0 Cd: 0.1–0.6 Ag: 1.0–3.0	Simeonov et al. (2003)
3	Southern Vietnam and neighboring Cambodia	Groundwater/drinking water	Geogenic activities and caused by natural anoxic conditions in the aquifers	Cd: BDL–5.0 Co: BDL–44.0 Cr: BDL–10.0 Cu: 0.1–480.0 Ni: 0.1–53.0 Se: BDL–64.0 Fe: BDL–56.0 Mn: BDL–34.0	Buschmann et al. (2008)
4	Adyar River, South India		Domestic and industrial sewage Surface water Urban runoff Atmospheric depositions Weathering of rocks	Cu: 11.0–98.0 Co: 2.0–74.0 Zn: 5.0-250.0 Fe: 65.0–4618.0 Pb: 8.0–618.0 Cr: 12.0–650.0	Venugopal et al. (2008)
5	Meghna floodplain aquifer, southeastern Bangladesh	Groundwater	Microbial oxidation of organic matter Reductive dissolution of Fe and Mn oxyhydroxide Natural geochemical processes Agricultural activity	Al: BDL–104.0 As: 2.0–585.0 Ba: BDL–354.0 B: 51.0–465.0 Co: BDL–1.0 Cr: BDL–12.0 Cu: BDL–11.0 Fe: 590.0–13100.0	Zahid et al. (2008)

(continued)

Table 11.1 (continued)

S. no.	Location	Water system	Source(s)	Concentrations (μg/L)	Reference (s)
				Mn: 19.0–1060.0 Mo: 0.5–1.8 Ni: BDL–33.0 Pb: BDL–11.0 Zn: 27.0–1379.0	
6	Sirajdikhan Upazila, Munshiganj district, Bangladesh	Groundwater	Fe-Mn oxyhydroxides with microbially mediated degradation of organic matter Weathering of sulfide minerals like arsenopyrite	As: 6.0–461.0 Fe: 570.0–5725.0 Mn: 51.0–1125.0	Halim et al. (2009)
7	Amik plain, Hatay, southern Turkey	Groundwater	Geogenic and agricultural activities	Cd: 3.30–19.40 Co: 8.10–21.10 Cr: 2.70–353.60 Cu: 1.10–54.90 Fe: 0.50–657.10 Mn: 14.4–1026.1 Ni: 3.50–161.80 Zn: BDL–193.90	Ağca et al. (2014)
8	Southern Hunan province, China	Groundwater	Airborne volatile particulates Mineral processing Mining activities Wastewater of smelters Leakage of tailings and waste residues Natural geologic sources	Pb: 17.0–365.0 Zn: 12.0–4170.0 As: BDL–36.0 Cd: BDL–21.0 Cr: BDL–4.0 Cu: BDL–25.0	Gong et al. (2014)
9	Al Hasa Oasis, Saudi Arabia	Groundwater	Phosphate fertilizer Wastewater mixing Industrial and domestic activities	Mn: 63.0–151.0 Fe: 101.0–145.0 Cu: 11.0–17.0 Zn: 10.0–18.0	Assubaie (2015)

(continued)

Table 11.1 (continued)

S. no.	Location	Water system	Source(s)	Concentrations (μg/L)	Reference (s)
				Cd: 9.0–13.0 Pb: 5.0–8.0	
10	Faridpur district, central Bangladesh	Groundwater	Rock-water interaction Surface runoff Agriculture fertilizers	As: 8.00–1460.0 Fe: 52.0–19600.0 Mn: 0.04–4.23 Ni: 1.10–18.80 Pb: 0.02–28.60 Zn: 2.0–58.0	Bodrud-Doza et al. (2016)
11	Langmaoshan, Wohushan, Jinxiuchuan, Queshan, and Yuqing reservoirs, northern China	Surface water	Atmospheric sedimentation Coal burning Domestic discharge Agricultural activities	As: 0.10–5.40 Hg: 0.02–0.50 Cd: 0.30–0.60 Cr: 2.00–34.00 Se: 0.10–0.90 Pb: 0.50–2.50	Hou et al. (2016)
12	Nile River, Egypt	Surface water	Agricultural runoff Industrial and municipal wastes	Fe: 199.0–2211.0 Mn: 30.0–298.0 Zn: 10.0–115.0 Ni: 1.0–33.10 Cu: 10.0–50.5 Pb: 5.0–51.40 Cd: 0.20–7.90	Abdel-Satar et al. (2017)
13	Aksu River and tributaries, Antalya basin, southwest Turkey	Surface water	Water-rock interaction Wastewater discharge from the leather industry and marble factories Domestic wastewater Agricultural runoff	Pb: 1.0–3.4 Cr: 1.0–19.2 Mn: 2.0–143.8	Şener et al. (2017)
14	Huaihe River, Anhui, China	Surface water	Anthropogenic activities and industrial wastes Coal combustion Vehicle exhaust Agrochemicals Weathered rocks and crustal materials	Cu: 1.07–315.11 Pb: 3.25–1165.42 Zn: 1.36–183529.44 Ni: 2.01–199.08 Co: 2.00–170.92 Cr: 1.13–	Wang et al. (2017)

(continued)

Table 11.1 (continued)

S. no.	Location	Water system	Source(s)	Concentrations (μg/L)	Reference (s)
				187.91 Cd: 4.69–300.02 B: 1.77–840.93 Mn: 0.03–224.51 Fe: 19.17–3903.02 Al: 1.53–9965.87 Ba: 14.84–455.87 Mg: 271.84–35273.45	
15	Ain Azel plain, Algeria	Groundwater	Polymetallic mines Agricultural activities Rock-water interaction	Al: 10–90 Cd: 9–165 Cu: 56–430 Fe: 55–499 Pb: 17–292 Zn: 45–276	Belkhiri et al. (2017)
16	Kalpakkam, Tamil Nadu, India	Groundwater	Seawater intrusion Corrosion of pipings and fittings Dissolution of minerals Agricultural sources	Fe: BDL–89.9 Cu: BDL–11.4 Ni: BDL–16.6 Zn: 1.8–245.6 Mn: BDL–587.3	Samantara et al. (2017)
17	Northern part of Vojvodina, northern Serbia	Groundwater	Landfill leachate	Fe: 50.0–740.0 Mn: 13.25–181.0 Zn: 9.7–332.0 Cr: 0.31–1.40 Cu: 0.31–5.66 Pb: 10.60–133 As: 0.97–87.60	Krčmar et al. (2018)
18	Muledane area of Vhembe district, Limpopo province, South Africa	Groundwater	Agricultural runoff Atmospheric deposition Agricultural and domestic sewage	Cr: 5.0–150.0 Fe: 150.0–1860.0 Mn: 10.0–1220.0 Cu: 10.0–410.0 Pb: 2.0–26.0	Edokpayi et al. (2018)
19	Qareh-Ziaeddin plain, Iran	Groundwater	Absorption and release by clays Weathering of volcanic formations	Fe: 18.3–191.0 Pb: 0.5–13.0 Cd: 2.0–4.7 Zn: 9.0–160.0 Mn: 2.9–14.6	Esmaeili et al. (2018)

(continued)

Table 11.1 (continued)

S. no.	Location	Water system	Source(s)	Concentrations (µg/L)	Reference (s)
			Agricultural sources	Cr: 5.8–26.4 Al: 61.0–26.0 As: 32–2	
20	Subarnarekha River basin, India	Groundwater	Mining activities Discharge of industrial and domestic effluents	Fe: 118.37–3721.28 Mn: 3.96–1691.03 As: 0.04–59.92 Zn: 9.22–12966.36 Cu: 0.28–15.43 Cr: 0.37–6.72 Ni: 2.9–20.43 Se: 0.02–2.99 Al: 4.54–686.39 Ba: 4.06–263.2	Chaturvedi et al. (2018, 2019)
21	Ije-ododo, Lagos state, Nigeria	Groundwater and surface water	Oil spill Natural geochemical sources	Cr: 3.0–170.0 Cd: 1.0–130.0 Pb: 1.0–180.0 Zn: 10.0–2650 Cu: 10.0–670.0 Mn: 50.0–350.0 Ni: 10.0–2850 V: 20.0–910.0	Ogunlaja et al. (2019)
22	Neyshabur plain, Ravazi Khorasan province, Iran	Drinking water	Storm runoff and metal disposal Infiltration of water and leachates from landfill and dumpsites Dissolution of iron minerals due to nitrate leaching in groundwater, oxidation, and a decrease in pH Chromite mines and bedrocks hosting chromite deposits	Cr: BDL–20.0 As: BDL–9.86 Pb: 1.0–12.0 Zn: BDL–11.38 Cu: 1.15–53.81 Fe: 1.09–29.70	Saleh et al. (2019)
23	Bazman basin, southeastern Iran	Groundwater	Chemical weathering of parent materials and ion exchange Arsenical pesticides	As: 1.9–74.60 Ba: 7.0–103.0 Cr: 5.0–17.0 Cu: 0.05–2.8 Fe: 5.0–50.0 Mn: 0.025–20	Rezaei et al. (2019)

(continued)

Table 11.1 (continued)

S. no.	Location	Water system	Source(s)	Concentrations (µg/L)	Reference (s)
			Wastewater efflu-ents, irrigation Fertilizer applica-tion Groundwater abstraction	Ni: 0.60–32.3 Se: 0.25–15.7 Zn: 0.25–11.3	
24	Emilio Portes Gil, Ocoyucan, Puebla, Mexico	Well water	Geogenic sources	Al: BDL–156.0 Fe: BDL–9.0 Zn: BDL–111.0 Ni: BDL–89.0 Pb: BDL–56.0 Cr: BDL–11.0 Cu: BDL–89.0	Castresana et al. (2019)
25	Semiarid region of Birbhum dis-trict, East India	Groundwater/ drinking water	Weathering sedi-mentary rocks Adsorption/ desorption of HMs at clay layer Extensive agricul-tural activities Leaching of HMs from the laterite and lateritic soil layers Combustion resi-dues of the ther-mal power plant Weathering of minerals.	Cd: BDL–64.0 Cr: BDL–1500.0 Cu: BDL–394.0 Mn: 5.0–1352.0 Ni: BDL–156.0 Pb: BDL–580.0 Zn: BDL–3249.0 Fe: BDL–4665.0 Sr: 25–1538.0	Mukherjee et al. (2020)

BDL below detection limit

observed in the exposed infant and children population along with reduced IQ level due to Pb exposure (Evens et al. 2015; Reuben et al. 2017).

11.4 Remediation of the HMs in Water Systems

Safe water is a global concern for better public health, environment, and healthy ecosystems as threatened by the several geogenic sources and man-made activities (Mukherjee and Singh 2018b). Avoiding release of the poorly treated effluent and sewage sludge into the river bodies or environment can considerably decrease the HM contamination of the surface and groundwater systems. Similarly, scientific waste management practices with leachate collection technique can reduce the HM pollution to the surrounding water environments. Adequate management of the residual combustion and avoiding unscientific bottom fly ash dumping at the thermal power plant sites could result in less water pollution with HMs. Furthermore, air quality management could result in less particulate deposition on the soil and reduce the contamination of the water systems. However, these are basically source reduction approaches, and several countries, especially the developing countries, are facing challenges to adopt proper management practices of the wastes/emissions caused by these sectors. Therefore, it is necessary to remove the HMs from water using removal techniques. HMs in water can be mitigated from water through several physical, chemical, biological, and ecological approaches. Among the HM treatment methods such as chemical precipitation, physical separation, ion exchange, membrane filtration, membrane distillation, and hybrid methods, adsorption-based methods are economic and extensively used.

11.4.1 Membrane Processes

Membrane processes are dependent on hydrostatic pressure which is used to remove suspended particles and high molecular weight solutes and allow water and low molecular weight solutes to pass through. Retention of pollutants is determined by membrane materials, membrane characteristics (such as pore size, surface charge, and hydrophobicity), specific micropollutant characteristics, and membrane fouling. Montmorillonite, kaolin, tobermorite, magnetite, silica gel, and alumina-based nanofilter membrane can remove up to 97% of Cd(II); chitosan-coated magnetic nanoparticles modified with α-ketoglutaric acid can remove 99.9% Cu(II); polymeric cation exchanger containing nano-$Zr(HPO_3\text{-}S)_2$ can remove up to 84% of Pb (II); acid-modified carbon can remove up to 93% of As(V); and polonite can remove up to 98% of Mn(II) (Khulbe and Matsuura 2018). Forward osmosis (FO) uses a semipermeable membrane and a highly concentrated draw solution (relative to that of the feed solution) to filter dissolved contaminants from the water. Osmotic pressure gradient is used as driving force for this separation. The draw solution induces a net flow of water through the membrane into the draw solution, thus effectively separating the water from its solutes. Reverse osmosis (RO) uses a high hydraulic pressure which is applied on the treating water, and a high-pressure-resistant membrane is used to filter the dissolved solutes. RO has a greater efficiency as it can remove particles as small as 10 Å (angstroms) and colloidal particles. Khulbe and Matsuura (2018) have proposed a FO process consisting of a thin-film composite (TFC) FO membrane made from interfacial polymerization on a macro void-free polyimide support and a novel bulky hydro-acid complex $Na_4[Co(C_6H_4O_7)_2].2H_2O$ (Na–Co–CA) which can remove Cr, As, Pb, Cd, Cu, and Hg from water. Moradi et al. (2020) analyzed the HM removal using the PES/B-Cur membranes prepared by integrating boehmite nanoparticles functionalized with curcumin (B-Cur) into PES membrane and observed an increased removal capability. The membrane showed the maximum adsorption capacity of 35.01 mg/g for Pb, 32.20 mg/g for Ni, 31.12 mg/g for Cu, 29.08 mg/g for Fe, 27.08 mg/g for Zn, and

25.32 mg/g for Mn. In another study, Zareei and Hosseini (2019) reported an increased performance of polyethersulfone-based nanofiltration membranes that were prepared by composite $CoFe_2O_4/CuO$ nanoparticles for Ni, Pb, and Cu removal.

11.4.2 Nanotechnology

Nanoparticles (NPs), nanomaterials (NMs), or nano-adsorbents (NAs) including nanofilms, nanowires, quantum dots, nanotubes, and various colloids are now gaining popularity in in situ and ex situ HM water treatment. HMs can be removed by adopting nanotechnologies where nanoscale biosensors detect the HMs in surface and groundwater and accelerate the efficiency of the chemical and photocatalytic processes. NMs such as nanomembranes, nanocatalysts, nanoscale metal oxides, graphenes, carbon nanotubes, nanobiological processes and iron oxides (FeO, Fe_2O_3, and Fe_3O_4), aluminum oxide (Al_2O_3), titanium dioxide (TiO_2), and silicon dioxide (SiO_2) are used to remediate the HMs from contaminated water systems. Several NAs such as clay materials, activated carbon, metal oxides, nano-titanates, silica, magnetic iron oxide nanoparticles (MNPs), and alginate biopolymer have also been employed to eliminate the HMs toxicity (Kumar et al. 2019).

Nanomaterial-induced phytoremediation techniques have also been studied to decontaminate the water systems where co-uptake of the NMs and the HMs occur by the particular plants and NMs enhance the accumulation capability of the plants by enhancing the cell wall permeability, regulating the transporter gene expression, and co-transportation. For example, graphene oxide NM can enhance the As uptake capability of *Triticum aestivum* (Hu et al. 2014), and TiO_2NPs can improve the Cd uptake capability of *Glycine max* (Singh and Lee 2016). A recent experiment by Arshad et al. (2019) reported that 1 gram of polyethylenimine-modified graphene oxide hydrogel composite (functionalized GOCA) can remove up to 602 mg of Pb, 374 mg of Hg, and 181 mg of Cd from wastewater at 25 °C. Li et al. (2015) reported maximum adsorption capacity of chitosan/sulfhydryl-functionalized graphene oxide composites for Pb, Cd, Cu, and Mn as 568.18 mg/g, 253.81 mg/g, 68.68 mg/g, and 18.29 mg/g. Amino siloxane oligomer-modified graphene oxide composite also exhibited the maximum adsorption capacity of 310.63 mg/g for U and 243.90 mg/g for Eu in aqueous solution (Zhao et al. 2017). In another study Najafi et al. (2012) reported the maximum adsorption capacity of amino-functionalized silica nanohollow sphere (NH_2-SNHS) as 96.79 mg/g for Pb, 40.73 mg/g for Cd, and 31.29 mg/g for Ni.

11.4.3 Phytoremediation

Phytoremediation has also been identified as one of the economic and effective techniques to remove/degrade the toxic HMs from water systems, and several invasive plants have been used for this purpose. Hyperaccumulators or hyperaccumulating plants can accumulate higher quantities of HMs in their above-ground tissues and are not sensitive to the HM toxicity and, therefore, used phytoremediation to treat contaminated water bodies. Water hyacinth (*Eichhornia crassipes*) is the commonly tested plant for phytoremediation to uptake several HMs from water bodies (Schneider et al. 1995). In a study, Goswami and Das (2018) reported 55–57% of Cu removal from the contaminated water by this plant. In a laboratory study conducted by Odjegba and Fasidi (2007), they observed the effective removal of Ag, Cd, Cr, Cu, Hg, Ni, Pb, and Zn from the contaminated water using water hyacinth. Alshaal et al. (2013) reported removal of Cd and Zn from water using giant reed (*Arundo donax*). Bulrush (*Typha latifolia*) was also reported to remove Fe, Mn, As, Au, Cu, Hg, and Zn from water (Newete et al. 2016). Pb from water can also be effectively removed by the water spinach (*Ipomoea aquatica*) plant (Bedabati Chanu and Gupta 2016), and Co can be removed by the Canadian pondweed (*Elodea canadensis*) plant (Mosoarca et al. 2018). Furthermore, frogbit (*Hydrocharis morsus-ranae*) can uptake and accumulate Co, Cu, Hg, K, Mn, and Ni from water (Polechońska and Samecka-Cymerman 2016).

11.4.4 Microbe-Assisted Bioremediation

Microbe-assisted bioremediation techniques are also adopted to reduce, degrade, or detoxify the toxic HMs from the contaminated water using different bacteria, fungi, and algae. Certain bacterial species are tolerant to HMs, and their cells can grow even at higher concentration of the toxic HMs. Such bacterial cells can uptake heavy metals by both ATP-dependent and ATP-independent processes and accumulate the HMs on cell wall, intra- as well as extracellular entrapment or through formation of complexes and redox reactions (Ahemad 2012). It has been reported that HM accumulation capability of the gram-positive bacteria is comparatively higher than the gram-negative bacteria (Rani and Goel 2009). *Staphylococcus aureus* and *Escherichia coli* bacteria can reduce As(V) to As(III) (Rosen 2002). *Cyanobacteria* can be used to remove cadmium from wastewater as it is tolerant to cadmium toxicity (Xu et al. 2018). Furthermore, *Pseudomonas* (Pepi et al. 2011), *Escherichia coli* (Bae et al. 2001; Deng and Wilson 2001; de Luca Rebello et al. 2013; LaVoie and Summers 2018), and *Staphylococcus aureus* (Monecke et al. 2016) are resistant to mercury and can be used to remove mercury from wastewater. Moreover, *Pseudomonas stutzeri* bacteria can also remove mercury from water systems. In a study, Zheng et al. (2018) reported that the marine bacterium *Pseudomonas stutzeri 273* is resistant to 50 μM Hg^{+2} and removed ~94% Hg^{+2} from culture. *Escherichia coli*

(Stoppel et al. 1995; Chen et al. 2009; Mengoni et al. 2010; Gallo et al. 2018) and *Klebsiella oxytoca* (Mulrooney and Hausinger 2003; Freeman et al. 2005) can uptake nickel from contaminated water systems. Several bacteria such as *Bacillus subtilis*, *Bacillus cereus*, *Bacillus licheniformis*, *Bacillus subtilis*, *Thiobacillus thiooxidans*, *Geobacillus thermodenitrificans*, *Geobacillus themocatenulatus*, *Staphylococcus* sp., *Enterobacter* sp., *Enterobacter cloacae*, *Pseudomonas* sp., *Pseudomonas aeruginosa*, *Pseudomonas putida*, and *Staphylococcus saprophyticus* are resistant to Cu and can be used for Cu-contaminated water treatment (Nanda et al. 2019).

Certain HMs from water can also be removed using electrocoagulation, photocatalysts, clays/layered double hydroxides (LDHs), ion exchange, and activated carbons. However, their performance is dependent on several factors such as pH and the presence of other ions in the treating water.

11.5 Conclusions

Occurrence of the toxic heavy metal(loid)s in water system is a threat to the public health, aquatic living organism, and dependent ecosystems. The water systems can be contaminated with the HMs due to geogenic processes, such as weathering of HM-containing minerals, sorption/desorption processes, and ion exchange process, and anthropogenic activities such as agricultural runoff, contaminated soil runoff, atmospheric deposition, release of the poorly treated industrial effluents, etc. Such contamination may cause several adverse effects to human health and other living organisms. Therefore, sustainable remediation methods need to be adopted to overcome the HM toxicity. Apart from adsorbent-based removal methods, phytoremediation and microbe-assisted bioremediation techniques are also effective, environment friendly, and economic. However, ion exchange, adsorption, membrane filtration, chemical precipitation, coagulation-flocculation, and electrochemical treatment methods can also be applied to treat the contaminated water up to a certain level.

References

Abdel-Satar AM, Ali MH, Goher ME (2017) Indices of water quality and metal pollution of Nile River, Egypt. Egypt J Aquat Res 43:21–29. https://doi.org/10.1016/j.ejar.2016.12.006

Ağca N, Karanlık S, Ödemiş B (2014) Assessment of ammonium, nitrate, phosphate, and heavy metal pollution in groundwater from Amik Plain, southern Turkey. Environ Monit Assess 186:5921–5934. https://doi.org/10.1007/s10661-014-3829-z

Ahemad M (2012) Implications of bacterial resistance against heavy metals in bioremediation: a review. IIOAB J 3:39–46

Ahmad A, Bhattacharya P (2019) Arsenic in drinking water: is 10 μg/L a safe limit? Curr Pollut Reports 5:1–3. https://doi.org/10.1007/s40726-019-0102-7

Ali W, Aslam MW, Feng C et al (2019) Unraveling prevalence and public health risks of arsenic, uranium and co-occurring trace metals in groundwater along riverine ecosystem in Sindh and Punjab, Pakistan. Environ Geochem Health 41:2223–2238. https://doi.org/10.1007/s10653-019-00278-7

Alissa EM, Ferns GA (2011) Heavy metal poisoning and cardiovascular disease. J Toxicol 2011:870125. https://doi.org/10.1155/2011/870125

Alshaal T, Domokos-Szabolcsy É, Márton L et al (2013) Phytoremediation of bauxite-derived red mud by giant reed. Environ Chem Lett 11:295–302. https://doi.org/10.1007/s10311-013-0406-6

Arshad F, Selvaraj M, Zain J et al (2019) Polyethylenimine modified graphene oxide hydrogel composite as an efficient adsorbent for heavy metal ions. Sep Purif Technol 209:870–880. https://doi.org/10.1016/j.seppur.2018.06.035

Assubaie FN (2015) Assessment of the levels of some heavy metals in water in Alahsa Oasis farms, Saudi Arabia, with analysis by atomic absorption spectrophotometry. Arab J Chem 8:240–245. https://doi.org/10.1016/j.arabjc.2011.08.018

Bae W, Mehra RK, Mulciiandani A, Chen W (2001) Genetic engineering of *Escherichia coli* for enhanced uptake and bioaccumulation of mercury. Appl Environ Microbiol 67:5335–5338. https://doi.org/10.1128/aem.67.11.5335-5338.2001

Barsova N, Yakimenko O, Tolpeshta I, Motuzova G (2019) Current state and dynamics of heavy metal soil pollution in Russian Federation—a review. Environ Pollut 249:200–207. https://doi.org/10.1016/j.envpol.2019.03.020

Bedabati Chanu L, Gupta A (2016) Phytoremediation of lead using *Ipomoea aquatica* Forsk. in hydroponic solution. Chemosphere 156:407–411. https://doi.org/10.1016/j.chemosphere.2016.05.001

Belkhiri L, Mouni L, Sheikhy Narany T, Tiri A (2017) Evaluation of potential health risk of heavy metals in groundwater using the integration of indicator kriging and multivariate statistical methods. Groundw Sustain Dev 4:12–22. https://doi.org/10.1016/j.gsd.2016.10.003

Boateng TK, Opoku F, Akoto O (2019) Heavy metal contamination assessment of groundwater quality: a case study of Oti landfill site, Kumasi. Appl Water Sci 9:1–15. https://doi.org/10.1007/s13201-019-0915-y

Bodrud-Doza M, Islam ARMT, Ahmed F et al (2016) Characterization of groundwater quality using water evaluation indices, multivariate statistics and geostatistics in Central Bangladesh. Water Sci 30:19–40. https://doi.org/10.1016/j.wsj.2016.05.001

Brown GE, Foster AL, Ostergren JD (1999) Mineral surfaces and bioavailability of heavy metals: a molecular-scale perspective. Proc Natl Acad Sci 96:3388–3395. https://doi.org/10.1073/pnas.96.7.3388

Buschmann J, Berg M, Stengel C et al (2008) Contamination of drinking water resources in the Mekong delta floodplains: arsenic and other trace metals pose serious health risks to population. Environ Int 34:756–764. https://doi.org/10.1016/j.envint.2007.12.025

Caporale AG, Violante A (2016) Chemical processes affecting the mobility of heavy metals and metalloids in soil environments. Curr Pollut Reports 2:15–27. https://doi.org/10.1007/s40726-015-0024-y

Castresana GP, Roldán EC, García Suastegui WA et al (2019) Evaluation of health risks due to heavy metals in a rural population exposed to Atoyac River pollution in Puebla, Mexico. Water 11:277. https://doi.org/10.3390/w11020277

Chakrabarty S, Sarma HP (2011) Heavy metal contamination of drinking water in Kamrup district, Assam, India. Environ Monit Assess 179:479–486. https://doi.org/10.1007/s10661-010-1750-7

Chaturvedi A, Bhattacharjee S, Singh AK, Kumar V (2018) A new approach for indexing groundwater heavy metal pollution. Ecol Indic 87:323–331. https://doi.org/10.1016/j.ecolind.2017.12.052

Chaturvedi A, Bhattacharjee S, Mondal GC et al (2019) Exploring new correlation between hazard index and heavy metal pollution index in groundwater. Ecol Indic 97:239–246. https://doi.org/10.1016/j.ecolind.2018.10.023

Chen CY, Kao CM, Chen SC, Chen TY (2009) Biodegradation of tetracyanonickelate by *Klebsiella oxytoca* under anaerobic conditions. Desalination 249:1212–1216. https://doi.org/10.1016/j.desal.2009.06.036

Chowdhary P, Raj A, Bharagava RN (2018) Environmental pollution and health hazards from distillery wastewater and treatment approaches to combat the environmental threats: a review. Chemosphere 194:229–246. https://doi.org/10.1016/j.chemosphere.2017.11.163

Chowdhury S, Mazumder MAJ, Al-Attas O, Husain T (2016) Heavy metals in drinking water: occurrences, implications, and future needs in developing countries. Sci Total Environ 569–570:476–488. https://doi.org/10.1016/j.scitotenv.2016.06.166

Cravotta CA (1993) Secondary iron-sulfate minerals as sources of sulfate and acidity: geochemical evolution of acidic ground water at a reclaimed surface coal mine in Pennsylvania. In: Alpers CN, Blowes DW (eds) Environmental geochemistry of sulfide oxidation. ACS symposium series, vol 550. Wiley, Washington, DC, pp 345–364

Craw D, Falconer D, Youngson JH (2003) Environmental arsenopyrite stability and dissolution: theory, experiment, and field observations. Chem Geol 199:71–82. https://doi.org/10.1016/S0009-2541(03)00117-7

Das Gupta A (2008) Implication of environmental flows in river basin management. Phys Chem Earth 33:298–303. https://doi.org/10.1016/j.pce.2008.02.004

de Luca Rebello RC, Machado Gomes K, Duarte RS et al (2013) Diversity of mercury resistant *Escherichia coli* strains isolated from aquatic systems in Rio de Janeiro, Brazil. Int J Biodivers 2013:1–8. https://doi.org/10.1155/2013/265356

Decharat S (2016) Heavy metals exposure and hygienic behaviors of workers in sanitary landfill areas in southern Thailand. Scientifica (Cairo) 2016:1–9. https://doi.org/10.1155/2016/9269210

Demková L, Jezný T, Bobulská L (2017) Assessment of soil heavy metal pollution in a former mining area-before and after the end of mining activities. Soil Water Res 12:229–236. https://doi.org/10.17221/107/2016-SWR

Deng X, Wilson DB (2001) Bioaccumulation of mercury from wastewater by genetically engineered *Escherichia coli*. Appl Microbiol Biotechnol 56:276–279. https://doi.org/10.1007/s002530100620

DeSisto SL, Jamieson HE, Parsons MB (2017) Arsenic mobility in weathered gold mine tailings under a low-organic soil cover. Environ Earth Sci 76:773. https://doi.org/10.1007/s12665-017-7041-7

Edelstein M, Ben-Hur M (2018) Heavy metals and metalloids: sources, risks and strategies to reduce their accumulation in horticultural crops. Sci Hortic (Amsterdam) 234:431–444. https://doi.org/10.1016/j.scienta.2017.12.039

Edokpayi JN, Enitan AM, Mutileni N, Odiyo JO (2018) Evaluation of water quality and human risk assessment due to heavy metals in groundwater around Muledane area of Vhembe District, Limpopo Province, South Africa. Chem Cent J 12:2. https://doi.org/10.1186/s13065-017-0369-y

Elgallal M, Fletcher L, Evans B (2016) Assessment of potential risks associated with chemicals in wastewater used for irrigation in arid and semiarid zones: a review. Agric Water Manag 177:419–431. https://doi.org/10.1016/j.agwat.2016.08.027

Esmaeili S, Asghari Moghaddam A, Barzegar R, Tziritis E (2018) Multivariate statistics and hydrogeochemical modeling for source identification of major elements and heavy metals in the groundwater of Qareh-Ziaeddin plain, NW Iran. Arab J Geosci 11:5. https://doi.org/10.1007/s12517-017-3317-1

Evens A, Hryhorczuk D, Lanphear BP et al (2015) The impact of low-level lead toxicity on school performance among children in the Chicago public schools: a population-based retrospective cohort study. Environ Heal A Glob Access Sci Source 14:1–9. https://doi.org/10.1186/s12940-015-0008-9

Fowler BA, Chou C-HSJ, Jones RL, Chen C-J (2007) Arsenic. In: Nordberg GF, Fowler BA, Nordberg M, Friberg LT (eds) Handbook on the toxicology of metals, 3rd edn. Elsevier, Amsterdam, pp 367–406

Fowler BA, Selene C-H, Chou RJ et al (2015) Arsenic. In: Nordberg GF, Fowler BA, Nordberg M (eds) Handbook on the toxicology of metals, 4th edn. Elsevier, Amsterdam, pp 581–624

Freeman JL, Persans MW, Nieman K, Salt DE (2005) Nickel and cobalt resistance engineered in *Escherichia coli* by overexpression of serine Acetyltransferase from the nickel hyperaccumulator plant *Thlaspi goesingense*. Appl Environ Microbiol 71:8627–8633. https://doi.org/10.1128/AEM.71.12.8627-8633.2005

Gabrielyan AV, Shahnazaryan GA, Minasyan SH (2018) Distribution and identification of sources of heavy metals in the Voghji River basin impacted by mining activities (Armenia). J Chem 2018:1–9. https://doi.org/10.1155/2018/7172426

Gaikwad P, Khare D, Mishra G (2009) Leaching studies and metal analysis of fly ash from thermal power plant of India. J Solid Waste Technol Manag 35:115–123. https://doi.org/10.5276/JSWTM.2009.115

Gallo G, Presta L, Perrin E et al (2018) Genomic traits of *Klebsiella oxytoca* DSM 29614, an uncommon metal-nanoparticle producer strain isolated from acid mine drainages. BMC Microbiol 18:198. https://doi.org/10.1186/s12866-018-1330-5

Gong X, Chen Z, Luo Z (2014) Spatial distribution, temporal variation, and sources of heavy metal pollution in groundwater of a century-old nonferrous metal mining and smelting area in China. Environ Monit Assess 186:9101–9116. https://doi.org/10.1007/s10661-014-4069-y

Goswami S, Das S (2018) Eichhornia crassipes mediated copper phytoremediation and its success using catfish bioassay. Chemosphere 210:440–448. https://doi.org/10.1016/j.chemosphere.2018.07.044

Haferburg G, Kothe E (2012) Biogeosciences in heavy metal-contaminated soils. In: Kothe E, Varma A (eds) Bio-geo interactions in metal-contaminated soils. Soil biology, vol 31. Springer, Berlin/Heidelberg, pp 17–34

Halim MA, Majumder RK, Nessa SA et al (2009) Groundwater contamination with arsenic in Sherajdikhan, Bangladesh: geochemical and hydrological implications. Environ Geol 58:73–84. https://doi.org/10.1007/s00254-008-1493-8

Hou W, Sun S, Wang M et al (2016) Assessing water quality of five typical reservoirs in lower reaches of Yellow River, China: using a water quality index method. Ecol Indic 61:309–316. https://doi.org/10.1016/j.ecolind.2015.09.030

Hu X, Kang J, Lu K et al (2014) Graphene oxide amplifies the phytotoxicity of arsenic in wheat. Sci Rep 4:1–10. https://doi.org/10.1038/srep06122

Hudson KM, Belcher SM, Cowley M (2019) Maternal cadmium exposure in the mouse leads to increased heart weight at birth and programs susceptibility to hypertension in adulthood. Sci Rep 9:1–13. https://doi.org/10.1038/s41598-019-49807-5

Jambor JL, Nordstrom DK, Alpers CN (2000) Metal-sulfate salts from sulfide mineral oxidation. Rev Mineral Geochem 40:303–350. https://doi.org/10.2138/rmg.2000.40.6

Järup L (2003) Hazards of heavy metal contamination. Br Med Bull 68:167–182. https://doi.org/10.1093/bmb/ldg032

Khdary NHM, Gassim AEH (2014) The distribution and accretion of some heavy metals in Makkah Wells. J Water Resour Prot 06:998–1010. https://doi.org/10.4236/jwarp.2014.611094

Khulbe KC, Matsuura T (2018) Removal of heavy metals and pollutants by membrane adsorption techniques. Appl Water Sci 8:1–30. https://doi.org/10.1007/s13201-018-0661-6

Kim JJ, Kim YS, Kumar V (2019) Heavy metal toxicity: an update of chelating therapeutic strategies. J Trace Elem Med Biol 54:226–231. https://doi.org/10.1016/j.jtemb.2019.05.003

Koda E, Miszkowska A, Sieczka A, Osiński P (2018) Heavy metals contamination within restored landfill site. Environ Geotech:1–32. https://doi.org/10.1680/jenge.18.00031

Krčmar D, Tenodi S, Grba N et al (2018) Preremedial assessment of the municipal landfill pollution impact on soil and shallow groundwater in Subotica, Serbia. Sci Total Environ 615:1341–1354. https://doi.org/10.1016/j.scitotenv.2017.09.283

Krishna AK, Satyanarayanan M, Govil PK (2009) Assessment of heavy metal pollution in water using multivariate statistical techniques in an industrial area: a case study from Patancheru,

Medak district, Andhra Pradesh, India. J Hazard Mater 167:366–373. https://doi.org/10.1016/j.jhazmat.2008.12.131
Kumar V, Chandra R (2020) Bioremediation of Melanoidins containing distillery waste for environmental safety. In: Saxena G, Bharagava RN (eds) Bioremediation of industrial waste for environmental safety. Springer Singapore, Singapore, pp 495–529
Kumar B, Singh UK (2018) Source apportionment of heavy metals and their ecological risk in a tropical river basin system. Environ Sci Pollut Res 25:25443–25457. https://doi.org/10.1007/s11356-018-2480-6
Kumar S, Prasad S, Yadav KK et al (2019) Hazardous heavy metals contamination of vegetables and food chain: role of sustainable remediation approaches – a review. Environ Res 179:108792. https://doi.org/10.1016/j.envres.2019.108792
LaVoie SP, Summers AO (2018) Transcriptional responses of Escherichia coli during recovery from inorganic or organic mercury exposure. BMC Genomics 19:1–24. https://doi.org/10.1186/s12864-017-4413-z
Li X, Zhou H, Wu W et al (2015) Studies of heavy metal ion adsorption on chitosan/sulfydryl-functionalized graphene oxide composites. J Colloid Interface Sci 448:389–397. https://doi.org/10.1016/j.jcis.2015.02.039
Liang N, Yang L, Dai J, Pang X (2011) Heavy metal pollution in surface water of Linglong gold mining area, China. Procedia Environ Sci 10:914–917. https://doi.org/10.1016/j.proenv.2011.09.146
Litter MI, Ingallinella AM, Olmos V et al (2019) Arsenic in Argentina: occurrence, human health, legislation and determination. Sci Total Environ 676:756–766. https://doi.org/10.1016/j.scitotenv.2019.04.262
Liu J, Zhang XH, Tran H et al (2011) Heavy metal contamination and risk assessment in water, paddy soil, and rice around an electroplating plant. Environ Sci Pollut Res 18:1623–1632. https://doi.org/10.1007/s11356-011-0523-3
Liu Q, Liu Y, Zhang M (2012) Mercury and cadmium contamination in traffic soil of Beijing, China. Bull Environ Contam Toxicol 88:154–157. https://doi.org/10.1007/s00128-011-0441-6
Mandour RA, Azab YA (2011) The prospective toxic effects of some heavy metals overload in surface drinking water of Dakahlia governorate, Egypt. Int J Occup Environ Med 2:245–253
Mengoni A, Schat H, Vangronsveld J (2010) Plants as extreme environments? Ni-resistant bacteria and Ni-hyperaccumulators of serpentine flora. Plant and Soil 331:5–16. https://doi.org/10.1007/s11104-009-0242-4
Mohammed AS, Kapri A, Goel R (2011) Heavy metal pollution: source, impact, and remedies. In: Khan MS, Zaidi A, Goel R, Musarrat J (eds) Biomanagement of metal-contaminated soils. Environmental pollution. Springer, Dordrecht, pp 1–28
Monecke S, Jatzwauk L, Müller E et al (2016) Diversity of SCCmec elements in *Staphylococcus aureus* as observed in South-Eastern Germany. PLoS One 11:1–24. https://doi.org/10.1371/journal.pone.0162654
Moradi G, Zinadini S, Rajabi L, Ashraf Derakhshan A (2020) Removal of heavy metal ions using a new high performance nanofiltration membrane modified with curcumin boehmite nanoparticles. Chem Eng J 390:124546. https://doi.org/10.1016/j.cej.2020.124546
Mosoarca G, Vancea C, Popa S, Boran S (2018) Adsorption, bioaccumulation and kinetics parameters of the phytoremediation of cobalt from wastewater using Elodea canadensis. Bull Environ Contam Toxicol 100:733–739. https://doi.org/10.1007/s00128-018-2327-3
Mukherjee I, Singh UK (2018a) Groundwater fluoride contamination, probable release, and containment mechanisms: a review on Indian context. Environ Geochem Health 40:2259–2301. https://doi.org/10.1007/s10653-018-0096-x
Mukherjee I, Singh UK (2018b) Groundwater quality assessment with special references to fluoride and its suitability for irrigation and drinking purposes in Birbhum district, West Bengal. In: Anshumali, Venkatesh M (eds) National conference on biogeochemical cycles and climate change (BCCC-2018). Department of Environmental Science and Engineering, Indian Institute of Technology (Indian School of Mines), Dhanbad, Jharkhand, India, pp. 365–380

Mukherjee I, Singh UK (2020) Fluoride abundance and their release mechanisms in groundwater along with associated human health risks in a geologically heterogeneous semi-arid region of East India. Microchem J 152:104304. https://doi.org/10.1016/j.microc.2019.104304

Mukherjee I, Singh UK, Patra PK (2019) Exploring a multi-exposure-pathway approach to assess human health risk associated with groundwater fluoride exposure in the semi-arid region of East India. Chemosphere 233:164–173. https://doi.org/10.1016/j.chemosphere.2019.05.278

Mukherjee I, Singh UK, Singh RP et al (2020) Characterization of heavy metal pollution in an anthropogenically and geologically influenced semi-arid region of East India and assessment of ecological and human health risks. Sci Total Environ 705:135801. https://doi.org/10.1016/j.scitotenv.2019.135801

Mulrooney SB, Hausinger RP (2003) Nickel uptake and utilization by microorganisms. FEMS Microbiol Rev 27:239–261. https://doi.org/10.1016/S0168-6445(03)00042-1

Najafi M, Yousefi Y, Rafati AA (2012) Synthesis, characterization and adsorption studies of several heavy metal ions on amino-functionalized silica nano hollow sphere and silica gel. Sep Purif Technol 85:193–205. https://doi.org/10.1016/j.seppur.2011.10.011

Nanda M, Kumar V, Sharma DK (2019) Multimetal tolerance mechanisms in bacteria: the resistance strategies acquired by bacteria that can be exploited to 'clean-up' heavy metal contaminants from water. Aquat Toxicol 212:1–10. https://doi.org/10.1016/j.aquatox.2019.04.011

Newete SW, Erasmus BFN, Weiersbye IM, Byrne MJ (2016) Sequestration of precious and pollutant metals in biomass of cultured water hyacinth (*Eichhornia crassipes*). Environ Sci Pollut Res 23:20805–20818. https://doi.org/10.1007/s11356-016-7292-y

Nordberg GF, Bernard A, Diamond GL et al (2018) Risk assessment of effects of cadmium on human health (IUPAC technical report). Pure Appl Chem 90:755–808. https://doi.org/10.1515/pac-2016-0910

Odjegba VJ, Fasidi IO (2007) Phytoremediation of heavy metals by *Eichhornia crassipes*. Environmentalist 27:349–355. https://doi.org/10.1007/s10669-007-9047-2

Ogunlaja A, Ogunlaja OO, Okewole DM, Morenikeji OA (2019) Risk assessment and source identification of heavy metal contamination by multivariate and hazard index analyses of a pipeline vandalised area in Lagos state, Nigeria. Sci Total Environ 651:2943–2952. https://doi.org/10.1016/j.scitotenv.2018.09.386

Pepi M, Gaggi C, Bernardini E et al (2011) Mercury-resistant bacterial strains *Pseudomonas* and *Psychrobacter* spp. isolated from sediments of Orbetello Lagoon (Italy) and their possible use in bioremediation processes. Int Biodeter Biodegr 65:85–91. https://doi.org/10.1016/j.ibiod.2010.09.006

Polechońska L, Samecka-Cymerman A (2016) Bioaccumulation of macro- and trace elements by European frogbit (*Hydrocharis morsus-ranae* L.) in relation to environmental pollution. Environ Sci Pollut Res 23:3469–3480. https://doi.org/10.1007/s11356-015-5550-z

Rani A, Goel R (2009) Strategies for crop improvement in contaminated soils using metal-tolerant bioinoculants. In: Khan M, Zaidi A, Musarrat J (eds) Microbial strategies for crop improvement. Springer, Berlin/Heidelberg, pp 85–104

Rawat M, Singh UK, Mishra AK, Subramanian V (2008) Methane emission and heavy metals quantification from selected landfill areas in India. Environ Monit Assess 137:67–74. https://doi.org/10.1007/s10661-007-9729-8

Ray RR (2016) Adverse hematological effects of hexavalent chromium: an overview. Interdiscip Toxicol 9:55–65. https://doi.org/10.1515/intox-2016-0007

Reuben A, Caspi A, Belsky DW et al (2017) Association of childhood blood lead levels with cognitive function and socioeconomic status at age 38 years and with IQ change and socioeconomic mobility between childhood and adulthood. JAMA – J Am Med Assoc 317:1244–1251. https://doi.org/10.1001/jama.2017.1712

Revan MK, Genç Y, Maslennikov VV et al (2014) Mineralogy and trace-element geochemistry of sulfide minerals in hydrothermal chimneys from the Upper-Cretaceous VMS deposits of the eastern Pontide orogenic belt (NE Turkey). Ore Geol Rev 63:129–149. https://doi.org/10.1016/j.oregeorev.2014.05.006

Rezaei A, Hassani H, Hassani S et al (2019) Evaluation of groundwater quality and heavy metal pollution indices in Bazman basin, southeastern Iran. Groundw Sustain Dev 9:100245. https://doi.org/10.1016/j.gsd.2019.100245

Rosen BP (2002) Biochemistry of arsenic detoxification. FEBS Lett 529:86–92. https://doi.org/10.1016/S0014-5793(02)03186-1

Saha R, Nandi R, Saha B (2011) Sources and toxicity of hexavalent chromium. J Coord Chem 64:1782–1806. https://doi.org/10.1080/00958972.2011.583646

Saleh MA, Ewane E, Jones J, Wilson BL (2001) Chemical evaluation of commercial bottled drinking water from Egypt. J Food Compos Anal 14:127–152. https://doi.org/10.1006/jfca.2000.0858

Saleh HN, Panahande M, Yousefi M et al (2019) Carcinogenic and non-carcinogenic risk assessment of heavy metals in groundwater wells in Neyshabur plain, Iran. Biol Trace Elem Res 190:251–261. https://doi.org/10.1007/s12011-018-1516-6

Samantara MK, Padhi RK, Sowmya M et al (2017) Heavy metal contamination, major ion chemistry and appraisal of the groundwater status in coastal aquifer, Kalpakkam, Tamil Nadu, India. Groundw Sustain Dev 5:49–58. https://doi.org/10.1016/j.gsd.2017.04.001

Schneider IAH, Rubio J, Misra M, Smith RW (1995) *Eichhornia crassipes* as biosorbent for heavy metal ions. Miner Eng 8:979–988. https://doi.org/10.1016/0892-6875(95)00061-T

Şener Ş, Şener E, Davraz A (2017) Evaluation of water quality using water quality index (WQI) method and GIS in Aksu River (SW-Turkey). Sci Total Environ 584–585:131–144. https://doi.org/10.1016/j.scitotenv.2017.01.102

Shengo ML, Kime MB, Mambwe MP, Nyembo TK (2019) A review of the beneficiation of copper-cobalt-bearing minerals in the Democratic Republic of Congo. J Sustain Min 18:226–246. https://doi.org/10.1016/j.jsm.2019.08.001

Simeonov V, Stratis JA, Samara C et al (2003) Assessment of the surface water quality in Northern Greece. Water Res 37:4119–4124. https://doi.org/10.1016/S0043-1354(03)00398-1

Singh UK, Kumar B (2017) Pathways of heavy metals contamination and associated human health risk in Ajay River basin, India. Chemosphere 174:183–199. https://doi.org/10.1016/j.chemosphere.2017.01.103

Singh J, Lee BK (2016) Influence of nano-TiO2 particles on the bioaccumulation of Cd in soybean plants (*Glycine max*): a possible mechanism for the removal of Cd from the contaminated soil. J Environ Manage 170:88–96. https://doi.org/10.1016/j.jenvman.2016.01.015

Singh UK, Ramanathan AL, Subramanian V (2018) Groundwater chemistry and human health risk assessment in the mining region of East Singhbhum, Jharkhand, India. Chemosphere 204:501–513. https://doi.org/10.1016/j.chemosphere.2018.04.060

Smith AH, Lingas EO, Rahman M (2000) Contamination of drinking water by arsenic in Bangladesh: a public health emergency. Bulletin of the World Health Organization 78: contamination of drinking-water by arsenic in Bangladesh: a public health emergency. World Heal Organ Bull World Heal Organ 78:1093. https://doi.org/10.1590/S0042-96862000000900005

Srinivasa Gowd S, Govil PK (2007) Distribution of heavy metals in surface water of Ranipet industrial area in Tamil Nadu, India. Environ Monit Assess 136:197–207. https://doi.org/10.1007/s10661-007-9675-5

Stoppel R-D, Meyer M, Schlegel H (1995) The nickel resistance determinant cloned from the enterobacterium Klebsiella oxytoca: conjugational transfer, expression, regulation and DNA homologies to various nickel-resistant bacteria. Biometals 8:70–79. https://doi.org/10.1007/BF00156161

Sushil S, Batra VS (2006) Analysis of fly ash heavy metal content and disposal in three thermal power plants in India. Fuel 85:2676–2679. https://doi.org/10.1016/j.fuel.2006.04.031

Tellez-Plaza M, Guallar E, Howard BV et al (2013) Cadmium exposure and incident cardiovascular disease. Epidemiology 24:421–429. https://doi.org/10.1097/EDE.0b013e31828b0631

Teta C, Hikwa T (2017) Heavy metal contamination of ground water from an unlined landfill in Bulawayo, Zimbabwe. J Heal Pollut 7:18–27. https://doi.org/10.5696/2156-9614-7.15.18

Venugopal T, Giridharan L, Jayaprakash M (2008) Groundwater quality assessment using chemometric analysis in the Adyar River, South India. Arch Environ Contam Toxicol 55:180–190. https://doi.org/10.1007/s00244-007-9117-y

Vongdala N, Tran HD, Xuan TD et al (2019) Heavy metal accumulation in water, soil, and plants of municipal solid waste landfill in Vientiane, Laos. Int J Environ Res Public Health 16:1–13. https://doi.org/10.3390/ijerph16010022

Walker FP, Schreiber ME, Rimstidt JD (2006) Kinetics of arsenopyrite oxidative dissolution by oxygen. Geochim Cosmochim Acta 70:1668–1676. https://doi.org/10.1016/j.gca.2005.12.010

Wang J, Liu G, Liu H, Lam PKS (2017) Multivariate statistical evaluation of dissolved trace elements and a water quality assessment in the middle reaches of Huaihe River, Anhui, China. Sci Total Environ 583:421–431. https://doi.org/10.1016/j.scitotenv.2017.01.088

Wei W, Ma R, Sun Z et al (2018) Effects of mining activities on the release of heavy metals (HMs) in a typical mountain headwater region, the Qinghai-Tibet plateau in China. Int J Environ Res Public Health 15. https://doi.org/10.3390/ijerph15091987

WHO (2001) Arsenic and arsenic compounds, vol 224. Environmental Health Criteria, Geneva

WHO (2010) Exposure to cadmium: a major public health concern. World Health Organization, Geneva

Wu H, Liao Q, Chillrud SN et al (2016) Environmental exposure to cadmium: health risk assessment and its associations with hypertension and impaired kidney function. Sci Rep 6:1–9. https://doi.org/10.1038/srep29989

Xu C, Sun T, Li S et al (2018) Adaptive laboratory evolution of cadmium tolerance in *Synechocystis* sp. PCC 6803. Biotechnol Biofuels 11:205. https://doi.org/10.1186/s13068-018-1205-x

Zahid A, Hassan MQ, Balke K-D et al (2008) Groundwater chemistry and occurrence of arsenic in the Meghna floodplain aquifer, southeastern Bangladesh. Environ Geol 54:1247–1260. https://doi.org/10.1007/s00254-007-0907-3

Zareei F, Hosseini SM (2019) A new type of polyethersulfone based composite nanofiltration membrane decorated by cobalt ferrite-copper oxide nanoparticles with enhanced performance and antifouling property. Sep Purif Technol 226:48–58. https://doi.org/10.1016/j.seppur.2019.05.077

Zhao D, Chen L, Xu M et al (2017) Amino siloxane oligomer modified graphene oxide composite for the efficient capture of U(VI) and Eu(III) from aqueous solution. ACS Sustain Chem Eng 5:10290–10297. https://doi.org/10.1021/acssuschemeng.7b02316

Zheng R, Wu S, Ma N, Sun C (2018) Genetic and physiological adaptations of marine bacterium *Pseudomonas stutzeri* 273 to mercury stress. Front Microbiol 9:1–14. https://doi.org/10.3389/fmicb.2018.00682

Chapter 12
An Overview of Sustainability of Textile Wastewater Management in Textile Sectors

Punyasloka Pattnaik and G. S. Dangayach

12.1 Introduction

In India, the textile sector is one of the oldest sectors for revenue generation as well as one of the most significant service providers. Most of the *ancillary organizations* are directly or indirectly dependent on this sector in the form of raw materials or semi-finished products, starting from the fashion industry to aerospace manufacturing organizations. Hence, the textile sector is the only sector where the production rate is quite higher as compared with other leading sectors (annual report on Ministry of Textiles 2017). Due to the availability of a sufficient quantity of raw materials and higher consumption rate, the production rate has grown significantly in the last few decades (Cohn 2000). With the increase in the production rate of textile products all over the world, the textile effluents also proportionally increased, which directly affect the soil, air, as well as water (Khan and Malik 2014). Nowadays, dyes are seriously being used for the development of textile products through different treatment techniques (Siddique et al. 2017). These chemicals are discharged to drain water, rivers, and dumping yards or, most of the time, evaporated to the open environment. These chemicals seriously affect human beings because of their toxic nature. Therefore, a sustainable waste management system is needed to improve the performance output of the sector. This may be brought through technology advancement, which in turn would reduce waste textile effluent generation during the fabrication of textile products. Hence, in the present day, it is a challenging task for the policymakers, who need to define the waste collection procedure as well as its uses in some other applications, simultaneously preparing a stringent regulatory document to help the textile manufacturers (Fischedick et al. 2014). Corsten et al. (2013) discussed sustainable solid waste management for municipality solid waste on how to improve energy uses, reuse products and materials, and reduce

P. Pattnaik (✉) · G. S. Dangayach
Department of Management Studies, MNIT Jaipur, Jaipur, Rajasthan, India

A. Singh et al. (eds.), *Water Pollution and Management Practices*,
https://doi.org/10.1007/978-981-15-8358-2_12

greenhouse emission in the Netherlands and, finally, recommended integrated waste management policies to improve high-quality recycling methodology.

12.2 Literature Review on Sustainable Development of Management System

The literature review on sustainable development of the management system is mainly focused on the current waste management scenario across different countries, assessment of present technologies and strategic innovations, decision-making approaches in STWWM, feasible approaches for sustainable recycling and utilization of solid wastes, sustainable consumption impact in textile sectors, innovation activities in textile wastewater management, operational capabilities for service in textile wastewater sectors, relationship between eco-innovations and the impact on business performance, identifying the role of MCDM approach, and nanotechnology applicability in industrial wastewater treatment, respectively. The major detailed criteria along with sub-criteria are tabulated in Table 12.1 for further development of the conceptual framework.

12.3 Conceptual Framework and Hypotheses

The proposed conceptual framework of sustainable development of the management system is to achieve sustainability in the textile sectors as shown in Fig. 12.1. In the subsequent stages, a hypothesis is developed that shows a better relationship among SDMS practices and output performances.

12.3.1 Current Waste Management Scenario Across Different Countries

In the last few decades, waste management scenarios across different countries played a significant role in improving the performance of the textile sector as well as the development of new technology to reduce the wastes obtained from the textile sectors, as the textile sector is one of the oldest sectors that support the improvement of economic condition in a country as well as create job opportunity, which balances the national economic growth. For most of the environment-related issues, specifically the sectors generating huge quantities of solid wastes, life cycle assessment (LCA) is the only potential tool to identify the real cause. Parkes et al. (2015) studied the life cycle assessment of ten integrated textile waste management systems to evaluate the effect of the emissions of municipal wastes during the treatment process.

Table 12.1 Categorization of sustainability of textile wastewater management system

Sl. no.	Major criteria	Sub-criteria	References
1.	Current waste management scenario across different countries	Integrated waste management systems are considered for each legacy scenario	Corsten et al. (2013)
		Energy-from-waste plant	Corsten et al. (2013) and Yılmaz and Abdulvahitoğlu (2019)
		Advanced thermal treatment process	Maqhuzu et al. (2019)
		Policymakers with crucial information required for the long-term planning of waste management solutions	Corsten et al. (2013), Horodytska et al. (2018), Heggelund et al. (2016), and Fuldauer et al. (2019)
		Implementation of modeling approaches instead of analytical techniques	Parkes et al. 2015 and Xiong et al. (2019)
		Estimation of the region-specific waste management of individual waste material fractions	Hossain et al. (2017), Landi et al. (2018), and Wigger et al. (2015)
2.	Assessment of present technologies and strategic innovations	Use of renewable energy technologies, development of pollution prevention systems, organic agriculture	Zhao and Lin (2019)
		Creation of green investment of funds and carbon emission technologies	Vajnhandl and Valh (2014)
		Development of innovative technologies to solve the environmental and social impacts of the textile industry	Tseng et al. (2019) and Prajogo (2016)
		Companies should adopt a comprehensive approach to the development and implementation of eco-innovation programs	de Oliveira Brasil et al. (2016)
3.	Policymaking approaches in STWWM	Analyzing the existing global emission of greenhouse gas evaluation indicators and developing new criteria and strategies to complement the connotation of the corporate management	Zhu et al. (2018) and Aydiner et al. (2016)
		Proving effectiveness of the system by assessing the relationship between existing performance	Shiwanthi et al. (2018) and Yukseler et al. (2017)

(continued)

Table 12.1 (continued)

Sl. no.	Major criteria	Sub-criteria	References
		Methods for quantifying and identifying microplastics	Henry et al. (2019)
		Measurement and sampling protocols	Shiwanthi et al. (2018)
4.	Feasible approaches for sustainable recycling and utilization of solid wastes	Sustainable water use in chemical, paper, textile, and food industry	Harijani et al. (2017)
		Possible waste stream separation and segregation concepts to increase wastewater treatability and reuse options	Zhang et al. (2019)
		Verifying applicability of small-scale tailor-made treatment technologies and their combinations	Kroll and Hoyer (2019) and Wang et al. (2019)
		Guidelines proposed within this project were validated through numerous laboratory scales	Das et al. (2019) and Ferronato et al. (2019)
5.	Sustainable consumption impact in textile sectors	Product choice	Niinimäki and Hassi (2011)
		Financial problems	Desore and Narula (2018)
		Product quality	Austgulen (2016) and Kang et al. (2013)
		Product benefits	Stål and Jansson (2017)
6.	Innovation activities in textile wastewater management	Product and service innovations	Vlyssides et al. (2000) and Cinperi et al. (2019)
		Management innovations	Bilińska et al. (2017) and Ye et al. (2020)
		Marketing innovations	Hubadillah et al. (2020)
7.	Operational capabilities for service in textile wastewater sectors	Service innovation competitiveness	Liu et al. (2020)
		Application of innovative knowledge	Wu et al. (2019)
		Number of new products/services applications	Pan et al. (2019)
		Self-generated innovative products/services	Ndubisi et al. (2019)
8.	Relationship between eco-innovations and the impact on business performance	Adopt a comprehensive approach to the development and implementation	de Oliveira Brasil et al. (2016)
		Socio-technical systems	Cheng et al. (2014)
		Culture and the organization management designs	Bossle et al. (2016) and Przychodzen and Przychodzen (2015)
		Social and technological aspects	Hojnik and Ruzzier (2016)

(continued)

Table 12.1 (continued)

Sl. no.	Major criteria	Sub-criteria	References
9.	Identifying the role of MCDM approach	Elimination or reduction of barriers to the growth of textile sector	Aragonés-Beltrán et al. (2009)
		Challenges to sustainable development in the textile and apparel sector	Ozturk and Cinperi (2018)
		Barriers to the sustainable development	Çelikbilek and Tüysüz (2016)
		Identifying the cause-effect relationship among the evaluation criteria	Cebeci (2009)
		Business development needs	Nayagam et al. (2011)
		Lack of support from industrial bodies and work associations	Çelikbilek and Tüysüz (2016)
		Governmental support	Ozturk and Cinperi (2018)
10.	Nanotechnology applicability in industrial wastewater treatment	Efficient treatment technologies to reach the standards for discharging the effluents	Babaei et al. (2017)
		Potential to recover by-products	Soltani et al. (2016)
		Treatment of wastewater using nanomaterials which assessed as acceptable	Hassanzadeh et al. (2017)
		Improvement of economic situation and treatment of industrial effluents using nanotechnology	Jorfi et al. (2016)
		The large-scale production of the nanomaterials, transportation and application of nanomaterials in the treatment process may produce new job opportunities	Babaei et al. (2017), Jorfi et al. (2016), Soltani et al. (2016) and Hassanzadeh et al. (2017)

They concluded that with the implementation of advanced technology, the rate of emission becomes reduced that directly affects the global warming potential. Hossain et al. (2017) also concluded that both off-site sorting and direct landfilling had shown a significant impact on the environment. Similarly, Landi et al. (2018) discussed the LCA technique to measure the reusing of textile fibers that directly benefit the environment. They maturely studied the textile material characterization in order to determine the technical feasibility to obtain a clean and compact fiber for further reuse instead of landfilling of the fiber. Horodytska et al. (2018) discussed

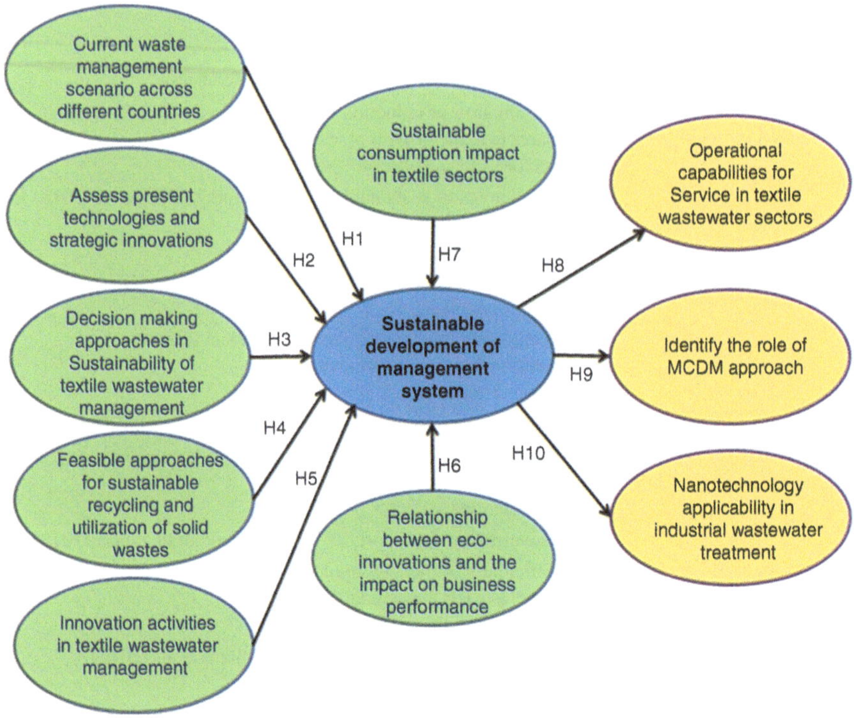

Fig. 12.1 A proposed conceptual framework for sustainable development of management system

postindustrial wastes as well as postconsumer of plastic film waste management technologies. They mainly differentiate between these two methods to analyze quality and uses of recycled materials. Waste management concerning engineered materials is a big issue across the globe in recent years. Although the bulk materials can be disposed of, the nano-sized engineered materials are a potential threat to the environment. To observe this, Heggelund et al. (2016) followed four steps to recover the nanoengineered materials from consumer products, across the markets of Europe, Denmark, and the United Kingdom. It was analyzed that silver and plastic are the two engineered nanomaterials which are commonly found in a variety of consumer goods. Wigger et al. (2015) have extracted silver from engineered nanomaterials used in textile industries and observed that the use of silver nanoparticles in the textile industry has been increased up to 350%. Although nanomaterials obtained from industries can be recycled, precautionary measurements should be undertaken before they are disposed into the environment.

Similarly municipal solid waste can also be recycled into different forms of energy (Maqhuzu et al. 2019). It was analyzed that 0.289 kg of fuel can be obtained from household waste, and with this process, the generation of SOx and greenhouse gases can be reduced up to 2.2–4.2%, respectively. Converting solid waste could also serve the purpose of alternative fuel production to minimize the usage of fossil fuels. Also, the waste from the household can be converted to carbon-free solid waste and is ready to be used directly in existing power plants that consequently reduced the usage of conventional coal by 18% annually. The authors also claimed that 289.3 gigagram (Gg) of alternative fuel can be saved annually from 1051.7 ± 270.7 Gg solid wastes collected from urban areas of Zimbabwe alone.

While the key issues associated with solid waste management pertaining to Turkey is invoked by Yılmaz et al. (2019) in their study, solid waste treatment in a country with a huge population seems to be a challenging issue with the prospect of economic growth and environmental and technological aspects. Improper solid waste management leads not only to shut down of plants and capital inflow but also becomes a severe threat due to uncontrolled gas emissions. A theoretical model was proposed in this study based on the energy recovery method adopted in Turkey. The theoretical model highlights the subsequent carbon mitigation as well as the financial strategies to follow European standards. The model presented in this article projected the electricity generation with the amount of solid waste generated from each city of Turkey for the next four decades through statistical analysis. The estimated electricity generation will be assumed up to 437–2236 TWh (terawatt hour) with the solid waste generated from 2023 to 2043 years. Besides, it was also about 35.2–66.7 billion dollars that can be saved with this proposed model of waste management.

Waste management in island developing states becomes the most significant challenge because of high investment, narrow resources, etc.; also, the framework considered in island developing states fails to cater to the needs of sustainable waste management. One such study on waste management in island developing states was carried out by Fuldauer et al. (2019). This study was in regard to the model developed to manage waste upon the guidelines of the United Nations on island developing states. The strategy to reuse the product, the importance of waste, and effective utilization of population within islands based on socioeconomic systems have been reported.

Toxic elements generated from solid waste necessitate the use of waste management around the globe. The impact of these toxic elements on humans and the environment is hazardous; special care needs to be taken during disposal, especially. Therefore, the study by Xiong et al. (2019) highlights an overview of efficient and effective methods to immobilize toxic elements by taking into consideration vital patterns of geographical distribution. The authors suggested that the use of the user-friendly practice to recover metals and rare elements from solid waste will benefit sustainability compared to conventional extraction of ore.

Hypothesis H1: Current waste management scenario across different countries is positively related to firm performance.

12.3.2 Assessment of Present Technologies and Strategic Innovations

An empirical survey was conducted by de Oliveira Brasil et al. (2016) to develop interrelationship among process, product, and organizational resource-based concept. The concept was started with data of 70 samples collected from different textile industries to expand business performance on the eco-innovation principle. Eco-innovation is the method that helps to develop its attributes and determinants to enhance performance. The results demonstrated that adopting the eco-innovation process helps in organizational and technological goals with ease. The important aspect of the firms working under product service relies on customer satisfaction. The best possible business strategy enables the firms having limited resources and helps not only in economic growth but also in social well-being. By keeping in mind the concept of sustainable product service for local firms, Tseng et al. (2019) proposed a model based on the fuzzy Delphi method. This model enables complex interrelationship between the supply chain and the business interest by considering four characteristic features such as sustainable consumption, followed by an advantage in collaboration and innovation in activities with innovation in services. It was also suggested that firms should maintain equilibrium between economic growth and high-quality products in business synergy.

Vajnhandl and Valh (2014) reported an in-depth survey taking into consideration the European textiles on wastewater treatment and management. Most commonly small- and medium-scale firms dominate the processing of fabric (fabric cutting and stitching) in the textile finishing industry. Adopting sustainable features for such small- and medium-scale firms becomes a severe issue as they have less resource. Ten years of data was carefully evaluated to frame the guidelines for reusing the wastewater (Vajnhandl and Valh 2014). In total, four sectors were targeted; they were textile, food, paper, and chemical industries; among them textile industry has the highest impact on the environment by discharging polluted water, having chemicals with coloring and other toxic agents. Reusing of such wastewater having tons of effluents becomes not feasible. Therefore, the reports highlight to use a separate treatment before it is discharged into the environment by using effective conventional methodologies.

Zhao and Lin (2019) investigated the methods to improve the energy efficiency of textile industries situated in china. The work was focused on implementing new policies to enhance the energy efficiency of industries. A regression model was adopted to draw the baseline, and it was expanded to identify factors associated with enhancing the energy efficiency of industrial agglomeration. The results were analyzed in three steps, the agglomeration to be affected as the value reaches the limit of the threshold; after that, the value moves toward the positive sign indicating its development related to economic growth, R&D, and drop in prices. In the third, the regression model enables to develop a nonlinear relationship connecting agglomeration and efficiency. Innovation is greatly affected by the organization's culture, human resources, and technological capabilities. This suggests that innovation is an internal activity of the firm and the management controls it.

On the contrary, the impact of innovation considering the external factors is a lesser known subject. A report by Prajogo (2016) examined the role dynamism and competitiveness in innovation. This report was based on the data obtained by nearly 207 firms in Australia. These reports suggest that dynamism helps in innovation, which in turn helps business performance. Besides, competitiveness weakens the product innovation but, on the other hand, strengthens the process innovation. Dynamism helps in establishing strategic fit, while competitiveness builds strategic mismatch.

Hypothesis H2: Assessment of present technologies and strategic innovations is positively related to production performance.

12.3.3 Decision-Making Approaches in Sustainability of Textile Wastewater Management

The rapid change in climatic conditions in the twenty-first century has been a challenge for mankind. This change accounts for the increased emission of greenhouse gases, such as CO_2, N_2O, HFC_5, etc., and was also witnessed by the Intergovernmental Panel on Climate Change. Emission of greenhouse gas (GHG) has an adverse effect not only on human life but also on food production. Of note, textile industries contribute 10% of the total GHG annually. Therefore, it is necessary to have an appraisal system to counter the challenges of GHG. Zhu et al. (2018) proposed a management system to counter greenhouse gas emissions from textile industries based in China. The evaluation methodology was divided into four steps. The first pertains to the policies and commitment, the second involves the implementation of program, the third is the expert review followed by a site visit, and finally the fourth was analyzing the data based on the Monte Carlo method. The methodology adopted by the team has led to draw accurate estimation of greenhouse gas emissions and its prevention.

In recent years, petroleum-based organic polymer dominates the textile industry. At least two-thirds of the item used in textiles has been switched over to polyamide, acrylic plastic, whose thickness varies from 100 nm to 5 mm. Plastic waste has become a potential threat to global warming and greenhouse gas. It was estimated that 35% of waste accumulated in the marine environment is of plastic, and a major portion can be seen in the coastal region. A critical review has been done by Henry et al. (2019) on factors that need to be focused on to control the discharge of fabric from textile industries. The report would provide an interim indication of sustainable management and strategies by adopting suitable analytical methods.

It is known that water is available abundantly in nature. Despite abundant availability, only $<1\%$ can be used directly due to the harsh discharge of effluents, rapid industrialization, and population. In a report, the current status of the water quality, as well as the necessary precautions to be taken in future, was studied by Kamali et al. (2019). The study was focused on the available treatments; further changes necessary to improvise the treatment methods have been discussed. The study was divided into three parts, namely, laboratory tests, pilot testing, and full-scale implementation. The initial stage of laboratory tests has been conducted, and it was in the stage to implement practically in industry. The sustainability treatment of nanoengineered materials was carried out using the conventional fuzzy Delphi method. The result is shown to be affected in the fields of economic, technological, social, and environmental segments.

Shiwanthi et al. (2018) in their study investigated the effect of textile industries (situated in Biyagama zone, Sri Lanka) on the environment. The study was started with an analysis for the cause of environmental pollution, wherein it was noticed that textile industries seem to be fifth in contributing CO_2 emissions. The growth of the textile market increased substantially from 7.8% in 2013 to 8.5% in 2014, and a total of 16% GDP of the Sri Lankan economy is based on textile industries. This increase in GDP is helpful for the economic growth of the country, but the emissions have an adverse effect on mankind as well as the environment, which has been depicted. The energy required, the quantity of material exported, and water consumption were regarded as input factors and emission and wastewater were regarded as output factors for a 5-year period that was considered in the study. The report also suggests that adopting eco-friendly treatment measures not only influences the revenue level but also can reduce the environmental burden.

Wastewaters generated from dairy plants contribute equally to emission as by the textile industries. An innovative sustainable approach in treating wastewater from dairy has been studied by Aydiner et al. (2016). The authors have evoked different MCDM approaches such as membrane distillation followed by forward osmosis. In this study, a total of five techniques were used to reclaim and reuse dairy wastewater. The results showed that the membrane distillation technique had been the most preferable novel solution for the treatment of wastewater from dairy. Further, the heat generated by such techniques can be utilized for other purposes; thereby, the process becomes cost-effective as well as an eco-innovative approach.

Yukseler et al. (2017), in an attempt, studied the best available techniques to reuse wastewater from textile industries present in Turkey. The study was undertaken not only to treat the wastewater but also to reduce the water consumption as requested by the textile mill. Before the pilot-scale implementation was processed, the best possible reuse opportunities were identified from the leading textile tycoon of denim mills. It was identified that wastewater from the dyeing and finishing process contributes more to environmental pollution by discharging rich effluents. Among the best available techniques, the membrane process was selected to reuse the wastewater, to recover the fabric discharged after the finishing process, and finally to reduce water consumption in dyeing as well as in finishing process.

Hypothesis H3: Decision-making approaches in STWWM are positively related to economic performance.

12.3.4 *Feasible Approaches for Sustainable Recycling and Utilization of Solid Wastes*

It was estimated that around 6.6 gigatons of cement was consumed in China alone from 2011 to 2013. The consumption was 4.5 gigatons more compared to the United States in nearly 100 years between 1901 and 2000, which took nearly 100 years to generate the same CO_2 by the United States, while China could generate the same amount from 2011 to 2013, respectively. This necessitates using more recycled materials in the construction industry to avoid depletion of the ozone layer and global warming. A combined Taguchi–response surface methodology (RSM) technique was adopted by Zhang et al. (2019) to synthesize the extracted recycled aggregates to be used in the construction industry. The study was focused on the recycling of different proportions of more elegant and coarse aggregates to be reused. It was noticed that 35% of recycled aggregates with 6% of finer aggregates can be combined in stringent conditions. On the other hand, 70% of recycled aggregates can be blended with 30% of finer aggregates in non-stringent conditions. The study also revealed that adopting such techniques would yield revenue of 12 billion dollars in 10 years and can reduce CO_2 emission up to 50.1×10^6 kg.

A study reveals not only the method to be adopted to recycle the municipal solid waste but also an innovative method for its disposal by utilizing an integrated network that would also earn a profit. A case study was carried out by Harijani et al. (2017) in Tehran regarding the disposal of solid waste by effectively utilizing innovative methods. The disposal of these solid wastes became necessary to avoid contamination of resources of water, soil, as well as air. A model based on mixed-integer programming was used ineffectively (i) to understand the capacity of each location, (ii) to allocate the solid waste according to their capacity, and (iii) to distribute solid waste based on capacity, followed by recycling. This model has been implemented in the city of Tehran, Iran, and it was estimated to earn a profit of 43.49 million USD in 5 years. The results of the model were also compared to analyze sustainability. The model 1 selected for comparison resulted in a loss of 308.6 USD, followed by 362.80 USD for model 2.

Kroll and Hoyer (2019) developed an in-house recycling machine having innovative technology. Sustainability can be implemented in those areas where waste product is available in abundance. However, waste product out of technical elastomers is nearly 20%; adopting such technology in the recycling of fewer waste products seems to be noneconomic. In-house recycling machine, renamed as Reaktruder, was based on the intense redesign of existing machines, and it is capable of converting a least 60% of waste into a reusable product with reduced power consumption. According to the authors, Reaktruder is best suitable for small- and medium-scale enterprises and has been installed across Europe. With the decreased investment, Reaktruder can bring profit in about 1500 h, as said by authors.

An article reported by Wang et al. (2019) highlights the recycling treatment involved in waste generated from the construction and demolition sector to transform that into ultrahigh-performance concrete. A technique based on Andreasen and

Andreasen was first used to the packed structure of ultrahigh-performance concrete, followed by aggregates that were prepared by using construction waste. Preparing aggregates out of ultrahigh-performance concrete is essential as it will reduce the negative impact on the packed structure. The results showed that heat accumulation during the process of concrete preparation had been reduced, in turn the shrinkage factor. Besides, consumption of energy was seen to be very low, and the best results were obtained for 50% of cement with around 19% of aggregates.

A project assisted by the World Bank in 2012 estimated that around 1.3 million tons of solid waste has been produced globally, and this will be doubled by the year 2025. Most commonly found solid waste was noticed to be plastic, paper, and organic waste. Therefore, innovative sustainable approaches would be useful and practical to be implemented to turn solid waste into a reusable product. Based upon the above discussion, Das et al. (2019) identified key objectives in their review that mainly focused on present scenario of technologies, innovations, and tools, methodologies adopted in different countries respectively. At the end, they analyzed the characteristic features of life cycle assessment tools and cumulative impact of all the above discussion that leads to recycling method according to the present scenario. This review describes the necessity of LCA models to facilitate future work among organizations with the help of algorithms.

The recent years have witnessed a solid waste management system employed in developing countries, seen to be a profitable organization, and the burden on the environment has also been significantly reduced. The scenario of waste management system of one such city of a developing country has been highlighted by Ferronato et al. (2019). This article presents the quantitative and qualitative approach of waste management effectively implemented in La Paz city of Bolivia, where no data of management systems were available. A total of six mega plans were carefully understood with the view of environmental, financial, transportation, and disposal methods and implemented in La Paz city by focusing on collecting organic waste to reduce global warming and to improve sanitary landfills.

Hypothesis H4: Feasible approaches for sustainable recycling and utilization of solid wastes are positively related to economic performance.

12.3.5 Sustainable Consumption Impact in Textile Sectors

The main strategy of the business is to change the design according to the trend that provides quick profit. The emphasis of recent textile industries has been to keep the final product as low as possible and improve the efficiency of production. Niinimäki and Hassi (2011) presented a report in which strategies adopted in the textile and clothing industry have been addressed. The author presented the concept of rethink and redesigns with the prospective to not compromise the production value. Further, the authors also proposed a new design concept that was based upon the adoption of sustainable concepts in niche markets.

The worth of textile industries across the globe is estimated to be 1 trillion USD, 35 million workers are associated with it, and 7% of the world's total export is from the clothing industry. Substantial increase in production is based on the demand from the western countries that have raised the concern of employment. Despite the considerable employment from the clothing industry, it is regarded to be the most polluting industry worldwide as it is based on the consumption of fuel, a variety of chemicals, and water. An article by Desore and Narula (2018) discusses the barrier, responses from the clothing industry to adopt sustainability in production, and final processing. The article highlights the significant gaps from the past literature that was to be addressed essentially to incorporate sustainability. Besides, the authors also suggest that the managerial unit should consider a few monetary steps.

An article reported by Stål and Jansson (2017) focused on analyzing the actual perception of the firms to adopt sustainability in practice. The authors considered case studies to find the gap among the Swedish firms. The report focused on pointing out two important aspects. Firstly, it is solely the producer who constructs the field of sustainability to satisfy their own needs, and consumers' perception of sustainability could not be considered. Secondly, the clothing industry focuses more on purchases and profit, while much attention has not been paid in recycling and disposal. The report also says that although the customer's choice should preferentially be satisfied, the gap between the attitude of the customer and the behavior of the producers obscures the business.

The ever-increasing concerns on environmental issues and sustainable consumption have been confined to the field of food, textile, electricity, and apparel. The past literature has enlightened the characteristic features of a consumer group to showcase its environmentally friendly nature. The products developed on a sustainability basis have been targeting a group of consumers only, and the major population has been left out. A report that focuses on customer behavior, their perception, and intentions toward the sustainable product has been comprehensively analyzed by Kang et al. (2013). The response of a group of 701 students among the universities in China, South Korea, and the United States has been analyzed. The response, as well as their behavior on sustainable products, was studied using a modeling approach. The results showed that consumers perceived knowledge, followed by personal relevance, and norms can significantly affect the interest of young consumers in purchasing sustainable textiles and apparel.

Hypothesis H5: Sustainable consumption impact in textile sectors is positively related to firm performance.

12.3.6 Innovation Activities in Textile Wastewater Management

The processing of wastewater, including dye, needs serious attention as it contains high pH with strong color, and it has low biodegradability in nature. The dyes used in recent years seem to be very stable in all weather conditions, and processing such dye is more complicated. The methods to treat dye that are in practice these days seem to be conventional, namely, biological oxidation, adsorption, and coagulation by iron salts. Sometimes a more effective conventional technique of ozonation has been used. However, reducing all the effluents from wastewater is a difficult task for any method. An innovative method of treating the early-stage dye and finishing-stage dye separately using a cellulosic reactive azo process was done by Vlyssides et al. (2000). They reported that electrolytic method gave satisfactory results after reduction of the organic load and the color of the effluent. At the end, they concluded that color reduction is of most importance factor for the textile industries. The results showed that chemical oxygen demand (COD), as well as biochemical oxygen demand (BOD), were substantially reduced.

A study by Bilinska et al. (2017) adopted the most advanced and innovative oxidation process to treat highly contaminated wastewater. An investigation was carried out on a simulated mixture that was prepared by blending the highly reactive agents of RY145, RR195, and RB221 at specific proportions. A variety of ozone-based advanced oxidation processes (AOPs) was compared with conventional techniques and assessed. Besides, the effect of hydrogen peroxide and UV during the process has also been investigated. As the mixture was composed of several dyes, the evaluation was carried out using the UV-Vis spectra technique. The addition of such coloring agents has turned to the decolorization of wastewater very rapidly. It was noticed that only 10% of decolorization was left behind after treatment for 10 min. Therefore, AOPs seem to be a promising technique and can readily be used in such a harsh environment very effectively.

A membrane modification technique with polydopamine (PDA) coating was used in the infiltration process by Ye et al. (2020) in their study. This modification technique has gained huge interest among researchers due to its intrinsic adhesion nature. Tight ultrafiltration membranes having a molecular weight between 1000 and 5000 Da were coated with polyethyleneimine ammonium persulfate. The use of ammonium persulfate is to oxidize the polydopamine rapidly along the surface of the substrate through a Schiff base reaction. By depositing such highly oxidizing polydopamine in ammonium persulfate atmosphere for nearly 1.5 h, a defect-free polydopamine with membranes of 1700 Da can be expected, which is capable of desalinating up to 99.95% and yielding 99.2% dye recovery. The results indicated that membrane modification using polydopamine is a better choice in dye desalination and recovery.

With an idea to find opportunities to reuse the separate streams of wastewater or composite wastewater from mill, Cinperi et al. (2019) prepared three separate wastewater streams. Moreover, five different concentrations of water were also characterized. A pilot-scale study was carried out using membrane bioreactor, nanofiltration, and reverse osmosis techniques. The addition of coloring agents in separate streams has resulted in peaks and oscillations depicting the mixing effects. After the membrane bioreactor test, the content of COD was 70%, while BOD was 74%, with TSS to be 86%, TN estimated to be 28%, and TP to be 43%. As the combined membrane bioreactor and nanofiltration resulted in 40–99% removal efficiency, whereas, 52–99% removal efficiency for membrane bioreactor with reverse osmosis technique. However, the combination of membrane bioreactors with nanofiltration and UV effluents had a great effect on product quality.

Another study by Hubadillaha et al. (2020) investigated the hollow fiber membranes made up of hydroxyapatite ceramic using the sintering technique. The properties of the membrane were analyzed to be significantly varied after using hydroxyapatite content. With the resulting least micro-sized pores (0.013 μm), it was capable of reducing 99.9% of color, 80.1% of COD, and 99.4% of turbidity followed by 30.1% of conductivity using sintered hydroxyapatite. It was also noticed that the sintering at a temperature range between 1000 and 1200 ° C demonstrated the capability to remove 100% metals.

Hypothesis H6: Innovation activities in textile wastewater management are positively related to firm performance.

12.3.7 Operational Capabilities for Service in Textile Wastewater Sectors

Product design is not only based on to look after the new design systems to attract consumers instead, it also establishing a design concept within the specific domain that is commonly termed to be facelift design, which is currently the scope of researchers worldwide for development of new design. To support this, cross-domain knowledge is essentially an influencing factor in innovative product design based on recombination and transformation technologies. Besides, knowledge of patent will also help in developing the conceptual design. In such a scenario, Liu et al. (2020) in their report depicted retrieving cross-domain patents and its importance to develop a retrieval tool based on patent knowledge necessary for innovative products. The report highlights a website that serves as a service provider in which technical terms and patent classification are extracted to provide a knowledge space. The specially built algorithms were used to transform technical terms to a specific function to which it has been extracted to define. The selected patent will then be offered to trigger the designers.

The issue of environmental concern arises when the firms pursue fast growth by ignoring environmental protection. As the environmental concern has risen, the firms have been forced to maintain a balance between profitability and economic responsibility. Wu et al. (2019) explored the opportunity to study whether innovative knowledge as well as firm's transparency acts to be an asset in building sustainability practices. This study was pertaining to the firms registered in China from 2006 to 2015. Both innovative knowledge and firm transparency have contributed to a positive impact on friendly practices of sustainability. Pan et al. (2019) reported that the firms work under the supply chain, and solutions have been falling toward collaboration and intelligence to meet the demands of consumers. A smart product service-based study was carried out especially to measure interoperability and sustainability. These product service-based tools are integrated in such a way that information would be processed autonomously with self-decision taking ability. A practical example of the physical Internet was also considered in this study. It was noticed that using smart product-service system (PSS) in association with physical Internet (PI) could be beneficial to address issues related to design and business models.

Service innovation also contributes to the economic growth of a country. It was estimated that the contribution of the service sector was around 48% in 1997 and has been increased to 57% by the year 2015, even among the countries having average and moderate income compared to countries with high income. It is because firms nowadays are not dependent on the resources as they have the opportunity to build a network between suppliers and consumers. Ndubisi et al. (2019) carried out intense research on firms' joint innovation capabilities, which are situated in the UAE. This work was based on the 302 responses collected from 151 firms across UAE that have been analyzed. This report also provides the gap between the knowledge and innovation capabilities required. It was also reported that the mediation effect in establishing a bridge between knowledge and service innovation is positive. Besides, the intensity of competitiveness to meet the demands of uncertainty has also been highlighted.

Hypothesis H7: Operational capabilities for service in textile wastewater sectors are positively related to operational performance.

12.3.8 Relationship Between Eco-innovations and the Impact on Business Performance

Firms with differentiated, innovative capabilities would have varied resources (heterogeneous and immobile) and are very difficult to imitate. Innovation in sustainability arises as a result of the confluence of discussion to meet the demands of a global society based on the conception of eco-innovation. Adapting to rapid changes in market strategies based on innovation would result in something new, reduce the burden on the environment, as well as influence factors such as social, cultural, and institutional values. A survey was carried out by de Oliveira Brasil et al. (2016) to build a relationship between eco-innovation and business performance based on data of 70 samples obtained from different companies. They analyzed that eco-innovation can influence the bridge between process and product. Every eco-innovation method has a different role in finding determinants, attributes, and contributions to expand the business. However, if a study is confined to the textile sector, eco-innovation should be based on holistic view and operation capabilities.

In recent years, the regulations imposed on textile industries to adopt innovative programs to reduce and reuse concept have brought down the burden on the environment. Eco-innovation refers to production, service, and business strategy, which are novel to the firms resulting in reduced pollution, environmental risk, and negative impact on the resources used. It has now become an integral part of the industries to adopt eco-innovation after government and environmental effects. Varied types of eco-innovation can be witnessed, such as innovation in product, innovation in process, innovation in organizational values, innovation in marketing with a different set of attributes, and contribution to reducing burden. Cheng et al. (2014) studied to create a link between eco-innovation and business, in which authors used structural modeling by considering almost 121 samples from TEMA (Taiwan Environmental Management Association). They noticed that the concept of eco-innovation has a strong influence on business performance. Besides, innovation in process and product conciliates organizational innovation. The three factors, such as eco-process, product, and organizational innovations, directly or indirectly affect business.

Bossle et al. (2016) reviewed the steps to be followed to integrate innovation based on sustainability. The highlights that were mentioned pertain to influencing factors and motivation necessary for the industries to adopt to eco-innovation systems. It was concluded that specific action would be necessary to make companies adopt such revolutionary changes in eco-innovation. Besides, the authors also recommend that the focus of companies to make a profit after sales should be diverted to adopt eco-innovations, which also results in profit. In finding out the driving force necessary to adopt eco-innovation in Slovenia, Hojnik and Ruzzier (2016) depict that variation in the economy of a particular country affects global trade and production. The same is true in the perception of pollution, management, and environmental phenomena. All of these have the potential to affect across geographic borders. Thus, the concept of sustainability shared by countries will also influence the others to adopt globally, because increasing population and limited resources become salient issues to be addressed. The study was focused on bringing a solution to technological and nontechnological issues in reducing the consumption of energy and material. Data of almost 223 companies across Slovenia was collected and analyzed. The study revealed that there are specific key issues that need to be deployed as they are conducive to eco-innovation. Empirical evidence revealed that competitive pressure is one such key factor that acts to be a driving force in eco-innovation. They concluded that eco-innovation results in growth and profitability and brings competitive culture, while profit has zero impact.

The analysis carried out by Przychodzen and Przychodzen (2015) enabled to provide a link between eco-innovation and financial aspects based on the company's public trading between 2006 and 2013 situated in Poland and Hungary, as the relationship between these two in central and western European companies remained unanswered. The work was to find the gaps in product and process innovation, followed by market and supply systems. The results indicated that eco-innovation was characterized by retention on low earnings, equity, and higher returns. Besides, it was noticed that more substantial companies could accommodate the concept of eco-innovation, as these companies possess lower financial risk compared to conventional firms. It was concluded that financial capabilities become relevant in the development of eco-innovative culture in an organization.

Hypothesis H8: Relationship between eco-innovations and the impact on business performance is positively related to firm performance.

12.3.9 Identifying the Role of MCDM Approach

Wastewater from the textile industry needs careful physical and chemical treatments as it is composed of industrial effluents. Treatment of such wastewater can yield 70–95% efficiency in removing solid suspension and COD, respectively. There are various treatment methods such as membrane distillation, hydrolysis, Fenton oxidation, and jar test. The difficulty lies in the optimum selection of chemical ratio as the solid suspension and organic matter in wastewater demand for optimal condition. Hence, multi-criteria decision tools help in better solving the design of treatment ratio. These MCDMs are based on a set of techniques that seem to help in making out the best decision. Beltran et al. (2009) selected a jar test to find the optimal chemical composition to treat wastewater. Jar test seems to be more efficient in the removal of suspended solids as well as organic matter. Despite the efficient MCDM technique used, coagulant selection is not an easy task as each coagulant shows different properties for the removal of solid suspension.

In recent years, the use of enterprise resource planning (ERP) systems is becoming more popular to introduce competitiveness in any organization. This acts to be the backbone of the firm, which can integrate and automate all business-related issues. The criticality remains in the selection of suitable ERP systems for a particular organization. Cebeci (2009) studied the suitability of ERP systems in association with the fuzzy AHP MCDM technique for the textile industry. As the textile industry involves a variety in production, products, as well as workers, it makes the implementation of ERP more complex. The results show that the ERP system's successful implementation needs initial screening based on analyzing the companies' vision and strategy. Once the specific requirements do not meet, then such ERP systems can be eliminated. After the screening phase, the methodology provides a recommendation before an ERP is selected, and then the suitable ERP systems that best fit to cater to the needs can be deployed in practice.

A different concept of fuzzy sets was proposed in 1965, while the generalization took place in 1983. Fuzzy sets are commonly employed in the field of decision-making and artificial intelligence, and in recent times, it is being tested in the socioeconomic system. The intuitionistic fuzzy set method was used to measure the interval value in the study by Nayagam et al. (2011). The reports illustrate the numerical examples, and analysis was done by comparing it with alternative methods as well. Based on the results, an innovative MCDM based on intuitionistic fuzzy sets has been used for the criterion of the values. The authors also claim that the proposed MCDM method provides substantially better results than the existing methods used in the real world. The energy sector is one of the fastest-growing economies and has a vital role in everyone's life, and improving such a domain has been a countries agenda all around the globe. Fossil fuel-based power accounts for almost 81% of total production. The demand for energy has been increasing by 1.8% every year. Different platforms are available to produce energy such as hydro and wind power, geothermal, and solar, and biomass contributes only 19% of the total energy production due to a lack of planning and failure to identify the appropriate energy sources. One such study depicts planning and identifying the most appropriate energy source by using MCDM technique of gray regression analysis. The study was focused on analyzing the appropriate source in the renewable sector. The MCDM selected was integrated into three steps, the process of analytical network, followed by optimization, and at the end solution based on compromisation. Initially, gray regression method was used to highlight the criteria, and then the obtained results were evaluated by formulating a network structure. A case study was also conducted to measure the effectiveness as well as the applicability by Çelikbilek and Tüysüz (2016).

The past studies so far that have been carried out were confined to reduce the effluents discharged to the environment, while the amount of wastewater generation and contamination loads were not depicted. Ozturk and Cinperi (2018) in their study used techno-economical minimization techniques to determine wastewater amount, including the pollutants in it, and also to reduce the water usage significantly in a woolen textile mill. The generation of wastewater, as well as water consumption for a specific application, was calculated during the on-site visit. Clean and wastewater samples have been collected at different periods and were analyzed to measure the specific pollutant. The reuse capability of the mill was then estimated. Also, to compare, reduction potentials of other mills were examined. MCDM technique was used to identify and evaluate a total of 82 such minimization techniques. Only 9 out of 82 decided to implement based on the results. Thus, a reduction of 41–69% in water consumption, 48–75% of wastewater reuse, and 28–63% of chemicals has been removed. It was also claimed that the potential payback period would be between 24 and 60 months.

Hypothesis H9: Identifying the role of MCDM approach is positively related to firm performance.

12.3.10 Nanotechnology Applicability in Industrial Wastewater Treatment

Dyes and pigments are the most commonly used organic compounds in the textile industry. It was estimated that around 7×10^5 tons of dyes were produced annually to meet the demand of the clothing industry worldwide. This shows the ever-increasing demand for apparel in the world market. It was also estimated that around 15–20% of these organic compounds are discharged in the effluents during dyeing operation. This will not only raise environmental pollution but also lead to nanoesthetic pollution. It was also noticed that dyes of nanoparticles in the waste body do not allow the sunrays to fall deep inside the water and, in turn, result in reduced photosynthesis. Despite having an eco-innovative process of water treatment, degradation of synthetic dyes from wastewater seems to be not amenable. Babaei et al. (2017) used nanoparticles of carbon to treat wastewater by enhancing adsorption and oxidation process. The results showed that effective adsorption was achieved when treated with H_2O_2 after screening through UV Fenton, US Fenton, and simple Fenton. It was observed that 94.8% of dye was removed by iron leaching.

Jorfi et al. (2016) synthesized MgO nanoparticles as a catalyst to treat the textile wastewater, especially to remove the acid red 73 dye. During the process, pH played an important role. Dyes were reduced to a greater extent from 200 to 31 mg/L, when pH and coagulants were set to 6 and 200 mg/L. The results also showed complete removal of acid 73 dye by using MgO nanoparticles with a concentration of 0.8 g/L with pH factor of 5 with reaction time of 60 mins.

While Soltani et al. (2016) studied the effect of ZnO nanoparticles in the removal of basic red 46 dyes from textile wastewater on catalytic decolorization, the results were compared between the ZnO nanoparticles in a suspended state with bentonite carrying ZnO nanoparticles. In the ZnO, when carried by bentonite, the size would vary between 5 and 40 nm (surface area, 80.6 m^2/g) compared to suspended particles of size 20–120 nm (surface area, 42.2 m^2/g). RSM technique with the central composite design was used, and the decolorization efficiency was achieved to be 89.92% when the initial dye concentration was 6 mg/L for 0.3 ZnO, and bentonite ratio and dosage were limited to 2.5 g/L. The by-product produced during decolorization was also studied.

It is necessary to treat with a high level of chemicals to address the wastewater issue from the woolen industry, as depicted by Hassanzadeh et al. (2017). The addition of chemicals to such extent yields another concern of high sludge production. FeSO4 as coagulant with a concentration range between 400 and 800 mg/L having pH value of 6–10 was used to reduce the COD to 200 mg/L and tributary to 25 NTU, respectively. A nanofiltration process with embedded polyamide nanofiber was used to reduce effluents from wastewater. Operating parameters were studied by using RSM technique. The optimal condition was obtained for pH, 8, and FeSO4, 600 mg/L, in the nanofiltration process. The increase in pressure and pH could yield removal efficiency to 98%. While with the addition of color concentration in the process, the efficiency decreased up to 90%.

Hypothesis H10: Nanotechnology applicability in industrial wastewater treatment is positively related to production performance.

12.4 Conclusions

The conceptual framework produces a benchmark to the textile manufacturing sectors and creates awareness for general consumers, for society, as well as for policymakers, respectively. The following significant points evolved to develop clean environment:

1. Current waste management scenarios across different countries try to adopt the proposed conceptual framework to improve the textile manufacturing technique.
2. Adopt the new technology for the development of better textile products, simultaneously to reduce waste effluents.
3. The decision-making approach is also one of the major criteria to implement as early as possible in a different part of the textile sectors.
4. A standard feasible approach may be required to develop an effective utilization of recycled wastewater as well as raw materials in the same textile processing applications or different sector applications.
5. The proposed ten criteria can also play a significant role in obtaining sustainable management policy and manufactures as well as improve the socioeconomic conditions.

References

Annual report, Ministry of Textiles (2017) Indian Textile Journal, Department of Industrial Policy and Promotion, Press Information Bureau, Union Budget

Aragonés-Beltrán P, Mendoza-Roca JA, Bes-Piá A, García-Melón M, Parra-Ruiz E (2009) Application of multicriteria decision analysis to jar-test results for chemicals selection in the physical–chemical treatment of textile wastewater. J Hazard Mater 164(1):288–295

Austgulen MH (2016) Environmentally sustainable textile consumption—what characterizes the political textile consumers? J Consum Policy 39(4):441–466

Aydiner C, Sen U, Koseoglu-Imer DY, Dogan EC (2016) Hierarchical prioritization of innovative treatment systems for sustainable dairy wastewater management. J Clean Prod 112:4605–4617

Babaei AA, Kakavandi B, Rafiee M, Kalantarhormizi F, Purkaram I, Ahmadi E, Esmaeili S (2017) Comparative treatment of textile wastewater by adsorption, Fenton, UV-Fenton and US-Fenton using magnetic nanoparticles-functionalized carbon (MNPs@ C). JIndus Eng Chem 56:163–174

Beltrán PA, Mendoza-Roca JA, Bes-Piá A, García-Melón M, Parra-Ruiz E (2009) Application of multicriteria decision analysis to jar-test results for chemicals selection in the physical-chemical treatment of textile wastewater. J Hazard Mater 15:288–295

Bilińska L, Gmurek M, Ledakowicz S (2017) Textile wastewater treatment by AOPs for brine reuse. Process Safe Environ Prot 109:420–428

Bossle MB, de Barcellos MD, Vieira LM, Sauvée L (2016) The drivers for adoption of eco-innovation. J Clean Prod 113:861–872

Cebeci U (2009) Fuzzy AHP-based decision support system for selecting ERP systems in textile industry by using balanced scorecard. Expert Syst Appl 36(5):8900–8909

Çelikbilek Y, Tüysüz F (2016) An integrated grey based multi-criteria decision making approach for the evaluation of renewable energy sources. Energy 115:1246–1258

Cheng CC, Yang CL, Sheu C (2014) The link between eco-innovation and business performance: a Taiwanese industry context. J Clean Prod 64:81–90

Cinperi NC, Ozturk E, Yigit NO, Kitis M (2019) Treatment of woolen textile wastewater using membrane bioreactor, nanofiltration and reverse osmosis for reuse in production processes. J Clean Prod 223:837–848

Cohn M (2000) The WTO: a New World government dedicated to the principle that property interests are more sacred than human rights. Guild Prac 57:134

Corsten M, Worrell E, Rouw M, Van Duin A (2013) The potential contribution of sustainable waste management to energy use and greenhouse gas emission reduction in the Netherlands. Resour Conserv Recycl 77:13–21

Das S, Lee SH, Kumar P, Kim KH, Lee SS, Bhattacharya SS (2019) Solid waste management: scope and the challenge of sustainability. J Clean Prod 228:658–678
de Oliveira Brasil MV, de Abreu MCS, da Silva Filho JCL, Leocádio AL (2016) Relationship between eco-innovations and the impact on business performance: an empirical survey research on the Brazilian textile industry. Rev Adm Empress 51(3):276–287
Desore A, Narula SA (2018) An overview on corporate response towards sustainability issues in textile industry. Environ Dev Sustain 20(4):1439–1459
Ferronato N, Ragazzi M, Portillo MAG, Lizarazu EGG, Viotti P, Torretta V (2019) How to improve recycling rate in developing big cities: an integrated approach for assessing municipal solid waste collection and treatment scenarios. Environ Dev 29:94–110
Fischedick M, Roy J, Acquaye A, Allwood J, Ceron JP, Geng Y, Kheshgi H, Lanza A, Perczyk D, Price L, Santalla E (2014) Industry in: climate change 2014: mitigation of climate change contribution of working group III to the fifth assessment report of the intergovernmental panel on climate change. Cambridge University Press, Cambridge/New York. Technical Report
Fuldauer LI, Ives MC, Adshead D, Thacker S, Hall JW (2019) Participatory planning of the future of waste management in small island developing states to deliver on the sustainable development goals. J Clean Prod 223:147–162
Harijani AM, Mansour S, Karimi B, Lee CG (2017) Multi-period sustainable and integrated recycling network for municipal solid waste – a case study in Tehran. J Clean Prod 151:96–108
Hassanzadeh E, Farhadian M, Razmjou A, Askari N (2017) An efficient wastewater treatment approach for a real woolen textile industry using a chemical assisted NF membrane process. Environ Nanotechnol Monit 8:92–96
Heggelund L, Hansen SF, Astrup TF, Boldrin A (2016) Semi-quantitative analysis of solid waste flows from nano-enabled consumer products in Europe, Denmark and the United Kingdom – abundance, distribution and management. Waste Manag 56:584–592
Henry B, Laitala K, Klepp IG (2019) Microfibres from apparel and home textiles: prospects for including microplastics in environmental sustainability assessment. Sci Total Environ 652:483–494
Hojnik J, Ruzzier M (2016) The driving forces of process eco-innovation and its impact on performance: insights from Slovenia. J Clean Prod 133:812–825
Horodytska O, Valdés FJ, Fullana A (2018) Plastic flexible films waste management – a state of art review. Waste Manag 77:413–425
Hossain MU, Wu Z, Poon CS (2017) Comparative environmental evaluation of construction waste management through different waste sorting systems in Hong Kong. Waste Manag 69:325–335
Hubadillah SK, Othman MHD, Tai ZS, Jamalludin MR, Yusuf NK, Ahmad A, Rahman MA, Jaafar J, Kadir SHSA, Harun Z (2020) Novel hydroxyapatite-based bio-ceramic hollow fiber membrane derived from waste cow bone for textile wastewater treatment. Chem Eng J 379:122396
Jorfi S, Barzegar G, Ahmadi M, Soltani RDC, Takdastan A, Saeedi R, Abtahi M (2016) Enhanced coagulation-photocatalytic treatment of acid red 73 dye and real textile wastewater using UVA/synthesized MgO nanoparticles. J Environ Manage 177:111–118
Kamali M, Persson KM, Costa ME, Capela I (2019) Sustainability criteria for assessing nanotechnology applicability in industrial wastewater treatment: current status and future outlook. Environ Int 125:261–276
Kang J, Liu C, Kim SH (2013) Environmentally sustainable textile and apparel consumption: the role of consumer knowledge, perceived consumer effectiveness and perceived personal relevance. Int J Consum 37(4):442–452
Khan S, Malik A (2014) Environmental and health effects of textile industry wastewater. In: Environmental deterioration and human health. Springer, Dordrecht, pp 55–71
Kroll L, Hoyer S (2019) Zero-waste production: technology for the in-house recycling of technical elastomers. Procedia Manuf 33:335–342
Landi D, Gigli S, Germani M, Marconi M (2018) Investigating the feasibility of a reuse scenario for textile fibres recovered from end-of-life tyres. Waste Manag 75:187–204

Liu L, Li Y, Xiong Y, Cavallucci D (2020) A new function-based patent knowledge retrieval tool for conceptual design of innovative products. Comput Ind 115:103–154

Maqhuzu AB, Yoshikawa K, Takahashi F (2019) The effect of coal alternative fuel from municipal solid wastes employing hydrothermal carbonization on atmospheric pollutant emissions in Zimbabwe. Sci Total Environ 668:743–759

Nayagam VLG, Muralikrishnan S, Sivaraman G (2011) Multi-criteria decision-making method based on interval-valued intuitionistic fuzzy sets. Expert Syst Appl 38(3):1464–1467

Ndubisi NO, Dayan M, Yeniaras V, Al-hawari M (2019) The effects of complementarity of knowledge and capabilities on joint innovation capabilities and service innovation: the role of competitive intensity and demand uncertainty. Ind Mark Manag. https://doi.org/10.1016/J. Indmaram.2019.05.011

Niinimäki K, Hassi L (2011) Emerging design strategies in sustainable production and consumption of textiles and clothing. J Clean Prod 19(16):1876–1883

Ozturk E, Cinperi NC (2018) Water efficiency and wastewater reduction in an integrated woolen textile mill. J Clean Prod 201:686–696

Pan S, Zhong RY, Qu T (2019) Smart product-service systems in interoperable logistics: design and implementation prospects. Adv Eng Inform 42:100996. https://doi.org/10.1016/j.aei.2019. 100996

Parkes O, Lettieri P, Bogle IDL (2015) Life cycle assessment of integrated waste management systems for alternative legacy scenarios of the London Olympic Park. Waste Manag 40:157–166

Prajogo DI (2016) The strategic fit between innovation strategies and business environment in delivering business performance. Int J Prod Econ 171:241–249

Przychodzen J, Przychodzen W (2015) Relationships between eco-innovation and financial performance – evidence from publicly traded companies in Poland and Hungary. J Clean Prod 90:253–263

Shiwanthi S, Lokupitiya E, Peiris S (2018) Evaluation of the environmental and economic performances of three selected textile factories in Biyagama Export Processing Zone Sri Lanka. Environ Dev 27:70–82

Siddique K, Rizwan M, Shahid MJ, Ali S, Ahmad R, Rizvi H (2017) Textile wastewater treatment options: a critical review. In: Enhancing cleanup of environmental pollutants. Springer, Cham, pp 183–207

Soltani RDC, Jorfi S, Safari M, Rajaei MS (2016) Enhanced sonocatalysis of textile wastewater using bentonite-supported ZnO nanoparticles: response surface methodological approach. J Environ Manage 179:47–57

Stål HI, Jansson J (2017) Sustainable consumption and value propositions: exploring product–service system practices among Swedish fashion firms. Sustain Dev 25(6):546–558

Tseng ML, Wu KJ, Chiu AS, Lim MK, Tan K (2019) Reprint of: service innovation in sustainable product service systems: improving performance under linguistic preferences. Int J Prod Econ 217:159–170

Vajnhandl S, Valh JV (2014) The status of water reuse in European textile sector. J Environ Manage 141:29–35

Vlyssides AG, Papaioannou D, Loizidoy M, Karlis PK, Zorpas AA (2000) Testing an electrochemical method for treatment of textile dye wastewater. Waste Manag 20(7):569–574

Wang X, Yu R, Shui Z, Song Q, Liu Z, Liu Z, Wu S (2019) Optimized treatment of recycled construction and demolition waste in developing sustainable ultra-high-performance concrete. J Clean Prod 221:805–816

Wigger H, Hackmann S, Zimmermann T, Köser J, Thöming J, von Gleich A (2015) Influences of use activities and waste management on environmental releases of engineered nanomaterials. Sci Total Environ 535:160–171

WuW LY, Zhang Q, Yu B (2019) How innovative knowledge assets and firm transparency affect sustainability-friendly practices. J Clean Prod 229:32–43

Xiong X, Liu X, Iris KM, Wang L, Zhou J, Sun X, Rinklebe J, Shaheen SM, Ok YS, Lin Z, Tsang DC (2019) Potentially toxic elements in solid waste streams: fate and management approaches. Environ Pollut 253:680–707

Ye W, Ye K, Lin F, Liu H, Jiang M, Wang J, Liu R, Lin J (2020) Enhanced fractionation of dye/salt mixtures by tight ultrafiltration membranes via fast bio-inspired co-deposition for sustainable textile wastewater management. Chem Eng J 379:122321

Yılmaz İH, Abdul vahitoğlu A (2019) Evaluation of municipal solid waste options in Turkey: scenarios for energy recovery, carbon mitigation and consequent financial strategies. Resour Conserv Recycl 147:95–110

Yukseler H, Uzal N, Sahinkaya E, Kitis M, Dilek FB, Yetis U (2017) Analysis of the best available techniques for wastewaters from a denim manufacturing textile mill. J Environ Manage 203:1118–1125

Zhang LW, Sojobi AO, Kodur VKR, Liew KM (2019) Effective utilization and recycling of mixed recycled aggregates for a greener environment. J Clean Prod 236:117600

Zhao H, Lin B (2019) Will agglomeration improve the energy efficiency in China's textile industry: evidence and policy implications. Appl Energy 237:326–337

Zhu L, Chen L, Wu X, Ding X (2018) Developing a greenhouse gas management evaluation system for Chinese textile enterprises. Ecol Indic 91:470–477

Chapter 13
Role of Coagulation/Flocculation Technology for the Treatment of Dye Wastewater: Trend and Future Aspects

Sonalika Sonal and Brijesh Kumar Mishra

13.1 Introduction

The word "coagulation" has been derived from the Latin word "coagulare" meaning to drive together and is known to be a process of destabilization of colloidal suspended particles along with some dissolved organic matters (Sawyer et al. 2003; Masters and Ela 2008). In the coagulation process, a coagulant is added along with rapid mixing to destabilize the non-settleable particles and to initiate their aggregation by neutralizing their electrostatic repulsive forces. Whereas flocculation is the process of aggregation of destabilized particles, through gentle agitation that leads to the formation of larger particles called flocs (Amuda and Amoo 2007; Fu and Wang 2011). The flocs formed as a result of the collision between the destabilized particles are transported through the flocculation process and get separated from liquid phase through settling or filtration (Masters and Ela 2008; Tchobanoglous et al. 2003). These two processes are generally encountered as an intact process in primary treatment process and, thus, referred as the most common treatment strategies used to remove colloidal particles and a portion of dissolved organic matter, in a combination form of coagulation/flocculation (Holkar et al. 2016; Patel and Vashi 2015; Renault et al. 2009).

Coagulation/flocculation is widely used primary treatment method in the wastewater treatment plant of the textile industry (Patel and Vashi 2015). Primary treatment includes the process of settlement of floating and suspended solid particles, present in the water, by either gravity or mechanical means. This system comprises of a grit chamber having fine screens that separate the large-sized particle pollutant and the sedimentation tank. The process of coagulation/flocculation occurs in a

S. Sonal · B. K. Mishra (✉)
Department of Environmental Science and Engineering, Indian Institute of Technology (Indian School of Mines), Dhanbad, India
e-mail: brijesh@iitism.ac.in

A. Singh et al. (eds.), *Water Pollution and Management Practices*,
https://doi.org/10.1007/978-981-15-8358-2_13

sedimentation tank where chemical coagulant is added to the water to facilitate the settling of suspended particles having a diameter of less than 50 μm (Baker 1948). The primary treatment will be beneficial in enhancing the efficiency of any treatment plant as it reduces the significant amount of pollutant prior to reaching the secondary treatment. The primary treatment generally reduces 90–98% of settleable solids and 60–80% of suspended solids along with 30–50% oxygen demand from the waste and hence reduces the load of other treatment subunits (Peavy et al. 1985). Coagulation is invariably achieved by using different metal salts that hydrolyses in water. These metal salts adsorb the colloidal particles and form larger flocs. These flocs, therefore, can easily be settled down after becoming larger particles under the effect of gravity and later known as settled sludge. However, the settling of these colloidal particles is not a facile task as they carry a net surface charge that causes to repel the particles. Thus, before understanding the coagulation/flocculation process, it is required to understand the colloids and their characteristics.

13.2 Colloidal Stability and Its Removal

Colloidal particles are an aggregate form of atoms/ions or molecules having density close to the water with a smaller diameter that prevents them not to settle down under the effect of gravity (Ghernaout et al. 2015). Thus, they are stable in nature if they are suspended and do not agglomerate naturally as their stable phase possesses randomness showing Brownian movement (Peavy et al. 1985). The most critical factors responsible for the stability of colloidal suspensions are surface charges and their excessive large surface to volume ratio resulting from their minute size (Sawyer et al. 2003). The surface charge plays a primary contribution to particle stability as surface phenomenon overruled mass phenomena. The surface phenomenon encompasses the accumulation of electrical charges around the particle surface (Peavy et al. 1985). These charges developed as a result of either the molecular arrangements within crystals or the loss of atoms due to abrasion of the surfaces or other different factors such as ionic strength of the surrounding medium, etc. Surface waters mostly possess negatively charged colloidal particles on their surface. The charged particles present in the solution phase cannot have a net charge, so the ions present in vicinity of the particles in solution will affect the surface charge of the particles. The particle charged must be counterbalanced by ions of opposite charge present in the solution. This electroneutrality results in the formation of an electrical double layer. The counter ions must be possibly configured as it is shown in Fig. 13.1.

The electrical double layer comprises of (1) a fixed or bound layer that contains oppositely charged ions to attract the negatively charged surfaces and (2) a diffuse layer, consisting of a mixture of charged ions as shown in Fig. 13.1. In between these two layers, a thin layer, named Shear surface, is present that separates the two layers. This arrangement of charged particles produces an electric potential between the surface of the colloidal particles and the bulk of the solution, which is stronger at the Stern layer and decreases exponentially with increase in distance from the colloids

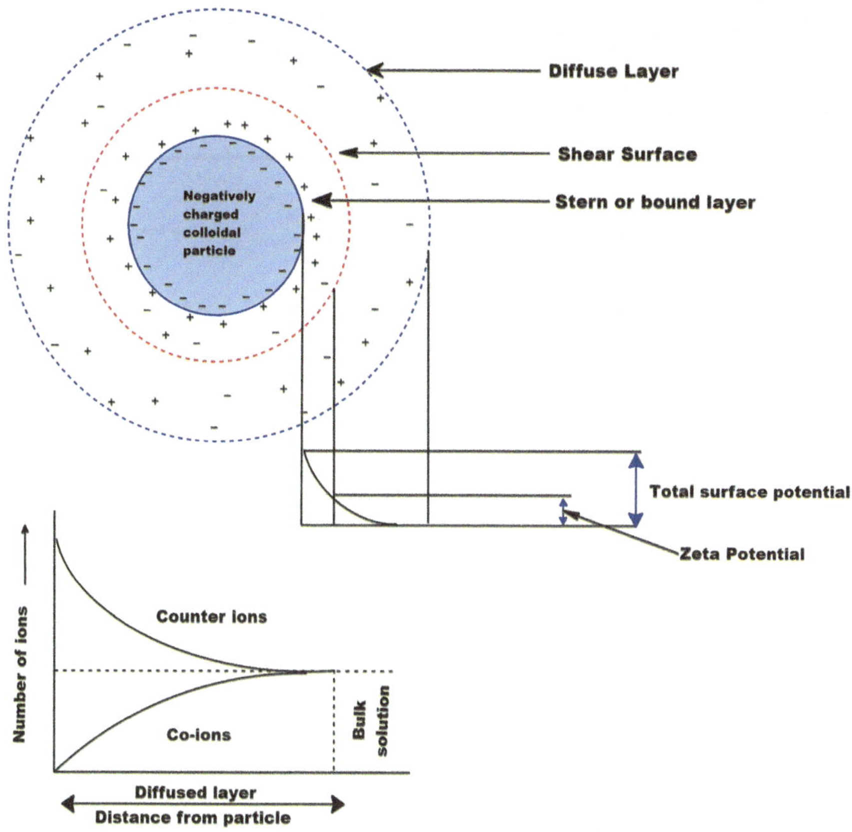

Fig. 13.1 Electrical potential around negatively charged colloidal particle. (Modified from Peavy et al. 1985; Sawyer et al. 2003; Ghernaout et al. 2015)

(Peavy et al. 1985; Asano et al. 2006; Crittenden et al. 2012). The counter ions in the bound layer are attracted electrostatically, but they might be loosely held and diffuse away or replaced by other ions in response to the thermal agitation.

When two similar charged particles come in close proximity, their electrical double layer begins to interact as a result, and two opposite forces act on them. As the two particles come closer, the electrostatic potential created by the halo of counter ions around the particles reacts to repel the particles, thus preventing aggregation of the particles. When the particles of similar charges come closer, then the repulsive forces will be stronger. The second force, an intermolecular attractive force called van der Waals force, supports attraction between the particles (Crittenden et al. 2012). This attractive force is possessed by all the molecules and colloidal particles regardless of their charge and composition. The magnitude of van der Waals force is inversely proportional to the sixth power of the distance between the particles and also decays exponentially with the distance (Sawyer et al. 2003, Mihelcic and Zimmerman 2014). Although this force decreases more rapidly than

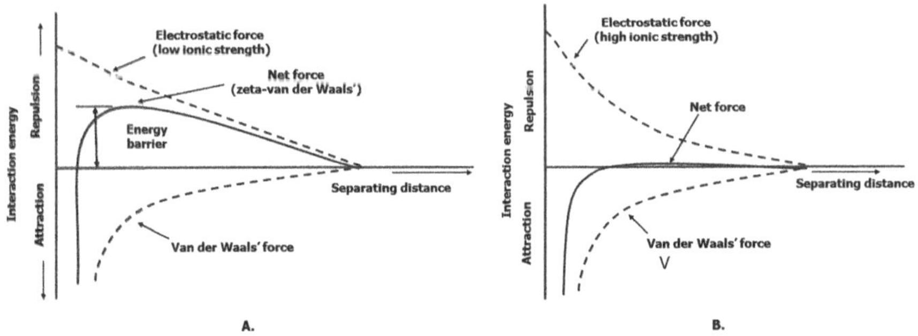

Fig. 13.2 Effect of interparticle forces on the stability of a colloidal system. (**a**) Stable system (**b**) Unstable system. (Modified from Peavy et al. 1985; Sawyer et al. 2003)

electrostatic potential, it is a stronger force at close distances. Figure 13.2a, b illustrates the sum of these two forces as the two colloidal particles come in close proximity to one another. As two similar charged particles come closer, the repulsive electrostatic forces increase to keep them apart. However, this repulsive force becomes attractive only if the particles can be brought sufficiently close together to get pass through maximum net repulsive force, called the energy barrier. If the particles do not surpass this energy barrier, then the colloidal particles remain stable and suspended in the solution. Once the particles pass the energy barrier, the attractive van der Waals force becomes a predominated one, and particles get attached.

If it is desired to destabilize and coagulate the colloidal particles, then a means for overcoming the energy barrier must be required or else the energy barrier must be lowered by some other means. Sometimes, Brownian movement, the random motion of colloidal particles because of molecular bombardment or mechanical agitation of the water, may produce enough momentum for particles to overcome the energy barrier and thus leads to collision. But these processes are too slow to occur; therefore, other means of agglomeration is used in water purification (Crittenden et al. 2012). This stable behaviour of colloidal particles due to attractive and repulsive forces was first examined by Derjagin, Landau, Verwey and Overbeek and is termed as DLVO theory after their work. Though their studies appear adequate to explain the colloidal stability, some recent work has shown that it does not sufficiently explain the kinetics of chemical destabilization of colloidal particles (Melia 1990; Elimelech and Melia 1990). To understand the colloidal stability-destability phenomenon, it is useful to know the different mechanisms of the coagulation/flocculation process that regulate the phenomenon, as explained in the next subsections below.

13.3 Mechanism of Coagulation/Flocculation Process

Although the coagulation/flocculation process follows a complicated method to be accomplished, four mechanisms have been found to more appropriately describe the whole process (Nharingo and Moyo 2016). These four mechanisms are as follows:

1. Electrical double layer (EDL) compression. The EDL plays an important role in the settling of the colloidal particles, primarily when the ionic strength of the solution is raised. The EDL constitutes a layer of cations bound around the negative charge particle surface and a diffuse layer encompassing a set of cations and anions that extend out into the solution. The ionic strength in surrounding water affects the decay function of the electrostatic potential (Sawyer et al. 2003). In other words, it can be inferred that high ionic concentration in surrounding water compresses the diffuse layer towards the surface of the colloid. As a result, the diffused layer is sufficiently compressed, and the net attractive force, van der Waals force, will predominantly prevail across the entire area of influence, and no energy barrier will exist. This ionic layer compression occurs in nature when water containing high turbidity increases the ionic content and thereby resulting in the settling of the particles (Ghernaout et al. 2015). Formation of delta around the oceans represents an excellent example of ionic compression occurring in nature (Miller et al. 2008).
2. Neutralization of charge and adsorption. Since most colloidal particles of surface water are negatively charged, the process of charge neutralization occurs by adsorption of positively charged cations or polymers. The metal salts or cationic organic polymers get ionized in water and produce cationic ions that interact with the negatively charged surface of the particles and neutralize them, leading to destabilization of the particles (Ghernaout et al. 2015). Here, the dose of the metallic coagulants and cationic polymers plays an essential role in the destabilization of colloidal particles. With the properly optimized dose, the particles come together, but if the dose exceeds the optimum level, the particles, instead of being neutralized, will attain the re-stabilization state (Sawyer et al. 2003).
3. Interparticle bridging. When non-ionic polymers or long-chain low-surface charge polymers are added alone or in addition with metallic salts, they dissociate and form larger molecules. These polymers might have a linear or branched structure having high surface reactivity. Thus, a particle gets adsorbed on the chain of one polymer, and other available active surface sites of other particulates may adsorb on the remainder of the polymers. As a result, the polymer-colloid groups may form enmeshment, leading into the formation of the bridge between the particles (Duan and Gregory 2003) and, thus, make the particles heavier that settles down the flocs. Again, the dose of the polymers plays a key role in the formation of the enmeshed polymer matrix. Overdosing of polymers may further lead to re-stabilization of the colloidal particles. Aggressive mixing or extended agitation can also break the interparticle bridging of polymer-colloid matrix and leads to re-stabilization of the colloids.

4. Entrapment of particles in the precipitate (sweep coagulation). When a high dose of metallic salts is added to the solution, they form heavier amorphous gelatinous flocs. These gelatinous flocs entrapped the particulate matters within themselves or enmeshed them by their sticky surface and swept them out from the water along with the settling flocs (Duan and Gregory 2003). This process of enmeshing and sweeping out of colloidal particles from the suspension is generally termed as sweep coagulation (Peavy et al. 1985). For this process, a high dose of metallic salts and neutral pH are required. Besides these, the presence of various anions in water and the higher concentration of colloidal particles enhance this coagulation mechanism.

13.4 Dyes and Its Classification

Dyes are synthetic complex aromatic and heterocyclic molecules that comprise of two groups, i.e. auxochrome (colour intensifier) and chromophores (colour-imparting compounds) (Sonal et al. 2018; Patel and Vashi; 2015; Verma et al. 2012). These aromatic compounds are more stable and difficult to biodegrade. The aromatic rings of dyes contain de-located electrons along with different functional groups (Ding et al. 2010). Their auxochrome groups are electron donor or withdrawing groups and are responsible for dyeing capacity of any dye, while colour-imparting group, chromogen-chromophores are electron acceptor. The chromogens are composed of cyclic aromatic structures such as benzene rings, anthracene or naphthalene which adhere to the chromophore radicals containing double-conjugated bonds with de-located electrons. The chromophore groups are ethenyl (–C≡C–), carbonyl (–C=O), imino (–C≡N–), nitro ($-NO_2$), azo group (–N=N–), ethylene group (–C=C–), methine group (–C=H), carbon-nitrogen (=C=NH; –CH=N–), carbon-sulphur (=C=S; ≡CS–S–C≡), nitro ($-NO_2$; –NO–OH), nitroso (–N=O; =N–OH) or quinoid groups, while auxochrome groups are amino ($-NH_2$), carboxylic (–COOH), sulphonyl ($-SO_3H$) and hydroxyl (–OH) group (Carmen and Daniela 2012; Welham 2000).

The textile dyes are classified in various ways depending on different criteria of origin, nature or reactivity (Fig. 13.3). On a general basis, the dyes are classified as natural and synthetic textile dyes. Since the beginning of 2600 BC, natural textile dyes extracted from vegetable and animal resources were used in textile processing units in China (Carmen and Daniela 2012). During the fifteenth century, Phoenicians used Tyrian purple produced from crushed sea snail's species, and since 3000 BC, an indigo dye extracted from the indigo plant was also used. Besides this, several other naturally derived dyes such as saffron (natural yellow 6) extracted from the stigmata of *Crocus sativus* and hematein (natural black 1) obtained from the heartwood of a tree, etc. were discovered and commonly used (Carmen and Daniela 2012).

In 1856, the first synthetic dye named "mauve dye" (aniline), a brilliant fuchsia colour, was first discovered by W.H. Perkin (UK), followed by some azo dyes synthesized by diazotization reaction discovered in 1958 by P. Gries (Germany)

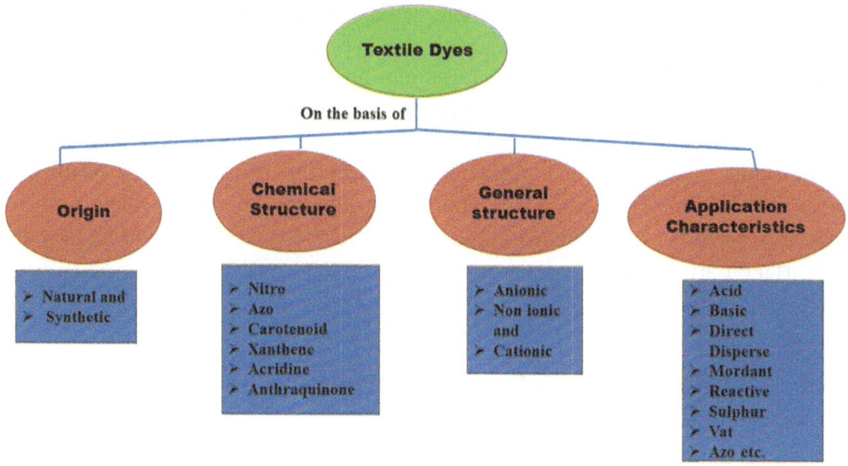

Fig. 13.3 Classification of dyes on different basis

(Welham 2000). Among all classes of dyes, azo class of dyes is the most widely used followed by anthraquinone and accounts for 65–75% of the total textile dye (Verma et al. 2012). The azo dyes comprise of characteristic reactive groups such as –OH, –NH or –SH that form covalent bonds with the fibres. These are commonly used for yellow, orange or red colours. Apart from this, about 60–70% of azo dyes are toxic and carcinogenic and are resistant to degradation. Due to this reason, in India, the MoEF has announced that the handling of 112 dyes (including 42 benzidine-based dyes and 70 azo dyes) that are capable of releasing harmful amines is prohibited (Nadigera 2001). After azo, the second most important class of textile dyes is anthraquinone group that are used for violet, blue and green colours mostly. Anthraquinone dyes are resistant to degradation due to fused aromatic ring structure (Fontenot et al. 2003). Based on general structure, the dyes are categorized into anionic (e.g. reactive, acid dyes, etc.), cationic (e.g. basic dyes) and non-ionic dyes (e.g. disperse dyes). Reactive dyes are one of the most popular classes of dyes that are abundantly used in textile industries to adhere colour on cotton, wool or nylon fibres. Their bright, vibrant gamut of colours, easy application and availability make them more acceptable for the use in textile industries (Robinson et al. 2001; Sonal et al. 2020). But its high solubility with water and high remains of unutilized dyes make this class of dye more problematic to treat. These dyes are released in copious amount along with wastewater containing several types of pollutants such as salts, acids, alkalis, dispersants, levelling agents, carriers, suspended solids and heavy metals that make the effluent more harmful and pernicious. The dyes possess toxicity and are likely known to be a carcinogen because of benzidine and other aromatic compounds they constitute off (Sonal et al. 2020). Dyes can potentially harm the aquatic as well as human life. The release of colour affects the photosynthesis function in plants and algae that impacts the aquatic life because of low light penetration and oxygen consumption. Metals and chlorine or other inorganic component are also found to be more lethal to specific forms of marine life. Suspended particles released along with the dye wastewater can choke fish gills and kill them (Wang et al. 2005). Dyes can also influence directly or indirectly human life, causing various abnormalities in the form of tumours, cancers and allergies (Carmen and Daniela 2012). The presence of these pollutants is also aesthetically undesirable and consequently alters the self-purification ability of streams and rivers. Hence, the decontamination of these dyes molecules from dye wastewater effluent has become of utmost importance to protect both the aquatic and terrestrial environment.

13.5 Characteristics of Dye Wastewater

The textile industries are one of the most growing sectors as industrialization and modern fashion world proliferate. These industries are considered as one of the biggest polluter industries demanding plenty of water and various complex chemicals during the processing stages. The textile industries are classified as dry processing mills and wet processing mills based on waste and effluent generation (Verma et al. 2012). The dry processing unit generates dry fabric rejects, while wet processing units generate a considerable amount of effluent because of different processing stages that are required for the production of textile fibres. The wet processing unit has been accomplished by various stages of sizing, de-sizing, scouring, bleaching, mercerizing, drying, printing and finishing stages. All these different stages generate distinct wastewater depending on the process and the chemical used. Dye wastewater released during these operational units in textile industries needs critical attention during coagulation because of the toxic/organic nature of colloids present in dye wastewater. Table 13.1 represents the various stages of wet processing unit and characteristics of effluents generated from them.

Consequently, the water discharge from the textile industries has a high environmental impact because of the tremendous amount of water requirement among other industries, globally, and each subunit effluent is highly polluted (Sonal and Mishra 2019). An average-sized textile industry consumes approximately 200 L per kg of fabric produced per day (Kant 2012). It has been estimated by the World Bank that textile dyeing and finishing treatment given to a fabric generate around 17–20% of total industrial wastewater (Kant 2012). In India, approximately 80% of the total production of 1,30,000 tons of dyestuff are consumed by the textile industry, due to high demand for cotton and polyester, globally (Naik et al. 2013).

On the whole, the dye wastewater released from textile and dyestuff industries is very complex and arduous to treat (Fewson 1988). The effluent discharged from the industry contains high biological and chemical oxygen demand, besides acids, alkalis, dyes, hydrogen peroxide, starch, surfactants, dispersing agents and soaps of metals (Paul et al. 2012). Still, the most notable consideration has been given to the dyes. These dyes are released in high concentration and are persistent even in low concentration. During manufacturing and textile dyeing-printing, complete

Table 13.1 Textile processing stages of the wet processing unit and generated wastewater

Stages	Process	Effluent characteristics
Sizing	First step that provides strength to the fibres and minimizes their breakage. Sizing agents such as polyvinyl alcohol (PVA), carboxymethyl cellulose and starch are added	Contains pH of 4–5, high biological oxygen demand (BOD) (range of 300–450 ppm) and medium chemical oxygen demand (COD), starch, enzymes and waxes
De-sizing	Process of removing sizing material before weaving	Effluents have a high temperature (70–80 °C), BOD and COD
Scouring	Impurities from the fibres such as fats, waxes, surfactant and natural oil and suspended contaminants in the scouring bath have been removed by using alkali (NaOH) solution	Oil, fats, high pH, alkali, temperature and strong colour
Bleaching	The unwanted colour from the fabric was removed by using bleaching agents such as peracetic acid and hydrogen peroxide	High pH, TDS, organic stabilizer and other organic compounds
Mercerization	It is the process of giving shine to the fabrics and enhances its dye uptake and absorbance. The process is carried by treating fabrics with high concentration of NaOH	High BOD content, TDS and pH
Dyeing and printing	In this process, fabrics or yarn is treated with dye, which requires a large volume of water typically not only in dye bath but also during the rinsing step. Other chemicals such as salts, surfactants, organic processing aids, etc. are added to improve dye adsorption depending on the dyeing process	High toxicity, pH, dissolved solids, metal concentration, BOD, COD and strong colour
Finishing	The process is used to improve the definite characteristics of the fabric. Specific features like softening, waterproofing, antibacterial and UV protection, etc. are imparted to the fabric	Low alkalinity, low BOD, high toxicity, solvents and softeners

Modified from Verma et al. (2011) and Patel and Vashi (2015)

utilization of the dyes does not occur, and some of them release into the environment. Notably, about 10% of dyes used by textile industries are discharged into the environment (Bazin et al. 2012).

Stringent environmental rules and regulation have grown up these days and awaken the textile industries to treat their effluents efficiently before discharge. For textile industries, aside from the degradation of dyes or its waste, the colour removal from water is also essential. The presence of colour in water even in low concentration will profoundly affect the aquatic ecosystem and bring aesthetically unpleasing view (Yagub et al. 2014). Different treatment techniques have been used to treat dye from wastewater such as coagulation/flocculation, adsorption, ion exchange, chemical reduction, advanced chemical oxidation, incineration, electrochemical treatment, membrane filtration, nanofiltration, biological degradation (aerobic, anaerobic) and ultrasonic mineralization. But among them, coagulation/flocculation gained more popularity because of its low cost, relatively simple operation and potential decolourization of the wastewater without degradation and by-product formation of dyes (Chen et al. 2010; Freitas et al. 2015). Coagulation/flocculation also favours the removal of colour from the dye bath but not by partial decomposition of dyes that results into formation of potentially more toxic and harmful aromatic compounds (Golob et al. 2005).

However, selection of appropriate coagulant needs a detailed investigation, including wastewater characteristics, the quantum of pollutant load and operational conditions during the coagulation process. A comprehensive list of different coagulant's performance for various dye molecules and the factors affecting the performance of selected coagulants has been described in the next sections.

13.6 Classified Coagulants and Its Application in Dye Wastewater (Current Trends)

Different chemicals have been used to fasten the natural destabilization process of colloidal particles along with some dissolved organic matter for accomplishing the process of coagulation/flocculation. Earlier, the most commonly practised method of coagulation was the addition of an adequate amount of electrolytes that sufficiently compresses the electrical double layer resulting in the settling of the flocs (Peavy et al. 1985). Later it has been noticed that the salts having trivalent ions of opposite charges are found more effective than divalent metallic salts. This significant relation between ionic charge and precipitating capacity of the metallic salts has been first established by Schulze and verified by Hardy. Thus, on their name, this finding is usually named as Schulze-Hardy rule. This rule states that "the coagulation of colloidal particles is effected by that ion of electrolytes that has a charge opposite in sign to that of the colloidal particles and the effect of such ion increases remarkably with the number of charges over these ions" (Jarvis et al. 2012; Peavy et al. 1985). Concomitantly, trivalent metallic salts such as alum, iron-based, titanium tetrachloride and zirconium oxychloride have been used for coagulation/flocculation process for removal of different organic compound such as natural organic matter (NOM), dyes, etc. (Jarvis et al. 2012; Hussain et al. 2014; Sonal et al. 2018).

Accordingly, in dye wastewater, these coagulants have been extensively used to remove the inorganic and organic components. In dye wastewater, various chemical and natural coagulants have been used to remove the targeted dye molecules, which are further classified in different groups, as shown in Fig. 13.4.

Some conventional alum and iron-based coagulants were practised widely, since a long time back (Gao et al. 2007a, b). But as the industries grow, new dyes are introduced, so these days some coagulant aid or synthetic/natural polymers have also

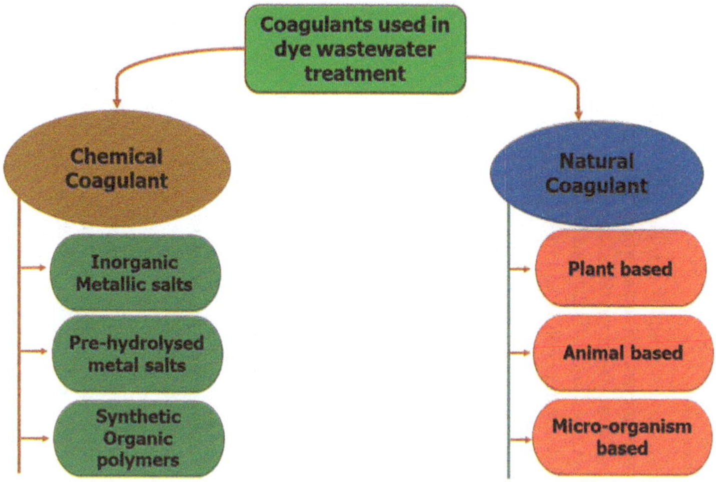

Fig. 13.4 Coagulants used in dye wastewater. (Modified from Verma et al. 2011)

been used along with different inorganic and organic coagulants, to treat the dye wastewater (Xue et al. 2018; Sanghi et al. 2006). Table 13.2 summarizes the effectiveness of different coagulants/coagulant aid for the treatment of dye wastewater. The chemistry of coagulants, when hydrolysed in solution, is quite multifaced. When aluminium or iron-based coagulants are added to the water, the pH is depressed due to the release of hydrogen ions, and they form aqua-metal complexes (Jarvis et al. 2012). The coagulation reactions comprise of an interaction between the hydrolysis product of metal salt and colloidal particles. Initially, metal salts come in contact with the solution; hydrolysis occurs followed by a series of hydrolytic reactions, where hydroxide ions from water molecules replace water molecules bound to the metals (Jarvis et al. 2012). This leads to a reduction in net charge that finally leads to the formation of monomeric then polymeric species that are responsible for the destabilization of colloidal particles. Likewise, other metals such as titanium tetrachloride and zirconium oxychloride also hydrolysed and formed polymeric species causing settling of colloidal particles and dye (Priya et al. 2017; Sonal et al. 2018). Thus, metallic coagulants are extensively used in the treatment plants based on their availability and performance.

Although the iron- and aluminium-based coagulants have been used commonly, some disadvantages urge the demand of new coagulant. Recent epidemiological, neuropathological and biochemical studies suggest that there is a possible link between the pathogenesis of Alzheimer's disease and neurotoxicity of aluminium (Banks et al. 2006; Polizzi et al. 2002). Also, iron-based coagulants are costly, and the residual iron may cause unpleasant taste, odour, colour, corrosion, staining or foaming (Li et al. 2008).

As an alternative, synthetic/natural organic polymers (such as starch, chitosan, polyethyleneimine, polyacrylamide, etc.) or pre-hydrolysed salts [such as

Table 13.2 Effectiveness of different coagulant studies for various dye molecules

S. no.	Coagulant and coagulant aid (if any) used in the coagulation/flocculation (CF) process	Coagulant type	Dye removed	Remarks	References
1.	*M. oleifera* Lam seeds extracted in NaCl and KCl sol. (1 Mol/L) and aluminium sulphate	Natural + chemical metallic coagulant	Textile dyes containing two reactive dyes	Removal was highly pH-dependent, most preferably occurs in acidic medium. *Moringa* extracted from KCl performs (82.2% and 83.05% for apparent colour and COD removal, resp.) well as compared with other two used coagulants	Dotto (2019)
2.	Polyaluminium ferric chloride-starch graft copolymer with acrylamide and dimethyl diallyl ammonium chloride	Synthetic polymer	Textile dyes and reactive dyes	Dose requirement is less as compared to a conventional coagulant	Zhou et al. (2019)
3.	Aluminium formate (AF) as a coagulant and polyamidine (PA) as coagulant aid	Chemical metallic coagulant + polymer	Reactive blue (K-GL)	In dye wastewater treatment, AF/PA showed better coagulation efficiency than PAC/PA	Xue et al. (2018)
4.	Zirconium oxychloride	Chemical metallic coagulant	Reactive blue 19	The dye removal efficiency was higher and the toxicity of the coagulant is negligible	Sonal et al. (2018)
5.	Three species (Al_a, Al_b and Al_c) based on a series of a hydrolysis reaction (as coagulant) and chitosan (CS) as a coagulant aid	Chemical metallic coagulant + animal-based natural coagulant	Disperse yellow S	Efficiency is significantly improved and larger flocs are generated with the addition of CS	Wang et al. (2017)
6.	Polysilicate-magnesium (PSM) modified with different acids	Synthetic inorganic coagulant	Congo red	The compact gel network structure and layer shape structure are more helpful to coagulate colloidal particles and form bridge aggregation among flocs. Thus, they exhibit better coagulation performance	Wei et al. (2017)

7.	Magnesium hydroxide as a coagulant and kaolin clay as coagulant aid	Chemical metallic coagulant	Reactive red and reactive yellow	The removal efficiency of dye tends to increase with the increase in initial turbidity of clay	Li et al. (2016)
8.	Aluminium sulphate (AS) and polyaluminium chloride (PAC) as coagulant and compound bioflocculant (CBF) as coagulant aid	Chemical metallic coagulant + pre-hydrolysed salt	Disperse yellow	The colour removal of AS and PAC was improved with the aid of CBF.	Huang et al. (2015)
				CBF also broadened the effective pH range	
9.	Ferric chloride as a coagulant and okra (*A. esculentus*) mucilage as coagulant aid	Chemical metallic coagulant + plant-based natural coagulant	Textile wastewater	A low dose of coagulant (88 mg/L Fe^{3+}) and coagulant aid (3.2 mg/L) is required to achieve 93.6% of colour removal, 97.2% of turbidity and 85.7% of COD at pH 6	Freitas et al. (2015)
10.	*Moringa oleifera* seed cake powder (MOSCP)	Plant-based natural coagulant	Congo red (CR)	Coagulation experiments indicated that MOSCP efficiently removed CR in a short time	Tie et al. (2015)
11.	*Ocimum basilicum L.* (basil) seed mucilage	Plant-based natural coagulant	Congo red	At dose 1.6 mg/L, 68.5% and 61.6% of dye and COD, removal was obtained at 50 mg/L dye conc	Shamsnejati (2015)
12.	*Plantago major L.* seed extract	Natural coagulant	Neutral red (basic dye)	At 297.6 mg/L of coagulant dose, at 49.6 min and pH 6.5, up to 92% dye removal and 82% COD reduction were achieved.	Chaibakhsh et al. (2014)
				But process was highly pH-dependent	
13.	An organic polymer flocculant (LA) and polyferric chloride (PFC) synthesized from papermaking sludge and spent pickling liquor, respectively	Organic polymer + pre-hydrolysed salt	Disperse yellow (DY) and reactive blue (RB)	Have excellent colour removal around 90% for the two dyes at optimal dosages and larger floc formation takes place	Rong et al. (2014)

(continued)

Table 13.2 (continued)

S. no.	Coagulant and coagulant aid (if any) used in the coagulation/flocculation (CF) process	Coagulant type	Dye removed	Remarks	References
14.	Aluminium sulphate (AS) as coagulant and sodium alginate (SA) as coagulant aid	Chemical metallic coagulant	Disperse yellow (DY)	AS plus SA exhibited a synergic effect on dye removal. Colour removal was more by SA at low AS doses than at higher ones	Wu et al. (2012)
15.	A lignin-based cationic polyelectrolyte (L-DAF)	Synthetic polymer	Acid black, reactive red and direct red (anionic azo dyes)	More than 95% dye removal was achieved in each case at different optimum dose. But longer settling time and lower pH increase the efficiency. COD removal was more than 89% with 5.4% less sludge generation	Fang et al. (2010)
16.	Steel industrial wastewater (SIWW)-rich $FeCl_3$	Chemical metallic coagulant	Disperse blue 79	SIWW is a very effective coagulant for disperse blue 79 removal	Anouzla et al. (2009)
17.	A tannin-based coagulant called ACQUAPOL C-1 and protein extract derived from *M. oleifera* seed as a coagulant	Plant-based natural coagulant	Alizarin violet 3R	Approx. 80% and 95% dye removal have been observed by tannin-based coagulant and seed extract, respectively. Dye conc. and pH have inverse impact on removal efficiency	Beltrán-Heredia et al. (2009)
18.	Chitosan	Animal-based coagulant	Acid black 1 (AB1), acid violet 5 (AV5) and reactive black 5 (RB5)	Settling time of AB1 is more than AV 5 and RB 5 as smaller flocs formed. High pH variations occur	Szyguła et al. (2008)
19.	Poly(dimethylaminoethyl methaacrylate)	Synthetic polymer	Congo red and direct blue 1	Congo red removal is 90–91% and direct blue 1 is 88–92% but at 24 h of sedimentation time	Dragan and Dinu (2008)
20.	Magnesium chloride	Chemical metallic coagulant	Reactive and disperse dye	Magnesium chloride was shown to be superior to $Ca(OH)_2$, $Al_2(SO_4)_3$, PAC and $FeSO_4/Ca(OH)_2$ for removing colour	Gao et al. (2007b)

21.	Polyaluminium chloride as a coagulant and polyacrylamide seed gum, as a coagulant aid	Pre-hydrolysed salts + natural polymer	Reactive, acid and direct	Removal efficiency is improved	Sanghi et al. (2006)
22.	Alum as coagulant and commercial cationic flocculant as coagulant aid	Chemical metallic coagulant + synthetic polymer	Reactive black 5 and acid dye	Combination yield effective for residual dye bath wastewaters.	Golob et al. (2005)
				TOC, COD, AOX, BOD and anionic surfactants were reduced and biodegradability increases	
23.	Ferric chloride	Chemical metallic coagulant	Reactive and disperse dye	–	Kim et al. (2004)
24.	Ferrous sulphate as coagulant and lime and cationic polymer as coagulant aid	Chemical metallic coagulant + polymer	Reactive dye	–	Georgiou et al. (2003)
25.	Cyanoguanidine, polyamine and polydiallyldimethylammonium chloride (polyDADMAC)	Inorganic salt + an organic polymer	Levafix brilliant red	For reactive dye coagulation, only cyanoguanidine polymer was appropriate but has an intense pungent smell and is not economical at large scale	Carvalho et al. (2002)
			ERN		

polyaluminium chloride (PAC) or polyferric sulphate (PFS)] have also been assessed as a coagulant aid to assist the conventional coagulants and to overcome their residual toxicity and dose requirement (Huang et al. 2015; Rong et al. 2014). The pre-hydrolysed salts assist the traditional additives as they get hydrolysed during the coagulant preparation, thus resulting into more control experimental conditions and very less impact on pH of the treated water (Renault et al. 2009). They also help in the formation of larger flocs as compared to alum (Huang et al. 2015). Despite this, their mode of action of floc formation, mechanism of targeting dye molecules, health and environmental safety of effluent are still lacking (Jeon et al. 2009). Unlike pre-hydrolysed salts, organic polymers are categorized as natural and synthetic polymers, having a large number of similar chemical units bonded together by covalent bonds. Natural polymers have advantages of non-toxicity and biodegradability but are not that effective as synthetic polymers (Zahrim et al. 2011). Also, their biodegradability influences their long storage life hence confined their applications (Dragan and Dinu 2008).

In contrast, synthetic polymers own more controlling specific properties of molecular weight and charge functionalities (Bratby 2006). Cationic polymers are found to be more effective for dye removal than anionic and non-ionic polymers (Salamone 1998). Despite this, these polymers are not found more effective in the removal of more diverse class of dyes, thus limiting their use (Table 13.2). Also, it was found in a study that the non-reactive monomers, such as acrylamide of synthetic polymers, are toxic and have carcinogenic potentials (Bratby 2006). Surmounting toxicity of chemical coagulants, natural coagulants (including plant- and animal-based coagulant) such as *Moringa* seeds (Tie et al. 2015), chitosan (Szyguła et al. 2008), okra mucilage (Freitas et al. 2015) and several other naturally derived polymers were also used for dye wastewater treatment, because of their advantages of no toxicity and bioavailability, over other chemical coagulants or polymers. Nevertheless, their use in practical treatment technology is still restricted because of short self-storage life (Dragan and Dinu 2008) and limited working pH range (Renault et al. 2009). Also, the extraction process of these natural coagulants varies because of their properties and thereby affects the coagulation/flocculation process (Renault et al. 2009; Guibal 2004) and rebuts their uses in industries. Moreover, one microorganism derived-coagulant named xanthan gum has been also observed. It is a high molecular weight polysaccharide derived from *Xanthomonas campestris* bacterial coat. But its use in the textile industry's wastewater has not been reported yet in the literature (Davidson 1980; Cohan 2010), might be because of its complex structure and higher molecular weight as compared to guar gum (Verma et al. 2011).

The current trends for the dye wastewater treatment clearly remark about the subsequent advancement in the use of metallic, inorganic and natural coagulants in the treatment process. It has been observed that initially alum and iron-based coagulants were extensively used, but as the dye developed, new coagulants have been introduced to overcome the limitations of the existing one. Different coagulant combinations and various coagulant aids have also been used for amelioration of the dye treatment process. Nevertheless, selection of these combinations is still an intricate task as many factors are responsible, which is discussed in the next section.

13.7 Selection Criteria for Coagulants

Selection of any specific coagulant used for the treatment of any target pollutants solely depends upon the characteristics of the coagulants, their hydrolysing properties, working pH range and several other factors. Thus, the factors responsible for coagulant selection are as follows.

13.7.1 Properties of the Coagulants

These properties include the hydrolysing characteristics of the coagulant and their flocs formation capacity in specific working pH. For example, alum, iron-based coagulant and zirconium oxychloride were more preferably used in coagulation/flocculation process over divalent metallic salts as they follow the Schulze-Hardy rule that says the higher the valence state of metal-based counter ions, higher will be the hydrolysis products, and more efficient will be destabilization of colloidal particles (Jarvis et al. 2012; Sonal et al. 2018).

13.7.2 Concentration and Type of Colloidal Particles

The presence of hydrophilic colloidal particles in water is found to be more challenging to precipitate as compared to that of hydrophobic particles. Hydrophilic particles form hydrogen bonding with the water molecules resulting in more dose requirement. The presence of turbidity sometimes plays a decisive role in the coagulation of colloidal particles (Peavy et al. 1985). Sufficient amount of colloidal particles promotes entrapment of particles in the precipitate during the coagulation process.

13.7.3 Quality of the Polluted Water

Pollutant quality also decides the coagulant and dose of the coagulant. The effluent characteristics such as pH, alkalinity, turbidity, presence of organic matters, etc. play an essential role in coagulant selection. For example, effective pH for alum is 6.5 to 8.5 only; beyond this pH, there must be an addition of external alkalis salts, thus rendering the overall cost of treatment. Similarly, the effective pH for ferric sulphate is in the range of 4 to 7 and above 9, whereas for ferric chloride, effective pH range is 3.5 to 6.5 and above 8.5 (Garg 2013).

13.7.4 Water Temperature

Higher temperature may lead to more collision of colloidal particle resulting in overcoming of electrostatic repulsive force by van der Waals force of attraction. This attractive force makes the colloidal particles to aggregate and settle down more quickly.

13.7.5 Cost and Availability

For the dye industry, treatment of effluent is a challenging task if stringent rules and cost of treatment have been considered in combination. Any coagulant can be used commercially only if that is readily available at low cost and efficient for target pollutants (Verma et al. 2011).

13.7.6 Dewatering Characteristics of the Solids that Are Produced

The flocs formed and settled down to form solid sludge. The treatment of sludge also requires huge economy and energy. So the dewatering capacity of sludge also impacts the use of any efficient coagulant.

The selection of coagulants and coagulant aids plays an important role in curbing the dye pollution load. Sometimes the use of these coagulants gets affected by various factors that govern their use in treating dyes and also makes them more specific for the particular pollutant (Fig. 13.5).

All these aforementioned parameters highly influence the selection of the coagulant for commencement of coagulation/flocculation. Along with the types of coagulants, other factors also affect the coagulation/flocculation process that are explained in detail by the following section.

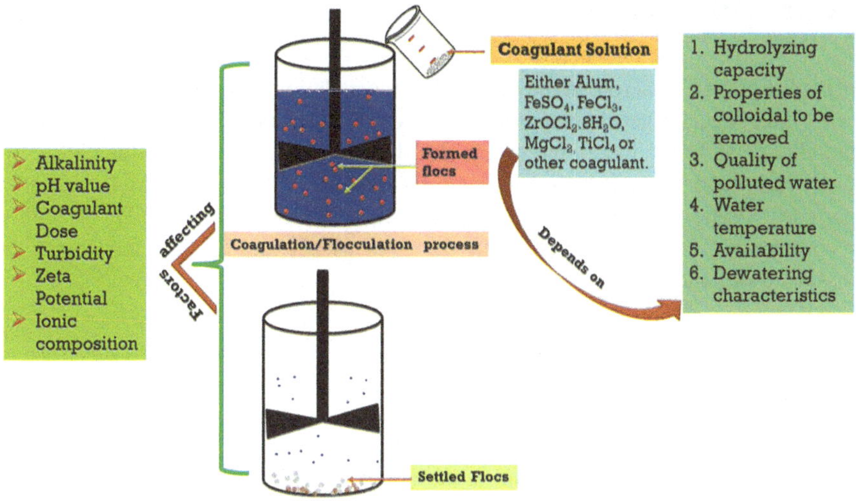

Fig. 13.5 Schematic representation for understanding the factors governing the coagulant selection and coagulation process

13.8 Factors Affecting Coagulation/Flocculation Practices in Dye Wastewater

The following discussion covers about the factors that affect the process of coagulation/flocculation in dye wastewater.

13.8.1 Alkalinity

Alkalinity refers to the acid neutralization capacity of water. In other words, it indicates the water buffering capacity. For agglomeration of colloidal particles, the alkalinity must be present in the medium. When the metallic salts get hydrolysed in water, the released hydrogen ions neutralize the alkalinity. If there is low alkalinity initially, further reduction in alkalinity will destroy its buffering capacity and leads to dropping down of pH (Wang et al. 2005). The latter may influence the floc formation as for specific coagulant optimum pH is required. Low alkalinity waters must be artificially buffered by addition of lime or soda ash (Sawyer et al. 2003).

13.8.2 pH Value

pH value plays a vital role in the coagulation/flocculation process as, at optimum pH only, the coagulant gets hydrolysed and the coagulation takes place. pH also affects the charge of NOM functional group and dissolved organic matter, the charge of dissolved-phase coagulant species and the surface charge of colloids and floc particles (Letterman 1999). In dye wastewater, the optimum pH is highly recommended for decolourization of the dyes as pH may change the charge of dissolved organics.

13.8.3 Coagulant Dose

The amount of coagulant addition is an essential factor, and the effectiveness of colloidal destabilization is highly correlated. An optimum dose is required for specific water chemistry. A lower dose of coagulant may cause insufficient destabilization of colloidal particles leading to cloudy sample with no flocs, while an overdose of coagulant results into re-stabilization of colloidal particles as the concentration of counter ions increases (Melia 1990; Elimelech and Melia 1990). Thus, an optimum dose of the coagulant is required to produce good flocs, and this can be determined by using the jar test apparatus.

13.8.4 Turbidity

The higher concentration of colloidal particles provides an ample chance for contact and building of good flocs and vice versa will occur if the colloidal concentration is low. For both cases, an optimum dose, pH and coagulant aid should be determined (Peavy et al. 1985).

13.8.5 Zeta Potential

The zeta potential measures the net charge of colloidal particles and is used to assess the stability of the suspension. The more negative the charge, the higher the zeta potential. As zeta potential increases, the repulsive forces between colloidal particles increase, and the colloidal suspension becomes more stable (Wang et al. 2005). While as the zeta potential approaches zero, the charge on the surface becomes so low that attractive forces overcome the repulsive force and destabilization take places. The magnitude of the zeta potential is measured by electrophoretic measurement of particles mobility in an electric field.

13.8.6 The Affinity of Colloids for Water

In turbid water usually, a mixture of hydrophilic-hydrophobic colloidal particles is present. Hydrophilic (water-loving) colloids seem very stable as their hydration shell prevents chemicals to readily replace the sorbed water molecules, thus inhibiting their destabilization from the suspension. The stability of hydrophilic colloidal particles mainly depends on their strong interaction with the water molecules than on their electrostatic charge (Hammer 1986). Dye molecules primarily are reactive, and soluble dyes have more affinity with the water molecules, thus rendering difficulties in the destabilization of the molecules (Patel and Vashi 2015). To settle down such particles, usually 10–20 times more coagulant dose has been required. Contrarily, hydrophobic (e.g. metal oxides) particles can be easily destabilized and coagulated.

13.8.7 Temperature

Lower temperature adversely affects the process of coagulation by altering the coagulant solubility. The lower temperature also increases the water viscosity, thereby hindering the kinetics of hydrolysis reactions and particle flocculation (Morris and Knocke 1984) and, thus, affects the sedimentation process. In contrast, higher temperature increases the rate and effectiveness of the process, (i) by increasing the kinetic energy of the molecules and rate of chemical reactions and (ii) by decreasing viscosity of water and altering the structure of the flocs, resulting in larger agglomerations.

13.8.8 Ionic Concentration and Composition (Cations and Anions in Solution)

The presence of higher ionic strength in the solution will enhance the destabilization behaviour of the colloidal particles. The presence of anions such as sulphate, silicate and phosphate suppressed the occurrence of charge reversal and re-stabilization of colloids (Boisvert et al. 1997). And the presence of divalent cations such as Ca^{2+} and Mg^{2+} also plays beneficial effects on coagulation of negatively charged particles by compressing the colloidal double layer and minimizing the repulsive potential (Black et al. 1965).

13.9 Merit and Demerits of Coagulation/Flocculation (C/F) in Dye Wastewater Treatment

The significant popularity of coagulation/flocculation process in dye wastewater treatment is because of its simple design and operational procedure. Apart from this, the process has the following advantages that make it an extensively used method for dye wastewater treatment.

13.9.1 Merits of the Coagulation/Flocculation Process

1. Provoke the destabilization of colloidal and dissolved particles. The coagulation process with the help of coagulant initiates the destabilization of colloidal particles which are difficult to settle under the effect of gravity (Baker 1948). Thus, this process fastens the natural settling of colloidal particles using coagulant/flocculants.
2. Reduces the turbidity of the water. The coagulation/flocculation process comprises of primary treatment that eliminates 90% of settleable particles, along with other dissolved and suspended particles (Peavy et al. 1985). As these particles sink down, the turbidity and dissolved organic compounds get significantly reduced.
3. Reduces the oxygen demand of water. The process substantially reduces the colloidal particles and dissolved organic compounds, thereby significantly curtailing the BOD and some COD of the wastewater.
4. Enhances the functioning of secondary and tertiary treatment units. This process carries a significant amount of organic and inorganic pollutants along with the settled colloidal particles and hence reduces the over-functioning of other secondary treatment by reducing maximum pollutants initially.
5. Curtail the total treatment cost. The coagulation/flocculation process is the cost-effective, simple and potential treatment process for decolourization of dyes (Chen et al. 2010). Hence this process reduces the total cost of treatment as compared to other advanced treatment methods used alone.
6. Prevent the formation of harmful by-products. Other advance oxidation processes such as photocatalytic oxidation, ozonation, etc. are used for the treatment of dye wastewater from toxic by-products that seem to be potential carcinogens. But the coagulation/flocculation process has advantage over them, as this process removes the colour but not by partial decomposition (Golob et al. 2005).

The efficiency of the primary treatment solely depends on the performance of other secondary treatment methods. Being the part of primary treatment, the coagulation/flocculation process, besides having some advantages, faces some flaws also which are described as follows.

13.9.2 Demerits of the Coagulation/Flocculation Process

1. Optimization of dose. For efficient C/F process, the dose of coagulant should be optimized regularly. A lower dose of coagulant may cause insufficient destabilization, while an overdose of coagulant results into re-stabilization of colloidal particles (Melia 1990; Elimelech and Melia 1990).
2. Regular monitoring on pH and metal solubility. The theoretical minimum solubility of different metals occurs at different pH. The solubility of metals must be controlled in a mixture of metals. Also, pH may vary with the variation in dose and characteristics of effluent. Thus, both of the parameters must be checked regularly on fixed intervals.
3. Massive sludge generation. Many coagulants such as alum, ferric chloride, etc. result in the generation of a tremendous amount of sludge that further requires another treatment before disposal. Also, the dewatering of sludge and its digestion require additional space and cost.
4. Cost and availability of coagulants. The cost of the coagulant and its easy availability are still a limiting factor on total treatment cost (Verma et al. 2011).
5. Residual impacts. If sulphide is present in the coagulant, scaling and corrosiveness will occur to the pipelines or tank accessories. Also, the concentration of alum in the effluent or sludge will cause Alzheimer's or other neurological diseases to human beings (Banks et al. 2006, Polizzi et al. 2002). Similarly, iron-based coagulant requires more skill as they are corrosive and deliquescent (Li et al. 2008).

13.10 Compatibility of the Coagulation/Flocculation Process with Other Treatment Options

To overcome the limitation of the coagulation/flocculation process for the treatment of dye wastewater, a number of researchers have investigated the combination of two different processes to reduce the demerits of the two processes and avail their advantages. The dye released from textile industries is recalcitrant and resistant to degradation; therefore, a single treatment process is not much reliable and economical to meet the regulatory discharge limits of the industries. Nowadays, combined or hybrid system (Fig. 13.6) has been developed to efficiently curb the environmental pollutants such as heavy metals or dyes (Su et al. 2016; Man et al. 2012; Lin and Lin 1993).

The combined treatment process includes sequential treatment method in which one process ends and another process starts simultaneously. In this process, the sequence of the treatment process is the same as written. Contrary to this, a hybrid method is the process of fusion of two or more treatment process into a single operation. Consequently, researchers have reported the performance and advantages of the combined or hybrid treatment process for dye wastewater, as summarized in Table 13.3.

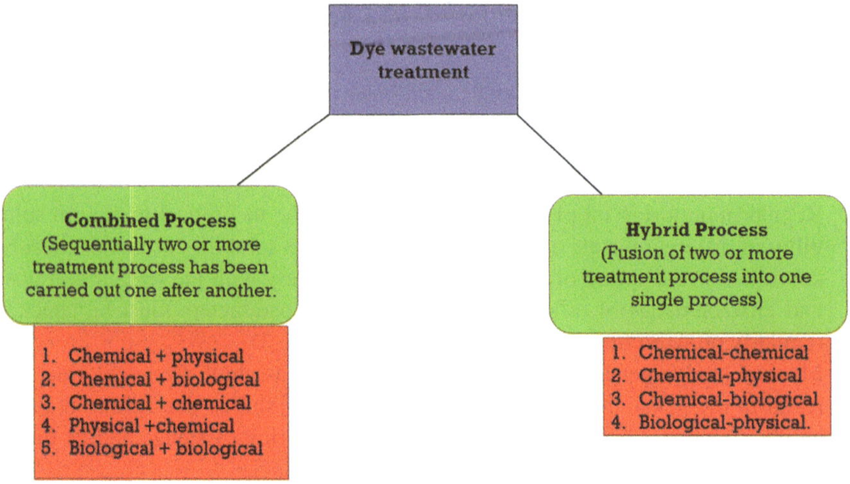

Fig. 13.6 Schematic representation of the combined treatment process along with coagulation/flocculation process

Table 13.3 Advantages of the combined treatment process to treat dye wastewater

Combined process	Advantages/remarks	References
Coagulation/flocculation and adsorption	Low sludge generation and low coagulant dose requirement	Papić et al. (2004)
Thermolysis and coagulation/flocculation	Able to decompose dyes and less sludge generation occurs	Kumar et al. (2008) and Verma et al. (2011)
Coagulation/flocculation and sequential batch reactor	Complete reduction of COD and BOD after the combined process	Lu et al. (2009)
Adsorption and coagulation/flocculation	Efficiently treat non-biodegradable dyes	Yang et al. (2007)
Coagulation/flocculation and nanofiltration	Less sludge generation, a low dose of coagulant and flocculant required, applicable to both anionic and cationic dyes. The NF treatment can altogether remove the intense colour left in CF treated dye solutions	Liang et al. (2014)
	Membrane fouling is abated and NF permeate flux is increased by applying the CF process as a pretreatment	
Coagulation/flocculation and Photocatalytic oxidation	Efficient treatment of textile dyes	Jorfi et al. (2016)
	Less sludge generation	
	Minimization of cost of the photocatalyst	
Forward osmosis (FO) and coagulation/flocculation	High water flux and recovery rate, well-controlled membrane fouling, high efficiency and minimal environmental impact	Han et al. (2016)
Coagulation/flocculation and ultrafiltration (UF)	Improve permeate fluxes and keep fouling of membrane at low levels	Beluci et al. (2019)

It has been suggested that for dye wastewater, single treatment restricts the effluent to meet the permissible regulatory limits. Thus, the combined treatment process overcomes the limitations of the advanced treatment process or secondary treatment, along with conventional coagulation/flocculation treatment.

13.11 Conclusion and Future Aspects

The stubborn and stable nature of organic dyes creates a problem during the treatment process, particularly when present in the form of complex matrix. The process of coagulation/flocculation plays a significant role in reducing the colour component of dye wastewater along with other pollutants. This process of treatment has been widely explored since years back, and subsequently, new technologies also came into existence. But the coagulation process is favoured because it does not form any toxic by-products after removal of dyes. However, the efficient working of this process depends on various factors that govern its functioning. The coagulant used in this process is highly selective concerning the targeting pollutant. Besides this, the process has some flaws also such as massive sludge generation, residual effects, etc., but these can be overcome by exploring new coagulants/coagulant aid combinations. Altogether, the process is still an efficient primary treatment method for the treatment of massively loaded dyeing unit effluent discharge. Further, the limitations of the process can be cured by fulfilling the following gaps in future.

The following points should be considered for coagulation/flocculation process in dye wastewater treatment:

1. As the textile industries grow, new emerging dyes and chemicals are introduced in the effluent discharge. So, the development of new efficient coagulants must be considered that will altogether overcome the availability, cost and toxicity problem.
2. Combined or hybrid technologies should be developed on a pilot-scale basis for better understanding and their practical use.
3. Reuse of coagulants should be focussed more.

Acknowledgement The authors are grateful to the Department of Environmental Science and Engineering, Indian Institute of Technology (ISM), Dhanbad, India, for providing technical and financial support.

References

Amuda OS, Amoo IA (2007) Coagulation/flocculation process and sludge conditioning in beverage industrial wastewater treatment. J Hazard Mater 141(3):778–783

Anouzla A, Abrouki Y, Souabi S, Safi M, Rhbal H (2009) Colour and COD removal of disperse dye solution by a novel coagulant: application of statistical design for the optimization and regression analysis. J Hazard Mater 166(2–3):1302–1306

Asano T, Burton FL, Leverenz HL, Tsuchihashi R, Tchobanoglous G (2006) Water reuse: issues, technologies, and applications. Metcalf and Eddy, Wakefield
Baker NM (1948) The quest for pure water. AWWA
Banks WA, Niehoff ML, Drago D, Zatta P (2006) Aluminum complexing enhances amyloid β protein penetration of blood–brain barrier. Brain Res 1116(1):215–221
Bazin I, Hassine AIH, Hamouda YH, Mnif W, Bartegi A, Lopez-Ferber M, Gonzalez C (2012) Estrogenic and anti-estrogenic activity of 23 commercial textile dyes. Ecotoxicol Environ Saf 85:131–136
Beltrán-Heredia J, Sánchez-Martín J, Delgado-Regalado A, Jurado-Bustos C (2009) Removal of Alizarin Violet 3R (anthraquinonic dye) from aqueous solutions by natural coagulants. J Hazard Mater 170(1):43–50
Beluci NDCL, Mateus GAP, Miyashiro CS, Homem NC, Gomes RG, Fagundes-Klen MR, Vieira AMS (2019) Hybrid treatment of coagulation/flocculation process followed by ultrafiltration in TIO_2-modified membranes to improve the removal of reactive black 5 dye. Sci Total Environ 664:222–229
Black AP, Birkner FB, Morgan JJ (1965) Destabilization of dilute clay suspensions with labeled polymers. J Am Water Works Assoc 57(12):1547–1560
Boisvert JP, To T C, Berrak A, Jolicoeur C (1997) Phosphate adsorption in flocculation processes of aluminium sulphate and poly-aluminium-silicate-sulphate. Water Res 31(8):1939–1946
Bratby J (2006) Coagulants. Coagulation and flocculation in water and wastewater treatment, 2nd edn. IWA Publishing, London, pp 50–68
Carmen Z, Daniela S (2012) Textile organic dyes–characteristics, polluting effects and separation/elimination procedures from industrial effluents – a critical overview. In: Organic pollutants ten years after the Stockholm convention-environmental and analytical update, Intech Open
Carvalho G, Delée W, Novais JM, Pinheiro HM (2002) A factorially-designed study of physico-chemical reactive dye colour removal from simulated cotton textile processing wastewaters. Coloration Technol 118(5):215–219
Chaibakhsh N, Ahmadi N, Zanjanchi MA (2014) Use of *Plantago major* L. as a natural coagulant for optimized decolorization of dye-containing wastewater. Ind Crops Prod 61:169–175
Chen T, Gao B, Yue Q (2010) Effect of dosing method and pH on color removal performance and floc aggregation of polyferric chloride–polyamine dual-coagulant in synthetic dyeing wastewater treatment. Colloid Surf A 355(1–3):121–129
Cohan W (2010) Could xanthan gum sensitivity be complicating your celiac disease recovery? Celiac.com. Retrieved 05–19
Crittenden JC, Trussell RR, Hand DW, Tchobanoglous G, Howe K (2012) Water treatment principles and design, 3rd edn. Wiley, New York
Davidson RL (1980) Handbook of water-soluble gums and resins. McGraw Hill
Ding S, Li Z, Wang R (2010) Overview of dyeing wastewater treatment technology. Water Resour Prot 26:73–78
Dotto J, Fagundes-Klen MR, Veit MT, Palácio SM, Bergamasco R (2019) Performance of different coagulants in the coagulation/flocculation process of textile wastewater. J Clean Prod 208:656–665
Dragan ES, Dinu IA (2008) Removal of azo dyes from aqueous solution by coagulation/flocculation with strong polycations. Res J Chem Environ 12(3):5
Duan J, Gregory J (2003) Coagulation by hydrolysing metal salts. Adv Colloid Int Sci 100:475–502
Elimelech M, O'Melia CR (1990) Kinetics of deposition of colloidal particles in porous media. Env Sci Tech 24:15–28
Fang R, Cheng X, Xu X (2010) Synthesis of lignin-base cationic flocculant and its application in removing anionic azo-dyes from simulated wastewater. Bioresour Technol 101(19):7323–7329
Fewson CA (1988) Biodegradation of xenobiotic and other persistent compounds: the causes of recalcitrance. Trends Biotecnol 6(7):148–153

Fontenot EJ, Lee YH, Matthews RD, Zhu G, Pavlostathi SG (2003) Reductive decolorization of a textile reactive dyebath under methanogenic conditions. Appl Biochem Biotechnol 109 (1–3):207–225

Freitas TKFS, Oliveira VM, De Souza MTF, Geraldino HCL, Almeida VC, Fávaro SL, Garcia JC (2015) Optimization of coagulation-flocculation process for treatment of industrial textile wastewater using okra (*A. esculentus*) mucilage as natural coagulant. Ind Crop Prod 76:538–544

Fu F, Wang Q (2011) Removal of heavy metal ions from wastewaters: a review. J Environ Manag 92(3):407–418

Gao BY, Wang Y, Yue QY, Wei JC, Li Q (2007a) Color removal from simulated dye water and actual textile wastewater using a composite coagulant prepared by polyferric chloride and poly dimethyl diallyl ammonium chloride. Sep Purif Technol 54(2):157–163

Gao BY, Yue QY, Wang Y, Zhou WZ (2007b) Color removal from dye-containing wastewater by magnesium chloride. J Environ Manag 82(2):167–172

Garg SK (2013) Water supply engineering, vol 1. Khanna Publishers

Georgiou D, Aivazidis A, Hatiras J, Gimouhopoulos K (2003) Treatment of cotton textile wastewater using lime and ferrous sulfate. Water Res 37(9):2248–2250

Ghernaout D, Al-Ghonamy AI, Boucherit A, Ghernaout B, Naceur MW, Messaoudene NA, Elboughdiri NA (2015) Brownian motion and coagulation process. Am J Environ Prot 4 (5–1):1–15

Golob V, Vinder A, Simonič M (2005) Efficiency of the coagulation/flocculation method for the treatment of dyebath effluents. Dyes Pig 67(2):93–97

Guibal E (2004) Interactions of metal ions with chitosan-based sorbents: a review. Sep Purif Technol 38(1):43–74

Hammer MJ (1986) Water and wastewater technology. Wiley, New York

Han G, Liang CZ, Chung TS, Weber M, Staudt C, Maletzk C (2016) Combination of forward osmosis (Water Res FO) process with coagulation/flocculation (CF) for potential treatment of textile wastewater. 91:361–370

Holkar CR, Jadhav AJ, Pinjari DV, Mahamuni NM, Pandit AB (2016) A critical review on textile wastewater treatments: possible approaches. J Environ Manag 182:351–366

Huang X, Gao B, Yue Q, Zhang Y, Sun S (2015) Compound bioflocculant used as a coagulation aid in synthetic dye wastewater treatment: the effect of solution pH. Sep Purif Technol 154:108–114

Hussain S, van Leeuwen J, Chow CW, Aryal R, Beecham S, Duan J, Drikas M (2014) Comparison of the coagulation performance of tetravalent titanium and zirconium salts with alum. Chem Eng J 254:635–646

Jarvis P, Sharp E, Pidou M, Molinder R, Parsons SA, Jefferson B (2012) Comparison of coagulation performance and floc properties using a novel zirconium coagulant against traditional ferric and alum coagulants. Water Res 46(13):4179–4187

Jeon JR, Kim EJ, Kim YM, Murugesan K, Kim JH, Chang YS (2009) Use of grape seed and its natural polyphenol extracts as a natural organic coagulant for removal of cationic dyes. Chemosphere 77(8):1090–1098

Jorfi S, Barzegar G, Ahmadi M, Soltani RDC, Takdastan A, Saeedi R, Abtahi M (2016) Enhanced coagulation-photocatalytic treatment of Acid red 73 dye and real textile wastewater using UVA/synthesized MgO nanoparticles. J Environ Manag 177:111–118

Kant R (2012) Textile dyeing industry an environmental hazard. Nat Sci 4:17027. 5 pages

Kim TH, Park C, Yang J, Kim S (2004) Comparison of disperse and reactive dye removals by chemical coagulation and Fenton oxidation. J Hazard Mater 112(1–2):95–103

Kumar P, Prasad B, Mishra IM, Chand S (2008) Decolorization and COD reduction of dyeing wastewater from a cotton textile mill using thermolysis and coagulation. J Hazard Mater 153 (1–2):635–645

Letterman RD (1999) American Water Works Association. Water quality and treatment. McGraw-Hill

Li WW, Zhou WZ, Zhang YZ, Wang J, Zhu XB (2008) Flocculation behavior and mechanism of an exopolysaccharide from the deep-sea psychrophilic bacterium *Pseudoalteromonas* sp. SM9913. Bioresour Technol 99(15):6893–6899

Li H, Liu S, Zhao J, Feng N (2016) Removal of reactive dyes from wastewater assisted with kaolin clay by magnesium hydroxide coagulation process. Colloids Surf A 494:222–227

Liang CZ, Sun SP, Li FY, Ong YK, Chung TS (2014) Treatment of highly concentrated wastewater containing multiple synthetic dyes by a combined process of coagulation/flocculation and nanofiltration. J Membr Sci 469:306–315

Lin SH, Lin CM (1993) Treatment of textile waste effluents by ozonation and chemical coagulation. Water Res 27(12):1743–1748

Lu X, Yang B, Chen J, Sun R (2009) Treatment of wastewater containing azo dye reactive brilliant red X-3B using sequential ozonation and upflow biological aerated filter process. J Hazard Mater 161(1):241–245

Man LW, Kumar P, Teng TT, Wasewar KL (2012) Design of experiments for Malachite Green dye removal from wastewater using thermolysis–coagulation–flocculation. Desalin Water Treat 40 (1–3):260–271

Masters GM, Ela WP (2008) Introduction to environmental engineering and science (No. 60457). Prentice Hall, Englewood Cliffs

Melia CR (1990) Kinetics of colloidal chemical processes in aquatic systems. In: Stumm W (ed) Aquatic chemical kinetics. Wiley, New York

Mihelcic RJ, Zimmerman BJ (2014) Environmental engineering: fundamentals, sustainability, design, 2nd edn. Wiley, New York

Miller SM, Fugate EJ, Craver VO, Smith JA, Zimmerman JB (2008) Toward understanding the efficacy and mechanism of *Opuntia* spp. as a natural coagulant for potential application in water treatment. Environ Sci Technol 42(12):4274–4279

Morris JK, Knocke WR (1984) Temperature effects on the use of metal-ion coagulants for water treatment. J Am Water Works Assoc 76(3):74–79

Nadigera GS (2001) Azo ban, eco-norms and testing. Corpus ID:36522979

Naik DJ, Desai HH, Desai TN (2013) Characterization and treatment of untreated wastewater generated from dyes and dye intermediates manufacturing industries of Sachin industrial area, Gujarat, India. J Environ Res Dev 7(4A):1602. New York

Nharingo T, Moyo M (2016) Application of *Opuntia ficus-indica* in bioremediation of wastewaters. A critical review. J Environ Manag 166:55–72

Papić S, Koprivanac N, Božić AL, Meteš A (2004) Removal of some reactive dyes from synthetic wastewater by combined Al (III) coagulation/carbon adsorption process. Dyes Pig 62 (3):291–298

Patel H, Vashi RT (2015) Characterization and treatment of textile wastewater. Elsevier

Paul SA, Chavan SK, Khambe SD (2012) Studies on characterization of textile industrial waste water in Solapur city. Int J Chem Sci 10(2):635–642

Peavy SH, Rowe RD, Tchobanoglous G (1985) Environmental engineering. Tata McGraw-Hill Education (India) Edition 2013

Polizzi S, Pira E, Ferrara M, Bugiani M, Papaleo A, Albera R, Palmi S (2002) Neurotoxic effects of aluminium among foundry workers and Alzheimer's disease. Neurotoxicology 23(6):761–774

Priya T, Mohanta VL, Mishra BK (2017) Performance evaluation of zirconium oxychloride for reduction of hydrophobic fractions of natural organic matter. Sep Purif Technol 174:104–108

Renault F, Sancey B, Badot PM, Crini G (2009) Chitosan for coagulation/flocculation processes–an eco-friendly approach. Eur Polym J 45(5):1337–1348

Robinson T, McMullan G, Marchant R, Nigam P (2001) Remediation of dyes in textile effluent: a critical review on current treatment technologies with a proposed alternative. Bioresour Technol 77(3):247–255

Rong H, Gao B, Li R, Wang Y, Yue Q, Li Q (2014) Effect of dose methods of a synthetic organic polymer and PFC on floc properties in dyeing wastewater coagulation process. Chem Eng J 243:169–175

Salamone JC (ed) (1998) Concise polymeric materials encyclopedia. CRC Press
Sanghi R, Bhattacharya B, Dixit A, Singh V (2006) *Ipomoea dasysperma* seed gum: an effective natural coagulant for the decolorization of textile dye solutions. J Environ Manag 81(1):36–41
Sawyer NC, McCarty LP, Parkin FG (2003) Chemistry for environmental engineering and science, 5th edn. Tata McGraw-Hill Edition
Shamsnejati S, Chaibakhsh N, Pendashteh AR, Hayeripour S (2015) Mucilaginous seed of *Ocimum basilicum* as a natural coagulant for textile wastewater treatment. Ind Crops Prod 69:40–47
Sonal S, Mishra BK (2019) Optimization of the operational conditions for the treatment of reactive dyes through a statistical tool: response surface methodology. Int J Environ Sci Dev 10(6)
Sonal S, Singh A, Mishra BK (2018) Decolorization of reactive dye Remazol Brilliant Blue R by zirconium oxychloride as a novel coagulant: optimization through response surface methodology. Water Sci Technol 78(2):379–389
Sonal S, Prakash P, Mishra BK, Nayak GC (2020) Synthesis, characterization and sorption studies of a zirconium (IV) impregnated highly functionalized mesoporous activated carbons. RSC Adv 10(23):13783–13798
Su CXH, Low LW, Teng TT, Wong YS (2016) Combination and hybridisation of treatments in dye wastewater treatment: a review. J Environ Chem Eng 4(3):3618–3631
Szyguła A, Guibal E, Ruiz M, Sastre AM (2008) The removal of sulphonated azo-dyes by coagulation with chitosan. Colloids Surf A 330(2–3):219–226
Tchobanoglous G, Burton FL, Stensel HD (2003) Wastewater engineering: treatment and reuse. Metcalf & Eddy Inc. McGraw-Hill, Inc, New York, USA. for Environmental Research Information. Cincinnati, Ohio, 45268, 27
Tie J, Li P, Xu Z, Zhou Y, Li C, Zhang X (2015) Removal of Congo red from aqueous solution using *Moringa oleifera* seed cake as natural coagulant. Desalin Water Treat 54(10):2817–2824
Verma S, Prasad B, Mishra IM (2011) Thermochemical treatment (thermolysis) of petrochemical wastewater: COD removal mechanism and floc formation. Ind Eng Chem Res 50(9):5352–5359
Verma AK, Dash RR, Bhunia P (2012) A review on chemical coagulation/flocculation technologies for removal of colour from textile wastewaters. J Environ Manag 93(1):154–168
Wang LK, Hung YT, Shammas NK (eds) (2005) Physicochemical treatment processes. Humana Press
Wang W, Yue Q, Li R, Song W, Gao B, Shen X (2017) Investigating coagulation behavior of chitosan with different Al species dual-coagulants in dye wastewater treatment. J Taiwan Inst Chem Eng 78:423–430
Wei Y, Ji Q, Chen L, Hao J, Yao C, Dong X (2017) Preparation of an inorganic coagulant-polysilicate–magnesium for dyeing wastewater treatment: effect of acid medium on the characterization and coagulation performance. J Taiwan Inst Chem Eng 72:142–148
Welham A (2000) The theory of dyeing (and the secret of life). J Soc Dye Colour 116(5):140–143
Wu C, Wang Y, Gao B, Zhao Y, Yue Q (2012) Coagulation performance and floc characteristics of aluminum sulfate using sodium alginate as coagulant aid for synthetic dying wastewater treatment. Sep Purif Technol 95:180–187
Xue M, Gao B, Li R, Sun J (2018) Aluminum formate (AF): synthesis, characterization and application in dye wastewater treatment. J Environ Sci 74:95–106
Yagub MT, Sen TK, Afroze S, Ang HM (2014) Dye and its removal from aqueous solution by adsorption: a review. Adv Colloid Interf Sci 209:172–184
Yang Zhou YWTL, Qiao QC, Shuo C (2007) Experimental study of wastewater treatment of reactive dye by phys-chemistry method. J China Univ Mining Technol 17(1):96–100
Zahrim AY, Tizaoui C, Hilal N (2011) Coagulation with polymers for nanofiltration pre-treatment of highly concentrated dyes: a review. Desalination 266(1–3):1–16
Zhou L, Zhou H, Yang X (2019) Preparation and performance of a novel starch-based inorganic/organic composite coagulant for textile wastewater treatment. Sep Purif Technol 210:93–99

The manufacturer's authorised representative in the EU is Springer Nature Customer Service Centre GmbH, Europaplatz 3, 69115 Heidelberg, Germany. If you have any concerns regarding our products, please contact ProductSafety@springernature.com

Printed and bound by CPI Group (UK) Ltd, Croydon, CR0 4YY
30/09/2026
02225297-0017